Interface Oral Health Science 2016

W0043631

Keiichi Sasaki • Osamu Suzuki
Nobuhiro Takahashi

Editors

Interface Oral Health Science 2016

Innovative Research on Biosis–Abiosis
Intelligent Interface

 Springer Open

Editors
Keiichi Sasaki
Division of Advanced Prosthetic Dentistry
Tohoku University Graduate School of
  Dentistry
Sendai, Japan

Osamu Suzuki
Division of Craniofacial Function
  Engineering
Tohoku University Graduate School
  of Dentistry
Sendai, Japan

Nobuhiro Takahashi
Division of Oral Ecology & Biochemistry
Tohoku University Graduate School
  of Dentistry
Sendai, Japan

ISBN 978-981-10-1559-5   ISBN 978-981-10-1560-1   (eBook)
DOI 10.1007/978-981-10-1560-1

Library of Congress Control Number: 2016959195

© The Editor(s) (if applicable) and The Author(s) 2017. This book is published open access.
Open Access This book is distributed under the terms of the Creative Commons Attribution 4.0 International License (http://creativecommons.org/licenses/by/4.0/), which permits use, duplication, adaptation, distribution and reproduction in any medium or format, as long as you give appropriate credit to the original author(s) and the source, provide a link to the Creative Commons license and indicate if changes were made.
The images or other third party material in this book are included in the work's Creative Commons license, unless indicated otherwise in the credit line; if such material is not included in the work's Creative Commons license and the respective action is not permitted by statutory regulation, users will need to obtain permission from the license holder to duplicate, adapt or reproduce the material.
The use of general descriptive names, registered names, trademarks, service marks, etc. in this publication does not imply, even in the absence of a specific statement, that such names are exempt from the relevant protective laws and regulations and therefore free for general use.
The publisher, the authors and the editors are safe to assume that the advice and information in this book are believed to be true and accurate at the date of publication. Neither the publisher nor the authors or the editors give a warranty, express or implied, with respect to the material contained herein or for any errors or omissions that may have been made.

Printed on acid-free paper

This Springer imprint is published by Springer Nature
The registered company is Springer Nature Singapore Pte Ltd.
The registered company address is: 152 Beach Road, #22-06/08 Gateway East, Singapore 189721, Singapore

Preface

The chapters in this book, which review the presentations at the 6th International Symposium for Interface Oral Health Science held January 18–19, 2016, at Tohoku University, Sendai, Japan, is being issued as a special publication. I am honored and pleased to deliver *Interface Oral Health Science 2016* to our publisher.

Interface Oral Health Science is a new concept in dentistry, established by the Tohoku University Graduate School of Dentistry in 2002, based on the knowledge that normal oral function is maintained through harmony between three biological and biomechanical systems: (1) structure of the mouth, including teeth, the mucous membrane, bones, and muscles; (2) microorganisms in the mouth (parasites); and (3) biomaterials. Tooth decay, periodontal disease, and other oral disorders can be recognized as "interface disorders", which are caused by the collapse of the interface between systems. This concept is shared not only by dentistry and dental medicine but also by a variety of disciplines, including medicine, material science, engineering, and others.

Since 2002, the Tohoku University Graduate School of Dentistry has regarded Interface Oral Health Science as the main theme of dental research in the twenty-first century. We are committed to advancing dental studies by implementing Interface Oral Health Science while promoting interdisciplinary research across a wide range of related fields. Based on this concept, we have successfully organized international symposiums for Interface Oral Health Science five times, in 2005, 2007, 2009, 2011, and 2014, which included inspiring special lectures, symposium sessions, poster presentations, and other discussions. These presentations were published in a series of English monographs titled *Interface Oral Health Science*.

These achievements were praised, and in 2007 the "Highly-functional Interface Science: Innovation of Biomaterials with Highly-functional Interface to Host and Parasite" was adopted as a program for research and education funding for an Inter-University Research Project 2007–2011, Ministry of Education, Culture, Sports, Science and Technology (MEXT), Japan. Subsequently, we have been developing a broader and more advanced concept: "Biosis–Abiosis Intelligent Interface", which also was adopted as the program for a Research Promotion Project 2012–2015, MEXT, Japan. It aims to create a highly functional and autonomic intelligent interface by combining the highly functional interface science established by the Tohoku University Graduate School of Dentistry and the Institute for Materials Research, Tohoku University. The accompanying technology for evaluation and control at the interface was provided by the Graduate School of Biomedical Engineering, Tohoku University. We firmly believe that the Biosis–Abiosis Intelligent Interface project can contribute to solving various problems not only in dentistry and medicine but also in other disciplines.

Therefore, the 6th International Symposium for Interface Oral Health Science was held in conjunction with the Innovative Research for Biosis–Abiosis Intelligent Interface Symposium, organized by the Tohoku University Graduate School of Dentistry with the aid of a Research Promotion Project Grant, MEXT, Japan. The symposium consisted of one special lecture, four focus sessions, and a poster session with 61 poster presentations. We hosted 22 distinguished invited keynote speakers for the focus sessions.

I hope that our project, including the symposium and the book, will accelerate the progress of dental science and point the way for dental research for future generation. Finally, I would like to thank all members of our school and the participants in the symposium for your contributions. Our thanks go especially to the authors of the excellent review chapters in this book.

Sendai, Japan   Keiichi Sasaki

Acknowledgment

The editors wish to acknowledge the following members of Tohoku University Graduate School of Dentistry; Institute for Materials Research, Tohoku University; and Tohoku University Graduate School of Biomedical Engineering, all of whom have contributed their valued expertise and time to hold the symposium and to edit manuscripts submitted to *Interface Oral Health Science 2016*. These colleagues have provided essential assistance, without which it wouldn't have been possible to publish this monograph in a timely manner.

Tohoku University Graduate School of Dentistry:

Satoshi FUKUMOTO
Guang HONG
Teruko YAMAMOTO
Minoru WAKAMORI
Shunji SUGAWARA
Yasuyuki SASANO
Masahiro SAITO
Koetsu OGASAWARA
Hidefumi FUKUSHIMA
Masatoshi TAKAHASHI
Yutaka HENMI
Akio MATSUMOTO
Yoshinori KONNO
Tsukasa SATO
Masashi SAITO
Hiroshi MIURA
Hiroaki OYAMADA
Satsuki HORITA
Teruko ANDO
Natsumi NIKAIDO
Yasuyuki ITAGAKI
Rika HASEBE

Yuki ONODERA
Genki FUKAYA
Naoko MIURA
Yoshinobu TAKEYAMA
Yuriko KIMURA
Maki HAMADA
Hidetomo KAWAUCHI
Toshikazu MICHIMUNE
Yasuhisa TAKEUCHI
Keiichi SAITO
Naoko SATO
Azusa FUKUSHIMA
Risa ISHIKO
Tomomi KIYAMA
Makiko MIYASHITA
Emika SATO
Misato ENDO
Tomoya SATO
Kenta SHOBARA
Ryo TAGAINO
Hiroaki TAKAHASHI
Satoko FURIYA
Takafumi YAMAMOTO
Jianlan LONG
Wei WANG
Ratri Maya SITALAKSMI
Chayanit CHAWEEWANNAKORN
Feng LUO
Yinying QU

Institute for Materials Research, Tohoku University:

Takashi GOTO
Takayuki NARUSHIMA
Hirokazu KATSUI

Tohoku University Graduate School of Biomedical Engineering:

Yoshifumi SAIJO
Ryoichi NAGATOMI

Contents

Part I   Symposium I: Biomaterials in Interface Science

1   **Low-Modulus Ti Alloys Suitable for Rods in Spinal Fixation Devices** 3
Mitsuo Niinomi

2   **Ceramic Coating of Ti and Its Alloys Using Dry Processes for Biomedical Applications** 23
Takatoshi Ueda, Natsumi Kondo, Shota Sado,
Ozkan Gokcekaya, Kyosuke Ueda, Kouetsu Ogasawara,
and Takayuki Narushima

3   **Dealloying Toxic Ni from SUS316L Surface** 35
Hidemi Kato, Takeshi Wada, and Sadeghilaridjani Maryam

4   **Bio-ceramic Coating of Ca–Ti–O System Compound by Laser Chemical Vapor Deposition** 47
Hirokazu Katsui and Takashi Goto

Part II   Symposium II: Innovation for Oral Science and Application

5   **Development of a Robot-Assisted Surgery System for Cranio-Maxillofacial Surgery** 65
Chuanbin Guo, Jiang Deng, Xingguang Duan, Li Chen,
Xiaojing Liu, Guangyan Yu, Chengtao Wang, and Guofang Shen

6   **Facilitating the Movement of Qualified Dental Graduates to Provide Dental Services Across ASEAN Member States** 73
Suchit Poolthong and Supachai Chuenjitwongsa

7   **Putting the Mouth into Health: The Importance of Oral Health for General Health** 81
Christopher C. Peck

8   Orofacial Stem Cells for Cell-Based Therapies of Local
      and Systemic Diseases 89
      Munira Xaymardan

9   Biomaterials in Caries Prevention and Treatment 101
      Lei Cheng, Yaling Jiang, Yao Hu, Jiyao Li,
      Hockin H.K. Xu, Libang He, Biao Ren, and Xuedong Zhou

Part III   Symposium III: Regenerative Oral Science

10   Efficacy of Calcium Phosphate-Based Scaffold Materials
      on Mineralized and Non-mineralized Tissue Regeneration 113
      Osamu Suzuki, Takahisa Anada, and Yukari Shiwaku

11   Gene Delivery and Expression Systems in Induced
      Pluripotent Stem Cells 121
      Maolin Zhang, Kunimichi Niibe, Takeru Kondo,
      Yuya Kamano, Makio Saeki, and Hiroshi Egusa

12   Emerging Regenerative Approaches for Periodontal
      Regeneration: The Future Perspective of Cytokine Therapy
      and Stem Cell Therapy 135
      Shinya Murakami

13   Molecular Mechanisms Regulating Tooth Number 147
      Maiko Kawasaki, Katsushige Kawasaki, James Blackburn,
      and Atsushi Ohazama

Part IV   Symposium IV: Medical Device Innovation
           for Diagnosis and Treatment of Biosis-Abiosis Interface

14   Open-Source Technologies and Workflows in Digital
      Dentistry 165
      Rong-Fu Kuo, Kwang-Ming Fang, and Fong-Chin Su

15   Detection of Early Caries by Laser-Induced Breakdown
      Spectroscopy 173
      Yuji Matsuura

16   Acoustic Diagnosis Device for Dentistry 181
      Kouki Hatori, Yoshifumi Saijo, Yoshihiro Hagiwara,
      Yukihiro Naganuma, Kazuko Igari, Masahiro Iikubo,
      Kazuto Kobayashi, and Keiichi Sasaki

Part V   Poster Presentation Award Winners

17   Activation of TLR3 Enhance Stemness and Immunomodulatory
      Properties of Periodontal Ligament Stem Cells (PDLSCs) 205
      Nuttha Klincumhom, Daneeya Chaikeawkaew,
      Supanniga Adulheem, and Prasit Pavasant

18 Influence of Exogenous IL-12 on Human Periodontal
   Ligament Cells.. 217
   Benjar Issaranggun Na Ayuthaya and Prasit Pavasant

19 Development and Performance of Low-Cost Beta-Type
   Ti-Based Alloys for Biomedical Applications Using
   Mn Additions.. 229
   Pedro F. Santos, Mitsuo Niinomi, Huihong Liu,
   Masaaki Nakai, Ken Cho, Takayuki Narushima, Kyosuke Ueda,
   Naofumi Ohtsu, Mitsuhiro Hirano, and Yoshinori Itoh

20 Effect of Titanium Surface Modifications of Dental
   Implants on Rapid Osseointegration.. 247
   Ting Ma, Xiyuan Ge, Yu Zhang, and Ye Lin

21 Development of Powder Jet Deposition Technique
   and New Treatment for Discolored Teeth 257
   Kuniyuki Izumita, Ryo Akatsuka, Akihiko Tomie,
   Chieko Kuji, Tsunemoto Kuriyagawa, and Keiichi Sasaki

22 Osteogenetic Effect of Low-Magnitude High-Frequency
   Loading and Parathyroid Hormone on Implant Interface
   in Osteoporosis 269
   Aya Shibamoto, Toru Ogawa, Masayoshi Yokoyama,
   Joke Duyck, Katleen Vandamme, Ignace Naert,
   and Keiichi Sasaki

Part I
Symposium I: Biomaterials in Interface Science

Chapter 1
Low-Modulus Ti Alloys Suitable for Rods in Spinal Fixation Devices

Mitsuo Niinomi

Abstract Low-Young's modulus Ti alloys are expected to be suitable for use in the rods of spinal fixation devices. However, in addition to a low Young's modulus, the rods must also exhibit a high Young's modulus at the deformed region to reduce springback when surgeons bend the rod inside the narrow bodies of patients. Ti–Mo and Ti–Cr alloys are potential candidates to satisfy these two conflicting demands. In these alloys, known as Young's modulus changeable or Young's modulus self-adjustable Ti alloys, the ω-phase with high Young's modulus is induced by deformation. Among these alloys, Ti–12Cr is judged to be the most suitable for use in the rods of spinal fixation devices. This alloy exhibits a high uniaxial fatigue strength, and its compressive fatigue strength is significantly improved by cavitation peening.

Keywords Low Young's modulus • Young's modulus self-adjustability • Ti–Cr alloy • Ti–Mo alloy • Spinal fixation device • Compressive fatigue strength • Cavitation peening

1.1 Introduction

To inhibit stress shielding, which is caused by the mismatch in Young's modulus between an implant and bone and leads to bone absorption and poor bone remodeling, low-Young's modulus titanium (Ti) alloys have been or are being developed for implant devices such as artificial hip joints, bone plates, and spinal fixation devices.

M. Niinomi (✉)
Institute for Materials Research, Tohoku University,
2-1-1 Katahira, Aoba-ku, Sendai, Miyagi 980-8577, Japan

Graduate School of Science and Technology, Meijyo University,
1-501, Shiogamaguchi, Tempaku-ku, Nagoya, Aichi 468-8502, Japan

Graduate School of Engineering, Osaka University, 2-1, Yamadagaoka,
Suita, Osaka 565-0871, Japan

Institute of Materials and Systems for Sustainability, Nagoya University,
Furo-cho, Chikusa-ku, Nagoya, Aichi 464-8603, Japan
e-mail: niinomi@imr.tohoku.ac.jp

© The Author(s) 2017   3
K. Sasaki et al. (eds.), *Interface Oral Health Science 2016*,
DOI 10.1007/978-981-10-1560-1_1

A novel β-type Ti alloy composed of nontoxic and allergy-free elements (Ti–29Nb–13Ta–4.6Zr mass%, hereafter abbreviated as TNTZ) is a low-Young's modulus Ti alloy developed by Niinomi et al. based on the d-electron design method [1]. This alloy exhibits excellent corrosion resistance and biocompatibility, with a low Young's modulus (~60 GPa), and has thus been investigated for use in many practical applications, including the production of implant rods for spinal fixtures [2]. In spinal fixation rods, the low Young's modulus and excellent biocompatibility of TNTZ is beneficial for patients; however, this alloy possesses only relatively moderate mechanical properties (ultimate tensile strength = ~510 MPa) [1], which are inadequate for practical application. However, according to surgeons specializing in spinal diseases, the implant rod should exhibit a small degree of springback to offer greater handling ability for surgeons, who are required to bend the rod to conform to the curvature of the spine within the limited space inside a patient's body [3]. The amount of springback is determined by both the strength and Young's modulus. Given the same strength, a rod with the higher Young's modulus will exhibit less springback, that is, a high Young's modulus is preferable to suppress springback from the viewpoint of surgeons. There is therefore a conflicting requirement concerning the Young's modulus from the viewpoint of patients and surgeons, which cannot be fully satisfied by TNTZ.

Recently, many researchers [4–7] have focused on the development of new β-type Ti alloys to solve this problem, such as Ti–Mo and Ti–Cr alloys. These alloys possess a novel property called "changeable Young's modulus" or "Young's modulus self-adjustability," in which the deformed region of the material exhibits a high Young's modulus, whereas the Young's modulus of the undeformed region remains low.

Therefore, changeable-Young's modulus Ti alloys suitable for the rods of spinal fixation devices are introduced with a focus on Ti–Mo and Ti–Cr alloys.

1.2   Mechanism of Increasing Young's Modulus in Deformed Region

The springback of the deformed rods of spinal fixation device can be reduced by increasing the Young's modulus in the deformed region. This reduction can be achieved if a secondary phase with a high Young's modulus is induced by deformation. In Ti alloys, metastable phases such as α' martensite with hcp structure, α'' martensite with orthorhombic structure, and ω-phase with hexagonal or trigonal structure can form between the stable α- and β-phases based on the amount of β-stabilizing element, as depicted in the schematic phase diagram of Ti alloys in Fig. 1.1. Therefore, various phases such as α'-, α''-, and ω-phases can be induced by deformation. ω-phase precipitation is known to increase the Young's modulus of the alloys, whereas α'- and α''-phase precipitations result in a decrease. Therefore, if the β-phase stability is controlled for the ω-phase to be induced by deformation, the

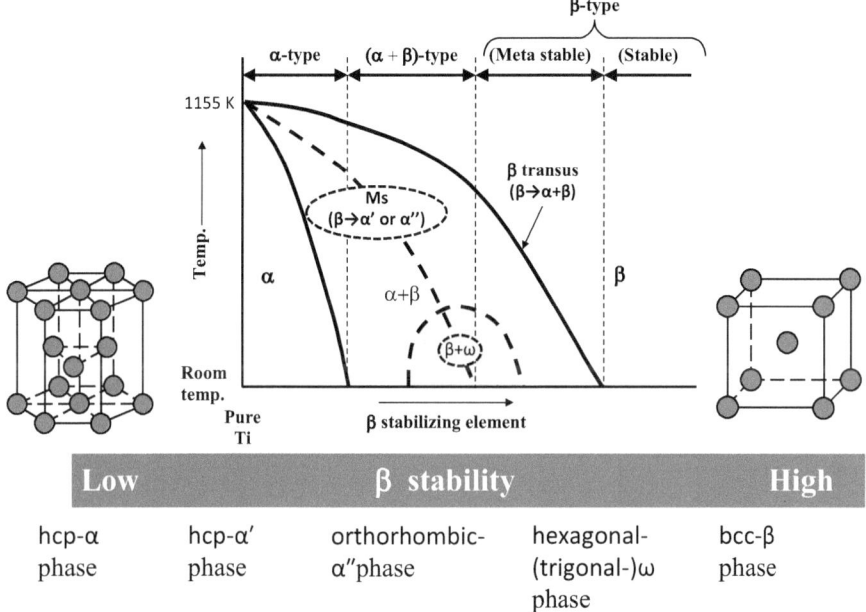

Fig. 1.1 Relationship between schematic phase diagram of titanium alloys and β stability

Young's modulus of the alloys is partially increased in the deformed region, leading to a decrease in springback of the alloy and thereby maintaining the deformed shape of the rod.

1.3   Possible Alloy System

The addition of molybdenum (Mo) results in β-stabilizing properties in Ti alloys [8], and, thus, it is an effective β stabilizer for designing β-type alloys. In addition, Hanada et al. reported that a ω-phase forms during the cold working of as-quenched Ti–Mo alloys within the composition range of 11–18 mass% Mo [9–11]. Thus, it is possible to realize a changeable Young's modulus in metastable Ti–Mo alloys via deformation-induced ω-phase transformation. Moreover, Mo is less toxic than Al and V [1]. Many studies have demonstrated the excellent mechanical compatibility and good cytocompatibility of Ti alloys containing Mo, such as Ti–Mo, Ti–Mo–Ta, and Ti–Mo–Zr–Fe [12–17]. Karthega et al. [18] reported that Ti–15Mo exhibits corrosion resistance similar to that of TNTZ in Hank's solution. Both Zhou et al. and Oliveira et al. [19] reported that Ti–Mo alloys exhibit excellent corrosion resistance.

Hanada et al. reported that the ω-phase forms during the cold working of as-quenched Ti–Cr alloys within the composition range of 8–11.5 mass% Cr [20]. Moreover, the ω-phase can be introduced by deformation at room temperature in

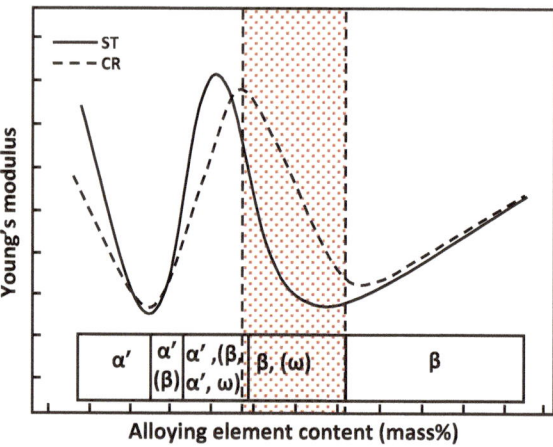

Fig. 1.2 Schematic variation of Young's modulus in Ti–M (M: β stabilizer) binary alloys as a function of alloying element content. ST and CR indicates solution treatment and cold rolling, respectively

Table 1.1 Nominal chemical compositions of Ti–Mo and T–Cr systems alloys examined to optimize their chemical compositions exhibiting the greatest increase in Young's modulus after deformation (the best Young's modulus adjustability)

Ti–Cr system	Ti–10Cr	Ti–11Cr	Ti–12Cr	Ti–13Cr
Ti–Mo system	Ti–13Mo	Ti–15Mo	Ti–16Mo	Ti–17Mo

Ti–V, Ti–Fe, and Ti–Mo metastable β-type alloys [10, 11, 21–23]. Furthermore, Cr is known to control the anodic activity of the alloy and increase the tendency of Ti to passivate [24], and the passive films of Ti alloys, in turn, allow them to maintain corrosion resistance [25–27].

Thus, Mo and Cr are suitable alloying elements to develop Ti-based biomaterials with Young's modulus adjustability.

However, as depicted in the schematic illustration of the relationship between temperature and alloying element content in Fig. 1.2 [23, 28], with each appearing phase, deformation-induced ω-phase precipitation occurs, and the lowest Young's modulus before deformation is highly expected in the relatively high β stability region.

To optimize the chemical compositions of Ti–Mo and Ti–Cr alloys exhibiting the greatest increase in Young's modulus after deformation (the best Young's modulus adjustability), the Mo equivalent, Mo_{eq}, was considered, and 1 mass%Cr is equivalent to 1.25 mass%Mo. Then, the Ti–Mo and Ti–Cr alloys, whose nominal chemical compositions are listed in Table 1.1 [29], were examined. The Young's moduli, microstructures, and deformation behaviors of both alloy systems were examined before and after deformation; deformation was simulated by cold rolling with a reduction of 10%.

1.4   Ti–Mo Alloys

1.4.1   Microstructures Before and After Deformation

The X-ray diffraction (XRD) profiles of the Ti–(13–18)Mo alloys subjected to solution treatment and cold rolling with a reduction of 10% (hereafter called as cold rolling (CR)) are shown in Fig. 1.3 [29, 30]. Only peaks corresponding to the β-phase are detected in each solution-treated (ST) specimen. No peaks corresponding to the ω-phase are detected by XRD. In addition, after cold rolling, only single β-phases are detected in each CR specimen. However, the enlarged XRD profiles of Ti–13Mo alloy subjected to solution treatment (ST) and CR in Fig. 1.4 [29] contain ω- and α″-phase peaks, respectively. For Ti–13Mo alloy, it is apparent that deformation-induced α″-phase transformation occurs.

Figure 1.5 [29, 30] shows selected area electron diffraction patterns of the Ti–(13–18)Mo alloys subjected to ST. The existence of a small amount of athermal ω-phase is recognized in the Ti–(13–16)Mo alloys subjected to ST. As the intensities of the ω reflections decrease with increasing Mo content, the amount of the athermal ω-phase in the ST specimens decreases. It is well known that the amount of the athermal ω-phase in β-type alloys depends on the stability of the β-phase. Specifically, the stability of the β-phase increases with increasing Mo content. Thus, the formation of the athermal ω-phase during water quenching (WQ) is suppressed in high-β-stable alloys. The amount of the athermal ω-phase in Ti–16Mo–ST is thus less than that in Ti–13Mo–ST, and the athermal ω-phase in Ti–17Mo–ST and Ti–18Mo–ST disappears.

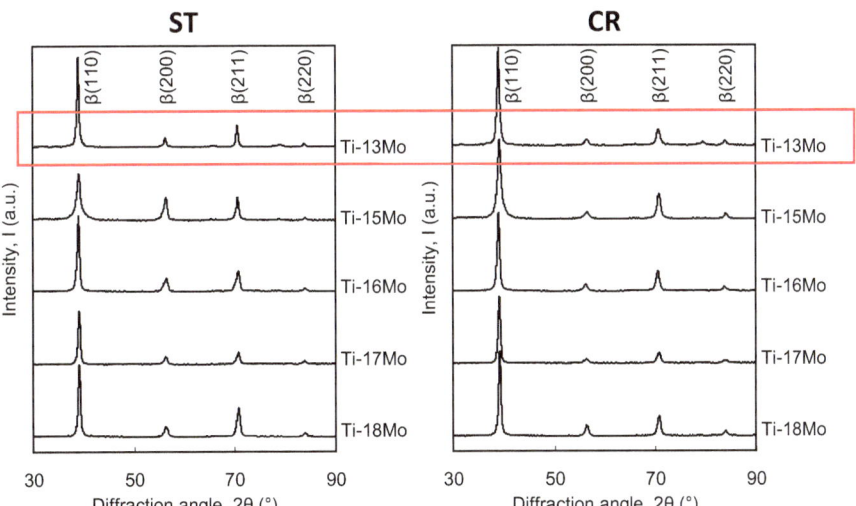

Fig. 1.3   XRD profiles of Ti–(13–18)Mo alloys subjected to solution treatment (ST) and cold rolling (CR)

Fig. 1.4 XRD profiles of
Ti–13Mo alloy subjected
to solution treatment (ST)
and cold rolling (CR)

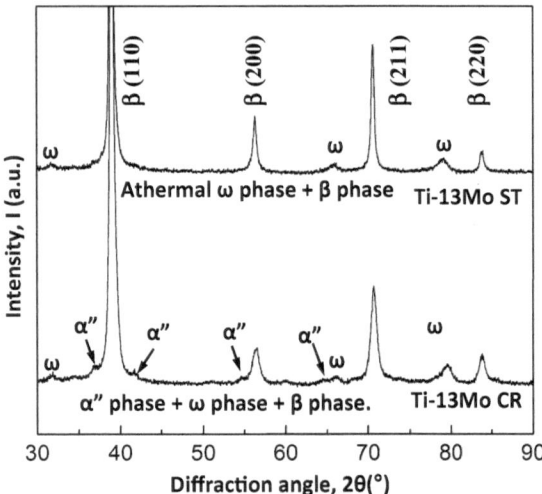

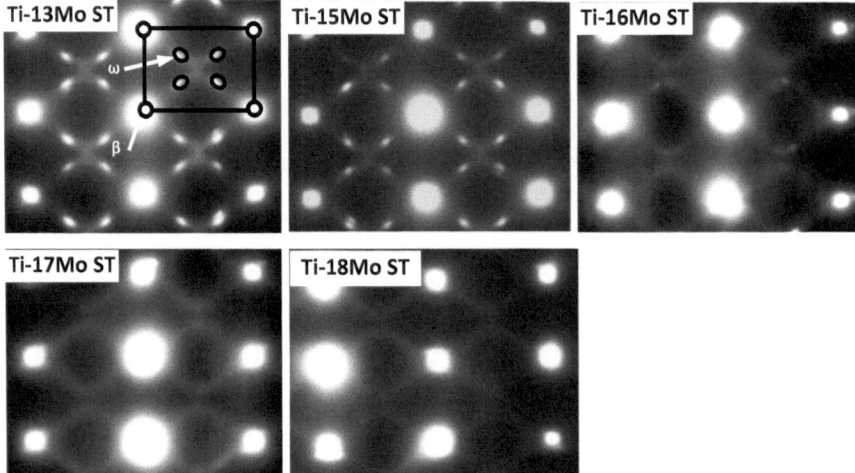

Fig. 1.5 Selected area electron diffraction patterns of Ti–(13–18)Mo alloys subjected to solution treatment (ST)

Figure 1.6 [29, 30] shows selected area electron diffraction patterns of the Ti–(13–18)Mo alloys subjected to CR. After cold rolling, the ω spots in all the alloys become sharp. This result confirms that deformation-induced ω-phase transformation occurs during CR. After CR, the ω reflections in Ti–13Mo–Cr, Ti–15Mo–CR, and Ti–16Mo–CR are much sharper than in Ti–13Mo, Ti–15Mo–ST, and Ti–16Mo–ST, indicating that the amount of ω-phase increases because of CR. Moreover, sharp extra spots corresponding to the ω-phase are observed in Ti–17Mo–CR and Ti–18Mo–CR, suggesting that a considerable amount of ω-phase is formed in these two alloys. These findings confirm that deformation-induced ω-phase transforma-

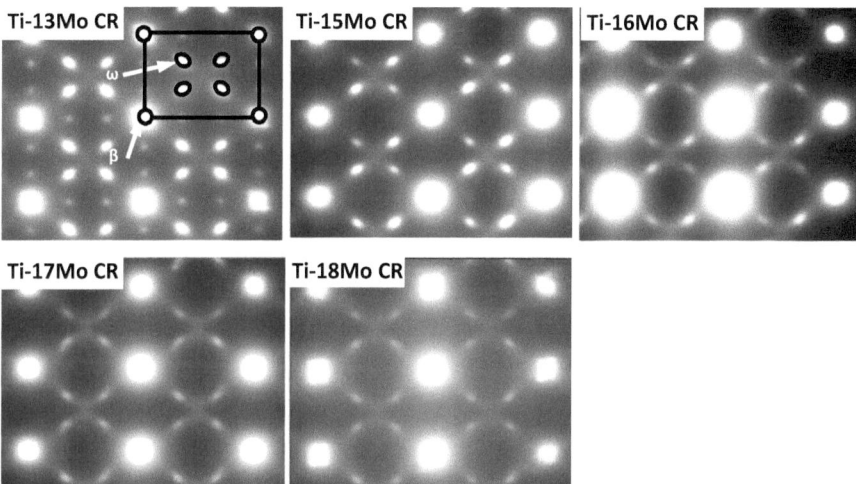

Fig. 1.6 Selected area electron diffraction patterns of Ti–(13–18)Cr alloys subjected to cold rolling (CR)

tion occurs in all the Ti–(13–18)Mo alloys during CR. With increasing Mo content, the intensities of the ω reflections decrease, indicating that deformation-induced ω-phase transformation is dependent on the stability of the β-phase. Specifically, as the Mo content increases, the stability of the β-phase increases and deformation-induced ω-phase transformation becomes difficult during CR. Thus, the amount of deformation-induced ω-phase decreases with increasing Mo content.

1.4.2 Young's Modulus Change by Deformation

Figure 1.7 [29, 30] shows the Young's moduli of the Ti–(13–18)Mo alloys subjected to solution treatment and CR. Except Ti–13Mo alloy, all the alloys subjected to ST exhibit low Young's moduli of <80 GPa, which is much lower than those of SUS 316 L stainless steel (SUS 316 L), commercially pure titanium (CP Ti), and Ti–6Al–4 V extra low interstitial (ELI) alloy (Ti64 ELI). For the Ti–(15–18)Mo alloys, the Young's moduli of the alloys subjected to CR are higher than those subjected to ST. For the ST specimens, with increasing Mo content, the Young's modulus first decreases slightly from 79 GPa for Ti–15Mo–ST to 73 GPa for Ti–17Mo–ST, after which the value slightly increases to 75 GPa for Ti–18Mo–ST. According to the TEM observation and XRD analysis results, the amount of athermal ω-phase in the alloys decreases with increasing Mo content. There is no athermal ω-phase in Ti–17Mo–ST or Ti–18Mo–ST. The ω-phase is known to have a significant effect on the mechanical properties of Ti alloys and is likely to increase the Young's modulus. In this case, the athermal ω-phase is the main factor contributing to the change in the

Fig. 1.7 Young's moduli of Ti–(13–18)Mo alloys subjected to solution treatment (ST) and cold rolling (CR)

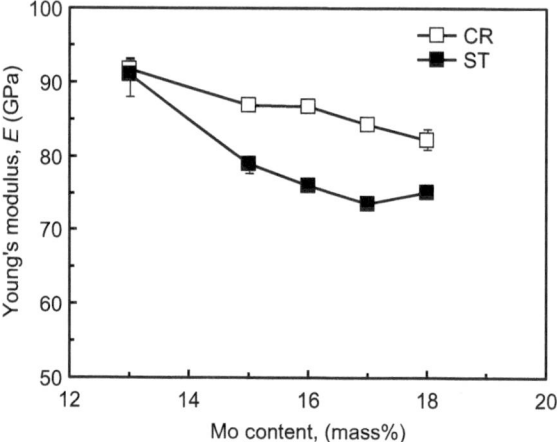

Young's modulus. Therefore, the Young's modulus of the alloys first decreases, and Ti–17Mo–ST exhibits the lowest Young's modulus among the designed alloys. With a further increase in the Mo content, the athermal ω-phase in the alloy disappears. Thus, the solid solution strengthening with Mo becomes the primary cause of the changes in the Young's modulus. The Young's modulus increases as the bcc lattice contracts with increasing Mo content. Therefore, the Young's modulus of Ti–18Mo–ST is slightly higher than that of Ti–17Mo–ST. The microstructural analysis indicates that deformation-induced ω-phase transformation occurs in all the alloys. Therefore, the increase in the Young's moduli of these four alloys following CR is attributed to the deformation-induced ω-phase transformation that occurred during CR. The lack of change in the Young's modulus of Ti–13Mo alloy is considered to result from the combined effect of deformation-induced ω-phase transformation and deformation-induced α''-phase transformation.

1.4.3 Deformation-Induced Products and Phase Constitutions

The deformation-induced products in Ti–(13–18)Mo alloys resulting from CR are summarized in Table 1.2 [29]. Except for Ti–13Mo alloy, the ω-phase as well as $\{332\}_\beta<113>_\beta$ and $\{112\}_\beta<111>_\beta$ twins are formed by CR. For Ti–13Mo alloy, the ω- and α''-phase as well as $\{332\}_\beta<113>_\beta$ and $\{112\}_\beta<111>_\beta$ twins are formed by CR.

The constituent phases of the Ti–(13–18)Mo alloys subjected to ST and CR are summarized in Table 1.3 [29]. The constituent phases of the Ti–(13–16)Mo alloys subjected to ST are the β- and athermal ω-phases, whereas that of the Ti–17Mo and Ti–18Mo alloys subjected to ST is ω-phase.

The constituent phases of the Ti–13Mo alloy subjected to CR are β-, athermal ω-, and deformation-induced ω-, and deformation-induced α''-phases. The constituent

Table 1.2   Summary of deformation-induced products by cold rolling in Ti–Mo alloys

Alloy	Deformation-induced ω	Deformation-induced α''	Mechanical twinning ($\{332\}_\beta<113>_\beta$ and $\{112\}_\beta<111>_\beta$)
Ti–13Mo	★	☆	★
Ti–15Mo	★	×	★
Ti–16Mo	★	×	☆
Ti–17Mo	★	×	☆
Ti–18Mo	★	×	☆

Table 1.3   Summary of phase constitutions of Ti–(13–18)Mo alloys subjected to solution treatment (ST) and cold rolling (CR)

Alloy	Subjected to solution treatment	Subjected to solution treatment and cold rolling
Ti–13Mo	β+athermal ω	β+athermal ω+deformation-induced ω+deformation-induced α''
Ti–15Mo	β+athermal ω	β+athermal ω+deformation-induced ω
Ti–16Mo	β+athermal ω	β+athermal ω+deformation-induced ω
Ti–17Mo	β	β+deformation-induced ω
Ti–18Mo	β	β+deformation-induced ω

phases of the Ti–15Mo and Ti–16Mo alloys subjected to CR are the β-, athermal ω-, and deformation-induced ω-phases. Further, the constituent phases of the Ti–16Mo and Ti–17Mo alloys subjected to CR are β- and deformation-induced ω-phases.

1.4.4   Springback

Ti–17Mo alloy exhibited the greatest increase in Young's modulus after CR, as stated above. The ratios of springback per unit load as a function of applied strain for Ti–17Mo alloy, Ti64 ELI, which is the most widely used Ti alloy for biomedical applications, and TNTZ, and strains for the calculation of the springback ratio are shown in Fig. 1.8 [29, 30]. The springback per unit load ratio of Ti–17Mo alloy is smaller than that of TNTZ and close to that of Ti64 ELI and reaches to a stable value for an applied strain greater than 2 %. The springback of a rod composed of Ti–17Mo alloy is expected to be significantly suppressed.

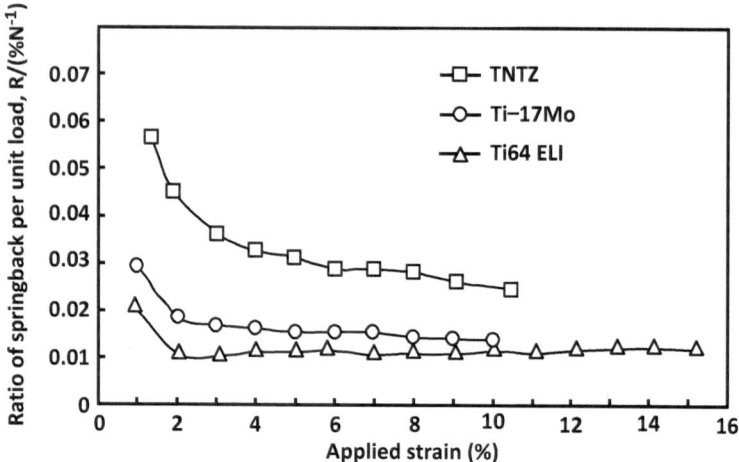

Fig. 1.8 Ratio of springback per unit load as a function of applied strain for Ti–17Mo alloy, Ti–6Al–4V ELI alloy (Ti64 ELI), and Ti–29Nb–13Ta–4.6Zr alloy (TNTZ) and strains for calculation of the springback ratio

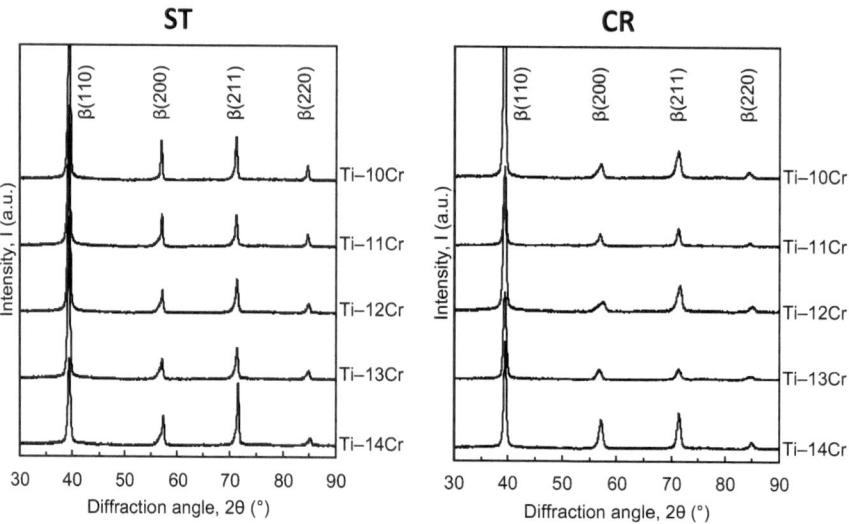

Fig. 1.9 XRD profiles of Ti–(10–14)Cr alloys subjected to solution treatment (ST) and cold rolling (CR)

1.5 Ti–Cr Alloys

1.5.1 Microstructures Before and After Deformation

Figure 1.9 [31] shows the XRD profiles of the Ti–xCr ($x = 10$, 11, 12, 13, and 14 mass%) alloys subjected to ST and CR. Only the peaks corresponding to the bcc β-phase were detected in every specimen under both ST and CR conditions. No peaks corresponding to the ω-phase were detected by XRD.

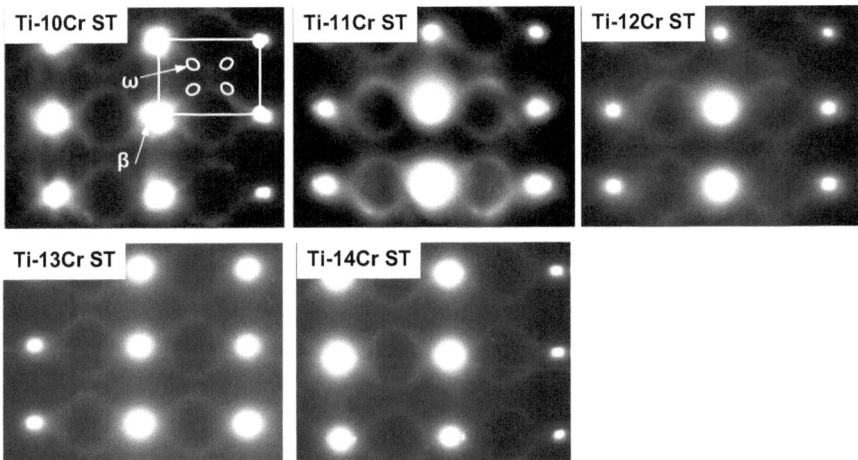

Fig. 1.10 Selected area electron diffraction patterns of Ti–(10–14)Cr alloys subjected to solution treatment (ST)

Figure 1.10 [31] shows the selected area electron diffraction patterns of the Ti–xCr ($x = 10, 11, 12, 13,$ and 14 mass%) alloys subjected to ST. The electron diffraction patterns of the $[110]_\beta$ zone of the Ti–10Cr–ST reveal weak extra spots in addition to spots derived from the β-phase, suggesting that a small amount of the athermal ω-phase is formed in Ti–10Cr–ST during WQ. For Ti–11Cr–ST, the intensities of the ω reflections decrease, which is reflected by the change in the circular diffuse streaks. As observed in Fig. 1.5b–e, as the Cr content continues to increase, the circular diffuse streaks are weakened. It is well known that the intensity of the ω reflection is related to the amount of the ω-phase. The amount of the athermal ω-phase in the designed alloys is observed to be dependent on the stability of the β-phase. Specifically, as the Cr content increased, the β-phase became more stable, leading to suppression of the formation of the athermal ω-phase during water quenching. Therefore, the amount of the athermal ω-phase in Ti–11Cr–ST is lower than that in Ti–10Cr–ST, and the athermal ω-phase disappears almost completely in Ti–12Cr–ST, Ti–13Cr–ST, and Ti–14Cr–ST.

Figure 1.11 [31] shows selected area electron diffraction patterns of the Ti–(1–14)Cr alloys subjected to CR. After CR, the ω reflections in Ti–10Cr–CR are much sharper than in Ti–10Cr–ST, indicating that the amount of the ω-phase increases in response to CR. Furthermore, the intensities of the ω reflections increase in Ti–11Cr–CR and Ti–12Cr–CR compared with those in Ti–11Cr–ST and Ti–12Cr–ST. These findings confirm that the deformation-induced ω-phase transformation occurs in the Ti–10Cr, Ti–11Cr, and Ti–12Cr alloys during CR. The amount of the ω-phase in Ti–10Cr–CR, Ti–11Cr–CR, and Ti–12Cr–CR is too small to be detected by XRD analysis, but the deformation-induced ω-phase is evident upon TEM observation. Specifically, as the Cr content increases, the intensities of the ω reflections

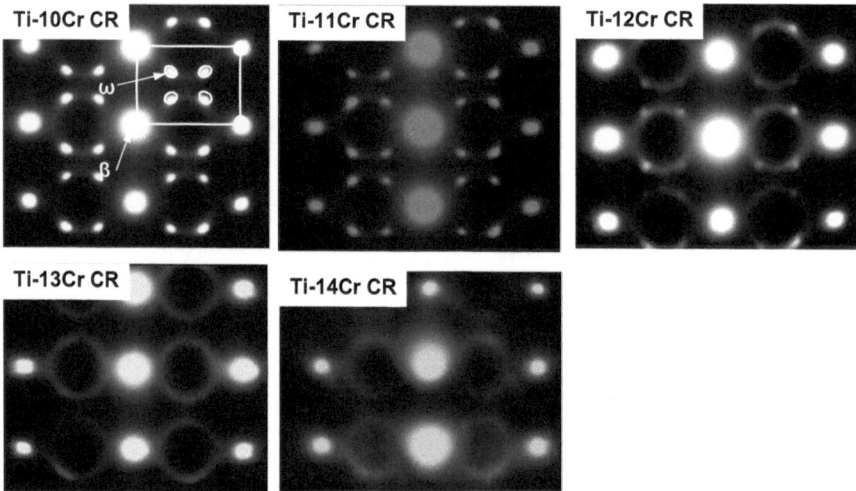

Fig. 1.11 Selected area electron diffraction patterns of Ti–(10–14)Cr alloys subjected to cold rolling (CR)

decrease. In addition, only diffuse streaks are observed in Ti–13Cr–CR and Ti–14Cr–CR, suggesting that the amount of the ω-phase does not change during cold rolling. These findings indicate that the deformation-induced ω-phase transformation does not occur in Ti–13Cr and Ti–14Cr alloys during CR. Thus, the deformation-induced ω-phase transformation is dependent on the β-phase stability. In lower-stability alloys, such as Ti–10Cr, Ti–11Cr, and Ti–12Cr, the deformation-induced ω-phase transformation may occur; however, with increasing Cr content, the β-phase can become more stable, thereby suppressing the deformation-induced ω-phase transformation. The combined EBSD analysis and TEM observation results indicate that the deformation-induced ω-phase transformation is accompanied by $\{332\}_\beta$ mechanical twinning.

1.5.2 Young's Modulus Change by Deformation

Figure 1.12 [31] shows the Young's moduli of the designed alloys subjected to ST and CR. In the ST specimens, as the Cr content increases, the Young's modulus decreases from 82 GPa for Ti–10Cr–ST to 68 GPa for Ti–12Cr–ST and then increases to 79 GPa for Ti–14Cr–ST. The changes in the Young's modulus can be attributed to two primary factors: the athermal ω-phase and solution strength of Cr. When the Cr content is less than 12 mass%, the amount of the athermal ω-phase in the designed alloys decreases drastically as the Cr content increases. In this case, the athermal ω-phase is the main factor responsible for the change in the Young's modulus; therefore, the Young's modulus initially decreases with increasing Cr content. However, when the Cr content increases beyond 12 mass%, the athermal ω-phase is no longer present in the designed alloys; thus, the solid solution strengthening

Fig. 1.12 Young's moduli of Ti–(10–14)Cr alloys subjected to solution treatment (ST) and cold rolling (CR)

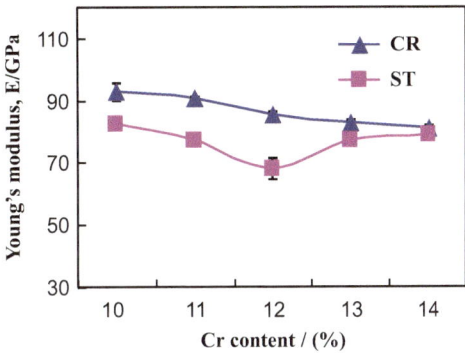

becomes the primary cause of the changes in the Young's modulus. As a result, the Young's modulus increases as the Cr content increases from 12 to 14 mass%. After CR, Young's moduli of Ti–10Cr, Ti–11Cr, and Ti–12Cr alloys increase remarkably compared with those after ST, whereas the changes in the moduli of Ti–13Cr and Ti–14Cr are negligible. Microstructural analysis indicates that the deformation-induced ω-phase transformation occurs in Ti–10Cr–CR, Ti–11Cr–CR, and Ti–12Cr–CR and that the ω-phase in β-type Ti alloys is likely to be the source of the increase in the Young's modulus. Therefore, the increase in the Young's moduli observed for these three alloys can be attributed to the deformation-induced ω-phase transformation that occurs during cold rolling. The degree of the increase in the Young's modulus is dependent on the amount of the deformation-induced ω-phase; thus, the Young's moduli of Ti–10Cr–CR, Ti–11Cr–CR, and Ti–12Cr–CR decrease as the Cr content increases. In this study, Ti–12Cr exhibits a low Young's modulus of 68 GPa after ST and a high Young's modulus of 85 GPa after CR.

1.5.3 Deformation-Induced Products and Phase Constitutions

The deformation-induced products formed by CR in the Ti–(10–14)Cr alloys are summarized in Table 1.4 [29]. For the Ti–(11–12)Cr alloys, the deformation-induced ω-phase and $\{332\}_\beta<113>_\beta$ mechanical twins are formed by CR. However, for Ti–13 and Ti–14 alloys, no deformation-induced products are formed by CR.

The constituent phases of the Ti–(10–14)Cr alloys subjected to ST and CR are summarized in Table 1.5 [29]. The constituent phases of the Ti–10 and Ti–11Cr alloys subjected to ST are β- and athermal ω-phases. The constituent phases that of the Ti–(12–14) alloys subjected to ST is the single β-phase.

The constituent phases of the Ti–10Cr and Ti–11Cr alloys subjected to CR are β-, athermal ω-, and deformation-induced ω-phases. The constituent phases of the Ti–12Cr alloy subjected to CR are β- and deformation-induced ω-phases. The constituent phase of the Ti–13Cr and Ti–14Cr alloys subjected to CR is the single β-phase.

Table 1.4 Summary of deformation-induced products by cold rolling in Ti–Cr alloys

Alloy	Deformation-induced ω	Deformation-induced α''	Mechanical twinning $(\{332\}_\beta<113>_\beta)$
Ti–10Cr	☆	×	☆
Ti–11Cr	☆	×	☆
Ti–12Cr	☆	×	☆
Ti–13Cr	×	×	×
Ti–14Cr	×	×	×

Table 1.5 Summary of phase constitutions of Ti–(1–14)Cr alloys subjected to solution treatment (ST) and cold rolling (CR)

Alloy	Subjected to solution treatment	Subjected to solution treatment and cold rolling
Ti–10Cr	β + athermal ω	β + athermal ω + deformation-induced ω
Ti–11Cr	β + athermal ω	β + athermal ω + deformation-induced ω
Ti–12Cr	β	β + deformation-induced ω
Ti–13Cr	β	β
Ti–14Cr	β	β

1.5.4 Springback

Ti–12Cr alloy exhibited the greatest increase in Young's modulus after CR as stated above. The ratios of springback per unit load as a function of applied strain for Ti–12Cr alloy, Ti64 ELI, and TNTZ, and strains for the calculation of the springback ratio are shown in Fig. 1.13 [32]. The springback per unit load ratio of Ti–12Cr alloy is less than that of TNTZ and close to that of Ti64 ELI and reaches to a stable value for an applied strain beyond 2%. The springback of a rod composed of Ti–13Cr alloy is expected to be significantly suppressed.

1.6 Young's Modulus Under Solution Treatment Conditions and Increment Ratio of Young's Modulus by Cold Rolling

To select of the optimal alloy for spinal fixation rods, the relationships between the Young's modulus under ST conditions and increment ratio, IR, of the Young's modulus by CR of Ti–Mo and Ti–Cr alloys are plotted in Fig. 1.14 [29]. IR of the Young's modulus by CR was calculated by the following equation:

$$\mathrm{IR} = \left(E_s - E_c \,/\, E_s \right) \times 100\% \qquad (1.1)$$

where E_s is the Young's modulus under ST conditions and E_c is the Young's modulus after CR.

Fig. 1.13 Ratio of springback per unit load as a function of applied strain for Ti–12Cr alloy, Ti–6Al–4V ELI alloy (Ti64 ELI), and Ti–29Nb–13Ta–4.6Zr alloy (TNTZ) and strains for calculation of the springback ratio

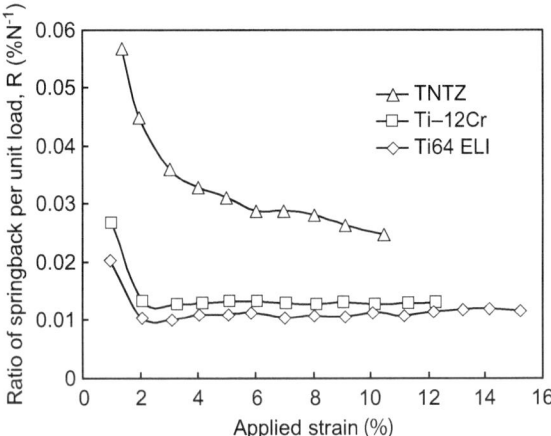

Fig. 1.14 Distribution of Ti–Cr and Ti–Mo alloys in a plot of Young's modulus under solution-treated (ST) conditions against increment ratio of Young's modulus by cold rolling (CR)

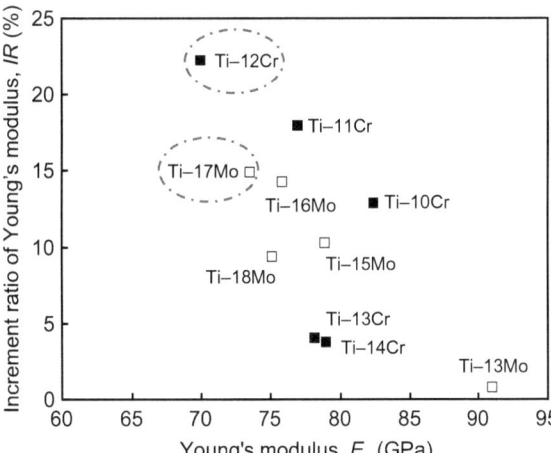

Ti–12Cr alloy exhibits the lowest Young's modulus and highest IR under ST conditions and is thus the most suitable for fixation device rods.

1.7 Toward Practical Applications

In the application of Ti alloys with changeable Young's modulus for spinal fixation device rods, their fatigue strengths are highly important. The uniaxial fatigue strength of Ti–12Cr alloy, a Ti alloy with changeable Young's modulus, was examined because it exhibited the lowest Young's modulus and highest IR under ST

Fig. 1.15 Fatigue limit of
Ti–12Cr alloy and
Ti–6Al–4V ELI alloy

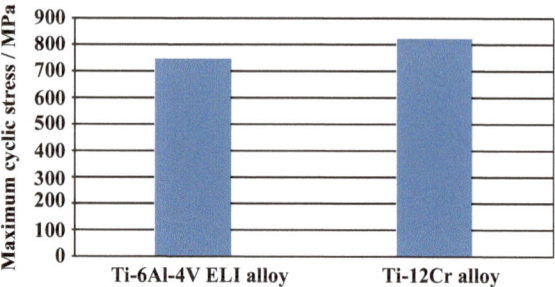

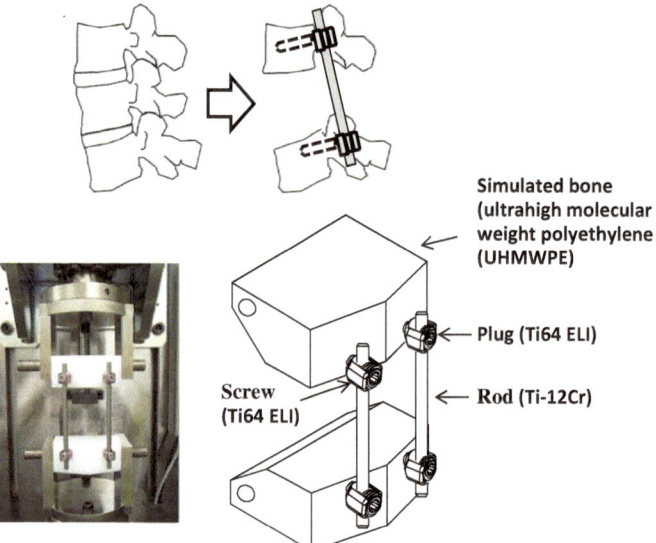

Fig. 1.16 Schematic drawing of compressive fatigue strength test method according to ASTM
F1717

conditions. The results are shown in Fig. 1.15 [33] and demonstrate excellent uniaxial fatigue strength of this alloy. The fatigue limits of Ti–12Cr alloy and Ti64 ELI
are shown for comparison. The fatigue ratio, which is the ratio of the fatigue limit
to tensile strength, of Ti–12Cr alloy is approximately 0.9, whereas that of Ti64 ELI
is approximately 0.6.

However, to employ the alloy in practical rod applications, its endurance must be
evaluated in a laboratory according to ASTM F1717, which describes a testing
method for evaluating the compressive fatigue strength of spinal fixation rods using
a simulated spinal fixation model, as shown in Fig. 1.16 [34]. The spinal fixation
device comprises a screw and plug made of Ti64 ELI and a rod made of Ti–12Cr
alloy. A Ti64 ELI rod was also used for comparison. Bone was simulated using
ultrahigh molecular weight polyethylene (UHMWPE). The compressive fatigue
limit of Ti–12Cr alloy subjected to ST is less than that of Ti64 ELI, as shown in

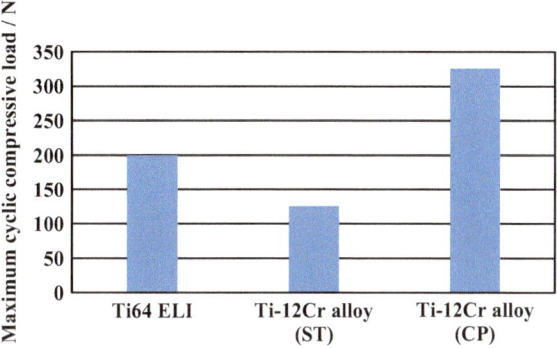

Fig. 1.17 Compressive fatigue limit of Ti–6Al–4V ELI alloy (Ti64 ELI) and Ti–12Cr alloy subjected to solution treatment (ST) and cavitation peening (CP) after solution treatment evaluated according to ASTM F 1717

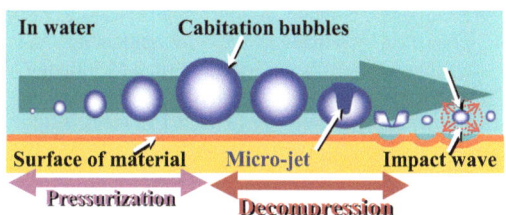

Fig. 1.18 Schematic drawings of development and crashing of cavitation

Fig. 1.17 [35]. In the ASTM F1717 compressive fatigue test, the rod typically fails at the contact area between the rod and plug. Therefore, fretting that occurs between the rod and plug is thought to reduce the compressive fatigue strength of the rod. An effective solution to such a problem is to improve the mechanical properties and tribological characteristics of the rod. The introduction of a hardened layer via compressive residual stress on the surface of the rod effectively prevents fretting fatigue. Peening techniques can introduce these hardened layers through plastic deformation, i.e., work hardening delivered by a large impact on the material's surface. Among the major peening techniques, cavitation peening, which is schematically illustrated in Fig. 1.18 [36], appears to be a highly promising method for improving the compressive fatigue strength of rods used in spinal fixation devices because it induces less surface damage than other peening techniques. Therefore, cavitation peening was performed on Ti–12Cr alloy rods to improve their compression fatigue strength as evaluated by ASTM F1717; the technique significantly increases the compressive fatigue strength of the rods, as demonstrated in Fig. 1.17.

1.8 Summary

Ti–17Mo and Ti–12Cr alloys with changeable Young's modulus are suitable for spinal fixation device rods, but Ti–12Cr alloy is more suitable because of its higher IR of the Young's modulus by deformation and lower Young's modulus. The fatigue strength of Ti–12Cr alloy is high, and its fatigue ratio is approximately 0.9. Ti–12Cr alloy is highly expected to be used for spinal fixation device rods because its compression fatigue strength, which is evaluated according to the ASTM 17171 standard, is significantly improved by cavitation peening.

Acknowledgments This work was partly supported by the interuniversity cooperative research program "Innovative Research for Biosis-Abiosis Intelligent Interface" of the Ministry of Sports, Culture, and Education in Japan and a Grant-in-Aid for Scientific Research (A) and Challenging Exploratory Research from the Japan Society for the Promotion of Science (JSPS).

References

1. Kuroda D, Niinomi M, Morinaga M, Kato Y, Yashiro T. Design and mechanical properties of new beta type titanium alloys for implant materials. Mater Sci Eng A. 1998;243:244–9.
2. Tamura T, Kozuka S, Oribe K, Nakai M, Niinomi M. Evaluation of Ti-Nb-Ta-Zr as new medical implants. In: Niinomi M, Akiyama S, Hagiwara M, Ikeda M, Maruyama K, editors. Ti-2007 science and technology. Sendai: The Japan Institute of Metals; 2007. p. 1441–4.
3. Steib JP, Dumas R, Skalli W. Surgical correction of scoliosis by in situ contouring – a distortion analysis. Spine. 2004;29:193–9.
4. Nakai M, Niinomi M, Zhao XF, Zhao XL. Self-adjustment of Young's modulus in biomedical titanium alloys during orthopaedic operation. Mater Lett. 2011;65:688–90.
5. Zhao XF, Niinomi M, Nakai M, Hieda J, Ishimoto T, Nakano T. Optimization of Cr content of metastable β-type Ti-Cr alloys with changeable Young's modulus for spinal fixation applications. Acta Biomater. 2012;8:2392–400.
6. Zhao XF, Niinomi M, Nakai M, Hieda J. Beta type Ti-Mo alloys with changeable Young's modulus for spinal fixation applications. Acta Biomater. 2012;8:1990–7.
7. Zhao XL, Niinomi M, Nakai M, Miyamoto G, Furuhara T. Microstructures and mechanical properties of metastable Ti-30Zr-(Cr, Mo) alloys with changeable Young's modulus for spinal fixation applications. Acta Biomater. 2011;7:3230–6.
8. Narushima N. Metallic biomaterials: titanium and its alloys. J Jpn Soc Biomater. 2005;23:86–95.
9. Hanada S, Yoshio T, Izumi O. Effect of plastic deformation modes on tensile properties of beta titanium alloys. Trans Jpn Inst Metals. 1986;27:496–503.
10. Hanada S, Izumi O. Correlation of tensile properties, deformation modes, and phase stability in commercial β phase titanium alloys. Metall Trans A. 1987;18:265–71.
11. Hanada S, Izumi O. Transmission electron microscopic observations of mechanical twinning in metastable beta titanium alloys. Metall Trans A. 1986;17:1409–20.
12. Ho WF, Ju CP, Chern Lin JH. Structure and properties of cast Ti–Mo alloys. Biomater. 1999;20:2115–22.
13. Wang K. The use of titanium for medical applications in the USA. Mater Sci Eng A. 1996;213:134–7.
14. Ho WF. A comparison of tensile properties and corrosion behavior of cast Ti–7.5Mo with c.p. Ti, Ti–15Mo and Ti–6Al–4V alloys. J Alloy Compd. 2008;464:580–3.
15. Lin DJ, Chuang CC, Chern Lin JH, Lee JW, Ju CP, Yin HS. Bone, formation at the surface of low modulus Ti–7.5Mo implants in rabbit femur. Biomater. 2007;28:2582–9.
16. Gordin DM, Gloriant T, Texier G, Thibon I, Ansel D, Duval JL, Nagel MD. Development of a β type Ti–12Mo–5Ta alloy for biomedical applications: cytocompatibility and metallurgical aspects. J Mater Sci Mater Med. 2004;15:885–91.
17. Trentania L, Pelilloa F, Pavesia FC, Ceciliania L, Cettab G, Forlino A. Evaluation of the TiMo12Zr6Fe2 alloy or orthopaedic implants: in vitro biocompatibility study by using primary human fibroblasts and osteoblasts. Biomater. 2002;23:2863–9.
18. Karthega M, Raman V, Rajendran N. Influence of potential on the electrochemical behavior of beta titanium. Acta Biomater. 2007;3:1019–23.

19. Zhou YL, Luo DM. Corrosion behavior of Ti–Mo alloys cold rolled and heat treated. J Alloy Compd. 2011;509:6267–72.
20. Hanad S, Izumi O. Deformation behaviour of retained β phase in β-eutectoid Ti-Cr alloys. J Mater Sci. 1986;21:4131–9.
21. Kuan TS, Ahrens RR, Sass SL. The stress-induced omega phase transformation in Ti-V alloys. Metall Trans A. 1975;6:1767–74.
22. Oka M, Taniguchi Y. {332} Deformation twins in a Ti–15.5 pct V alloy. Metall Trans A. 1979;10:651–3.
23. Matsumoto H, Watanabe H, Masahashi N, Handa S. Composition dependence of Young's modulus in Ti–V, Ti–Nb, and Ti–V–Sn alloys. Metall Trans A. 2006;37:3239–49.
24. Donachie MJ. Titanium: a technical guide. 2nd ed. Materials Park: ASM International; 2000. p. 126.
25. Hanawa T, Ota M. Characterization of surface film formed on titanium in electrolyte using XPS. Appl Surf Sci. 1992;55:269–76.
26. Ong JL, Lucas LC, Raikar GN, Connatser R, Gregory JC. Spectroscopic characterization of passivated titanium in a physiologic solution. J Mater Sci Mater Med. 1995;6:113–9.
27. Hanawa T, Asami K, Asaoka K. Repassivation of titanium and surface oxide film regenerated in simulated bioliquid. J Biomed Mater Res. 1998;40:530–8.
28. Hao YL, Li SJ, Sun SY, Zheng CY, Yang R. Elastic deformation behavior of Ti-24Nb-4Zr-7.9Sn for biomedical applications. Acta Biomater. 2007;3:277–86.
29. Zhao XF. Research and development of binary β-type titanium alloys with changeable Young's modulus for spinal fixation applications. PhD thesis. Tohoku University. (2012).
30. Zhao XF, Niinomi M, Nakai M, Hieda J. Beta-type Ti-Mo alloys with changeable Young's modulus for spinal fixation applications. Acta Biomater. 2012;8:1990–7.
31. Zhao XF, Niinomi M, Nakai M, Hieda J. Optimization of Cr content of metastable β-type Ti-Cr alloys with changeable Young's modulus for spinal fixation applications. Acta Biomater. 2012;8:2392–400.
32. Liu HH, Niinomi M, Nakai M, Hieda J, Cho K. Deformation-induced changeable Young's modulus with high strength in β-type Ti-Cr-O alloys for spinal fixture. J Mech Behav Biomed. 2014;30:205–13.
33. Nakai M, Niinomi M, Liu H et al. Fatigue properties of Ti-12Cr alloy with changeable elastic modulus, collected abstracts of 2015 autumn meeting of the Japan Institute of Metals and Materials. (2015). J21.
34. ASTM F1717-15: standard test methods for spinal implant constructs in a vertebrectomy model, ASTM standard. West Conshohocken, PA, ASTM Int.
35. Narita K, Niinomi M, Nakai M, Suyalatu T et al. Improvement of endurance of β-type titanium alloys by cavitation peening treatment. Collected abstracts of 2014 autumn meeting of the Japan Institute of Metals and Materials. (2014). p. 380.
36. Niinomi M, Li H, Nakai M, Liu H, Liu Y. Biomedical titanium alloys with Young's moduli close to that of cortical bone. Regen Biomater. 2016;3:1–13.

Open Access This chapter is distributed under the terms of the Creative Commons Attribution 4.0 International License (http://creativecommons.org/licenses/by/4.0/), which permits use, duplication, adaptation, distribution and reproduction in any medium or format, as long as you give appropriate credit to the original author(s) and the source, provide a link to the Creative Commons license and indicate if changes were made.

The images or other third party material in this chapter are included in the work's Creative Commons license, unless indicated otherwise in the credit line; if such material is not included in the work's Creative Commons license and the respective action is not permitted by statutory regulation, users will need to obtain permission from the license holder to duplicate, adapt or reproduce the material.

Chapter 2
Ceramic Coating of Ti and Its Alloys Using Dry Processes for Biomedical Applications

Takatoshi Ueda, Natsumi Kondo, Shota Sado, Ozkan Gokcekaya,
Kyosuke Ueda, Kouetsu Ogasawara, and Takayuki Narushima

Abstract In this chapter, bioceramic coatings on Ti and its alloys are examined. Surface modification processes of metallic biomaterials are reviewed, and the formation and evaluation of amorphous calcium phosphate (ACP) and anatase-rich TiO_2 coatings are described. Dry processes such as radio-frequency (RF) magnetron sputtering and thermal oxidation are employed for coating. Ag-containing ACP coating films exhibited antibacterial activity through the continuous release of Ag ions, caused by the resorbability of ACP. Anatase-rich TiO_2 layers fabricated by a two-step thermal oxidation process showed photodegradation of organic compounds under both UV- and visible-light irradiation. The introduction of Au into TiO_2 layers from Ti-Au alloy substrates by thermal oxidation contributed to the expression of visible-light response.

Keywords Calcium phosphate • Silver • Antibacterial activity • Anatase • Photocatalytic activity

2.1 Introduction

Ti and its alloys (along with stainless steels and Co-Cr alloys) are recognized as primary metallic biomaterials because of their excellent balance of strength and ductility, along with their high corrosion resistance when forming passivation films, low allergenic nature, and bone compatibility (osseointegration) [1, 2]. However,

---

T. Ueda • N. Kondo • S. Sado • O. Gokcekaya • K. Ueda • T. Narushima (✉)
Department of Materials Processing, Tohoku University,
6-6-02 Aza Aoba, Aramaki, Aoba-ku, Sendai, Miyagi 980-8579, Japan
e-mail: narut@material.tohoku.ac.jp

K. Ogasawara
Department of Immunobiology, Institute of Development, Aging and Cancer,
Tohoku University, 4-1 Seiryo-machi, Aoba-ku, Sendai 980-8579, Japan

Department of Intractable Diseases and Immunology, Tohoku University,
4-1 Seiryo-machi, Aobaku, Sendai 980-8575, Japan

© The Author(s) 2017   23
K. Sasaki et al. (eds.), *Interface Oral Health Science 2016*,
DOI 10.1007/978-981-10-1560-1_2

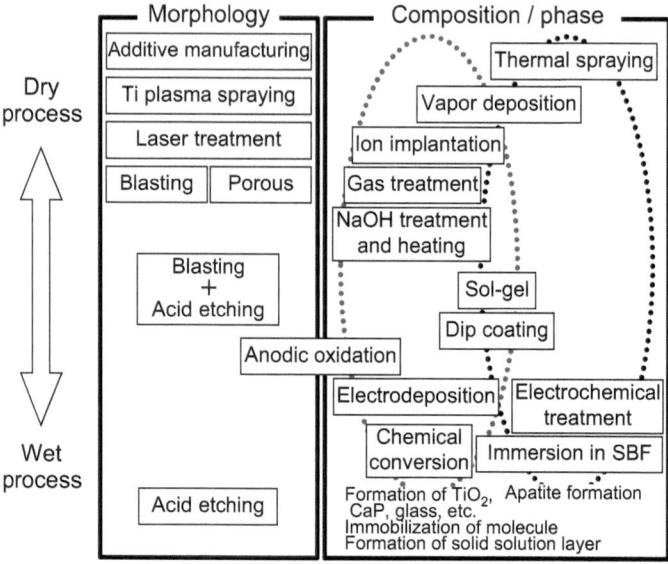

Fig. 2.1 Surface modification processes of metallic biomaterials

biofunctions of metallic biomaterials are insufficient compared to those of ceramic and polymer biomaterials. Some ceramics have chemical compositions close to that of hard tissue and exhibit bioactivity and bioresorbability, while polymers can polymerize to form functional groups and segments for biomedical applications. Therefore, surface modification is necessary to improve the biofunctionality of Ti and its alloys [3, 4]. Metallic biomaterial microstructures can generally be controlled to exhibit desired mechanical properties seen in the bulk material, an advantage they hold over ceramic and polymer biomaterials. Surface modification can in turn impart biofunctionality without impacting the bulk properties.

The different processes used for surface modification of metallic biomaterials are summarized in Fig. 2.1 [5]. These techniques are classified into two groups of wet and dry processes. Wet processes use aqueous solution for steps such as immersion and electrifying, while dry processes are based on the reactions of the surface with gases, ions, and plasma. Surface modification can control the morphology, composition, and/or phase of a given material. Controlling the surface morphology allows for high mechanical bonding between metallic biomaterials and hard tissues (mechanical anchoring) and for control of biological reactions of cells and proteins on the modified surface. Acid etching, blasting, laser treatment, additive manufacturing (AM), and porous treatments using beads, Ti plasma spraying, and fiber mesh coating are various examples of these processes.

Controlling the composition/phase of a surface can be conducted through physical, chemical, and biological processes. Since the primary application of metallic biomaterials is for hard-tissue substitution, bioceramics with excellent bone compatibility can be utilized as agents for surface composition/phase control. Processes such as thermal spraying, vapor deposition, electrochemical treatment, and immersion in simulated body fluid (SBF) have been used to coat apatite, which is an analogue for the inorganic components of hard tissues. In addition, TiO_2, calcium phosphate, sodium titanate, and $CaTiO_3$ coatings formed by anodic oxidation, gas treatment, NaOH treatment and heating, ion implantation, and chemical conversion can improve bone compatibility of implant substrates under biological conditions. Anodic oxidation changes both the chemical composition and roughness of a surface. Hybrid coatings of ceramics and polymers are candidates for further improvements in the biofunctionality of metallic biomaterials. To this end, hybrids such as calcium phosphate+collagen and calcium phosphate+bone morphogenetic protein have been seen in recent study [6–8].

In this chapter, the formation and evaluation of ceramic coatings of Ti and its alloys are described with a focus on their antibacterial potential. Radio-frequency (RF) magnetron sputtering and thermal oxidation are the main processes discussed here for controlling composition/phase of surfaces, which are classified as dry processes. Surgical site infection (SSI) related to implants occurs at rates of 2–30 % depending on the surgical site and the procedure, and SSI is even more frequent after revision surgery [9, 10]. Several approaches are known to lower the risk of SSI, such as the use of antibiotics. However, prolonged use of antibiotics at higher doses can lead to systematic drug resistance and local toxicity [11]. The use of ceramic coatings with antibacterial activity is a powerful method to mitigate SSI.

2.2 Antibacterial Properties of Ag-Containing Amorphous Calcium Phosphate Coating Films

Ag, Cu, and Zn ions are known antibacterial agents. Ag ions are effective against many types of bacteria, are less likely to cause bacterial resistance, and express antibacterial activity at lower concentrations than Cu and Zn ions [12–14]. Here, amorphous calcium phosphate (ACP) coating films on Ti and its alloys are fabricated by RF magnetron sputtering and evaluated in vivo and in vitro [15–17]. Coating of implants with Ag-containing ACP is one possible technique for preventing infection; ACP possesses bioresorbability under biological conditions, which causes the continuous release of Ag ions from such Ag-containing films, creating antibacterial activity for a desired duration. The addition of Ag to HAp (hydroxyapatite, $Ca_{10}(PO_4)_6(OH)_2$) has been studied previously [18–21], but HAp is less resorbable than ACP. Both antibacterial activity and improved bone forming are expected from implants coated with Ag-containing ACP.

Ag-containing β-TCP (β-type tricalcium phosphate, $Ca_3P_2O_8$) sintered body was fabricated by hot-pressing Ag and β-TCP raw powders as targets in RF magnetron

Table 2.1 Notation and composition of sputtering targets (mass%)

	Composition	
Notation	Ag	β-TCP
0AgTCP	0	100
15AgTCP	15	85
30AgTCP	30	70

sputtering. The Ag content in Ag-containing β-TCP targets was 0, 15, and 30 mass% as listed in Table 2.1. Plates ($10 \times 10 \times 1$ mm) of mirror-polished commercially pure (CP) Ti (grade 2) and blasted Ti-6mass%-Al-4mass%V (Ti-6Al-4V) alloy were used as substrates. The thickness of the coating films for analysis and evaluation of antibacterial properties was fixed at 0.5 μm, as controlled by adjusting the deposition time.

Figure 2.2 shows X-ray diffraction (XRD) patterns of the coating films formed on mirror-polished CP Ti substrates using a 15mass%-Ag-containing TCP (15AgTCP) target. A broad peak is present at a 2θ value of $25–35°$, which is a characteristic of ACP coating films [15]. Cross-sectional scanning electron microscopy (SEM) images of coating films on mirror-polished CP Ti substrates are present in Fig. 2.3. The coating films are dense and uniform, with good adhesion to the substrates. The deposition rates were $0.02–0.1$ nm·s^{-1}, depending on the target composition and RF power.

The Ag contents of the coating films formed using 15AgTCP and 30AgTCP targets were 2 and 15 mass%, respectively, lower than those of the targets. Lower Ag contents compared to target compositions have been previously reported in Ag-containing HAp coating films [22]. These discrepancies may be linked to differences in the ionization rates between elements during sputtering.

E. coli and *S. aureus*, Gram-negative and Gram-positive bacteria, respectively, were used for antibacterial testing. Ag-containing ACP coating films on blasted Ti-6Al-4V substrates were shaking cultured at 200 rpm in 1/500 nutrient broth (NB) solution with an initial bacterial concentration of 1×10^7 CFU (colony-forming unit)·mL^{-1} for *E. coli* and 1×10^5 CFU·mL^{-1} for *S. aureus*. Solution temperature was maintained at 310 K and incubation was carried out for 10.8 and 86.4 ks. Afterward, the numbers of viable bacteria colonies were counted using a smear-plate culture method.

Figure 2.4 shows the relationship between the incubation time and the number of viable bacteria for coating films formed on blasted Ti-6Al-4V substrates using a 15AgTCP target. After incubation for 10.8 ks, the number of viable *E. coli* was less than 1, which was plotted as 10^0 CFU. The number of viable *S. aureus* decreased after incubation for 10.8 ks and became less than 1 after incubation for 86.4 ks.

The amounts of Ca, P-related, and Ag ions after immersion of Ag-containing ACP coating films in the 1/500 NB solution with shaking at 200 rpm are depicted in Fig. 2.5. Ag ions elute from the Ag-containing ACP coating films and exhibit antibacterial activity. The Ag ion concentration was almost constant over all measured immersion times, caused by the formation of AgCl between the eluted Ag$^+$ and Cl$^-$ present in the 1/500 NB solution.

Fig. 2.2 XRD patterns of coating films fabricated on mirror-polished CP Ti substrates at RF powers of 50, 100, and 150 W using a 15AgTCP target

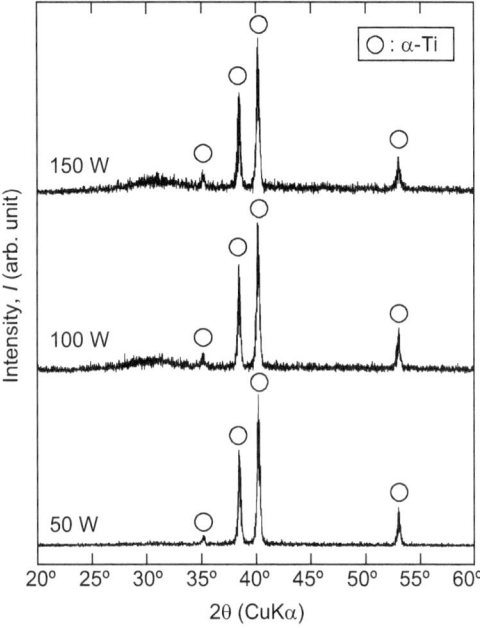

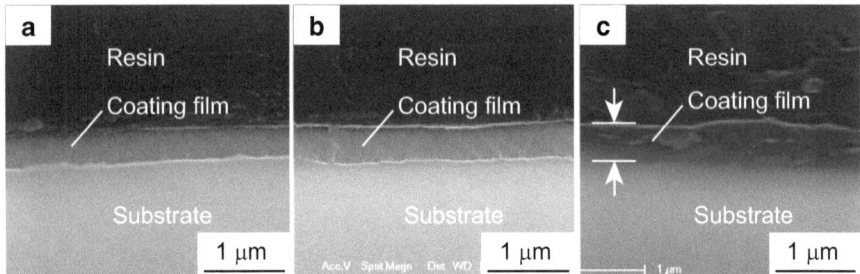

Fig. 2.3 Cross-sectional SEM image of coating films fabricated on mirror-polished CP Ti substrates using (**a**) 0AgTCP, (**b**) 15AgTCP, and (**c**) 30AgTCP targets

2.3   Photocatalytic Activity of TiO₂ Layers Formed by Two-Step Thermal Oxidation

2.3.1   UV (Ultraviolet) Response

TiO₂ layers have been used as surface coatings of Ti implants [23], as TiO₂ exhibits photocatalytic activity under UV-light irradiation, such as photoinduced superhydrophilicity and photodegradation of organic compounds [24]. The photoinduced superhydrophilicity of TiO₂ layers on Ti improves bone compatibility [25, 26], and photodegradation contributes to the antibacterial activity of Ti implants by killing

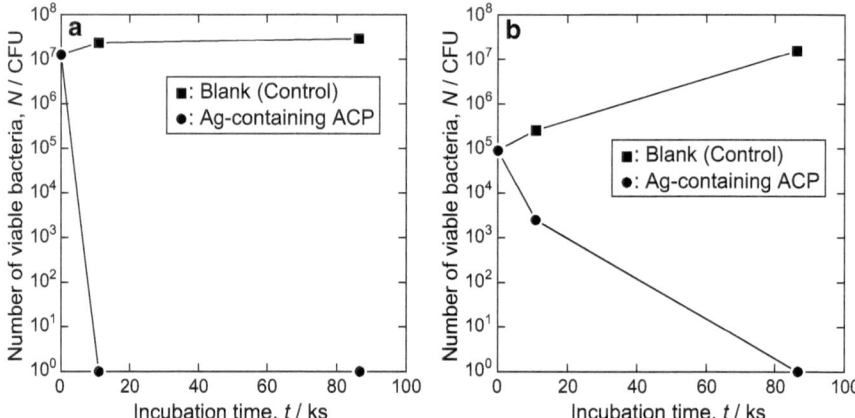

Fig. 2.4 Number of viable (**a**) *E. coli* and (**b**) *S. aureus* after incubation with Ag-containing ACP coating films fabricated on blasted Ti-6Al-4V substrates using a 15AgTCP target

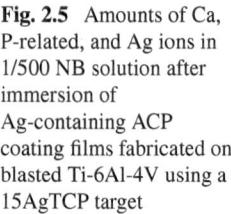 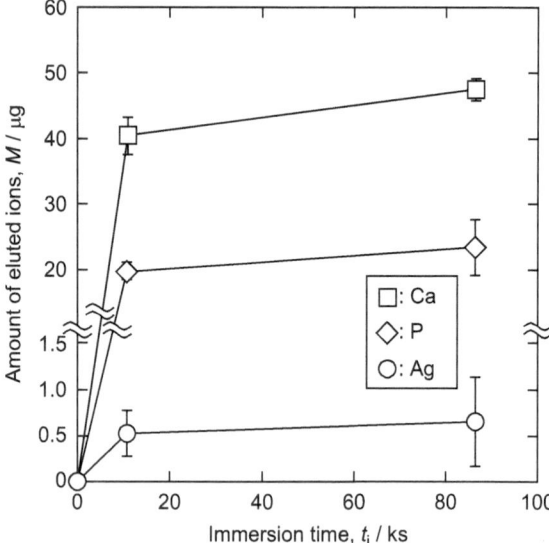

Fig. 2.5 Amounts of Ca, P-related, and Ag ions in 1/500 NB solution after immersion of Ag-containing ACP coating films fabricated on blasted Ti-6Al-4V using a 15AgTCP target

nearby bacteria through radical formation [27, 28]. Rutile is a thermodynamically stable phase of TiO_2 with a band gap of 3.0 eV, while anatase is a metastable phase of TiO_2 having a band gap of 3.2 eV. The lifetimes of these electronic excitations are an order of magnitude larger for anatase as compared to rutile [29], and as such the higher photocatalytic activity is expected in anatase.

The present authors presented a two-step thermal oxidation process for the fabrication of anatase-rich TiO_2 layer on Ti and its alloys [30–32], consisting of initial

Fig. 2.6 Effects of the anatase fraction in the TiO₂ layers on the rate constant of methylene blue degradation under UV-light irradiation for CP Ti and the Ti-25mass%Mo and Ti-25mass%Nb alloys [Reprinted from Ref. [32] with permission from Elsevier]

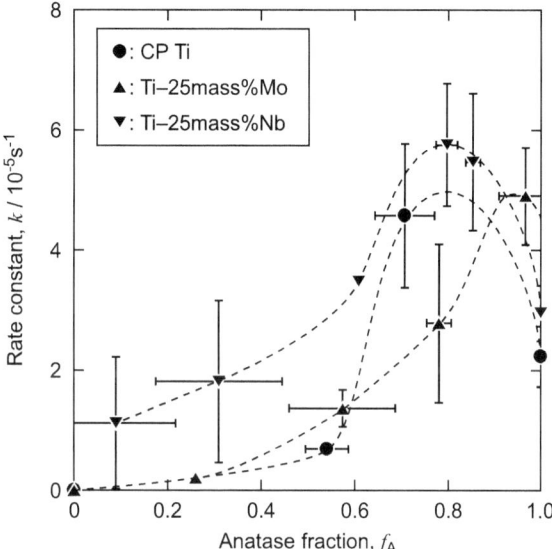

treatment under CO-containing atmosphere and subsequent treatment in air, and evaluated the photocatalytic activity of anatase-containing TiO₂ layers on CP Ti and Ti-25mass%Mo and Ti-25mass%Nb alloys [32]. The anatase fraction in the TiO₂ layer can be controlled during this process by controlling the temperature of the second step. By using TiO₂ layers with various anatase fractions, the degradation rate of methylene blue (MB) under UV-light irradiation was measured. Illumination was performed using a UV lamp with a central wavelength of 351 nm. The UV-light intensity was 1.0 mW·cm⁻² at the surface of the specimen. The effects of anatase fraction on the MB degradation rates are summarized in Fig. 2.6 [32], where the maximum rate was obtained at an anatase fraction of approximately 0.8 for both CP Ti and its alloys. Su et al. [33] measured the MB degradation rates of TiO₂ films on CP Ti formed by anodic oxidation, where the grain sizes, surface areas, and crystallinity of the TiO₂ films were fixed, but the anatase fraction was varied. They reported a maximum decomposition rate at an anatase fraction of approximately 0.6 and indicated that electrons that are photoexcited in rutile can migrate to the conduction band of anatase to leave electron holes in the rutile; as a result, recombination is effectively suppressed in anatase-rich TiO₂ layers. In our case, the second step temperature was altered to control the resulting anatase fraction, but the thickness and surface roughness were not controlled. Nevertheless, a similar dependence of decomposition rate upon the anatase fraction was obtained compared to Su et al.; these anatase-rich TiO₂ layers show excellent photodegradation of organic compounds.

2.3.2 Visible-Light Response

The visible-light-responsive photocatalytic activity of implants is useful for anti-bacterial activity during operation and in the reactivation of implants. Because SSI is caused by endogenous flora in the patient's skin, mucous membranes, or hollow viscera, SSI cannot simply be prevented by sterilization before operation [34]. In order to express the photocatalytic activity of TiO_2, UV-light irradiation is generally required, which is harmful to the human body. Therefore, visible-light response is more desirable for preventing infection during operation. Reactivation of implants is required during usage for some periods in a human body. While UV-light irradiation is known to reactivate Ti implant surfaces [35], reactivation using visible-light irradiation is more favorable for the human body.

Au solutes in TiO_2 [36] and the addition of metallic Au nanoparticles to TiO_2 [36, 37] can lead to visible-light response. Therefore, two-step thermal oxidation was applied to a Ti-Au alloy in order to fabricate Au-containing TiO_2 layers with visible-light-responsive photocatalytic activity. Ti-4.2at%Au alloy was employed as a substrate material. The first step treatment was conducted under an Ar-1%CO atmosphere at 1073 K for 3.6 ks, and the subsequent second step treatment was conducted in air at 673 K for 10.8 ks. An anatase-rich TiO_2 layer was obtained on the Ti-4.2at%Au alloy substrate after this process. Figure 2.7 depicts the cross-sectional microstructure of the anatase-rich TiO_2 layer as observed via transmission electron microscopy (TEM). Au nanoparticles can be seen here as small black particles approximately 5 nm in diameter.

The photocatalytic activity of the Au-containing TiO_2 layer was evaluated by self-cleaning tests (JIS R 1753: 2013) based on the decomposition of a layer of steric acid on the Au-containing TiO_2 layer under visible-light irradiation, where the steric acid layer was prepared by dip coating. Substrates with this steric acid layer were irradiated with visible light with an intensity of 10 mW·cm^{-2} at the surface of the steric acid layer, for as long as 86.4 ks. The light source was a Xe lamp with a UV filter to produce light of $\lambda > 400$ nm. Changes in the water contact angles under visible-light irradiation were measured to determine the extent of steric acid photo-degradation, and the TiO_2 layer expressed visible-light-responsive photocatalytic activity.

Figure 2.8 shows the changes in water contact angles on as-polished and two-step (i.e., after two-step thermal oxidation) CP Ti and Ti-4.2at%Au substrates under visible-light irradiation [38]. The water contact angles on the as-polished specimens remained almost constant, indicating no visible-light response. On the other hand, the two-step specimens showed decreased water contact angles. For the two-step Ti-4.2at%Au alloy, the water contact angle decreased from 80 to 10°. Silva et al. [37] reported that electrons in Au nanoparticles (1.87–6.40 nm in diameter) attached on TiO_2 particles were excited by surface plasmon resonance and migrated to TiO_2 particles, producing H_2 by the photoreduction of H_2O. The Au particle size recommended in their study is close to that of our study (Fig. 2.7). One possible mechanism of visible-light response in the Au-containing TiO_2 layer formed by two-step thermal oxidation involves the decrease in band gap due to Au solutes in the TiO_2 layer [36] and/or the surface plasmon resonance of the Au nanoparticles in the TiO_2 layer [37].

Fig. 2.7 Cross-sectional TEM image of Ti-4.2at%Au alloy after two-step thermal oxidation

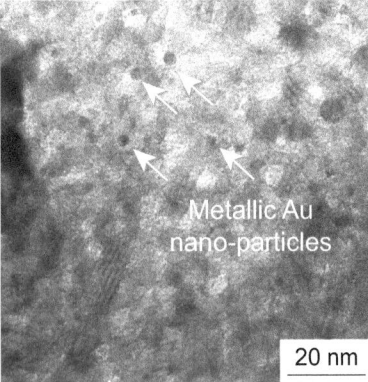

Fig. 2.8 Measured water contact angles on the as-polished and two-step CP Ti and Ti-4.2at%Au alloy substrates during visible-light irradiation

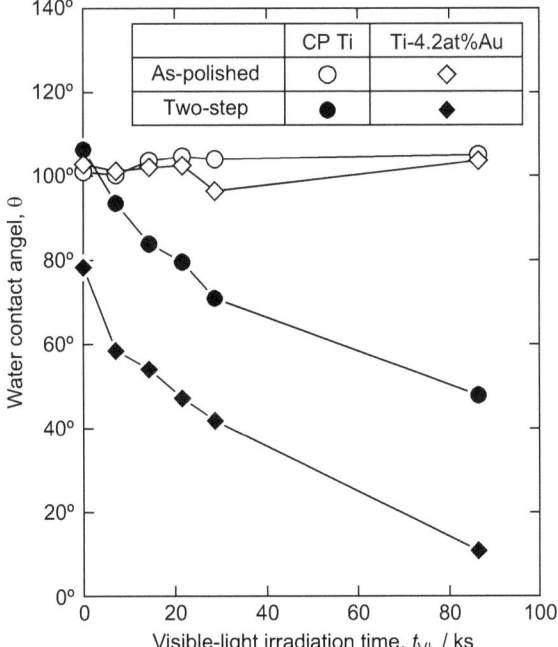

2.4 Summary

The surface modification of Ti and its alloys using ceramic coatings and their antibacterial activity were described in this chapter. Both Ag-containing ACP coating films and Au-containing TiO_2 layers, fabricated by RF magnetron sputtering and two-step thermal oxidation, respectively, indicated antibacterial activity. Control over Ag ion elution from Ag-containing ACP coating films and the improved visible-light response of Au-containing TiO_2 layers will be discussed in future publications.

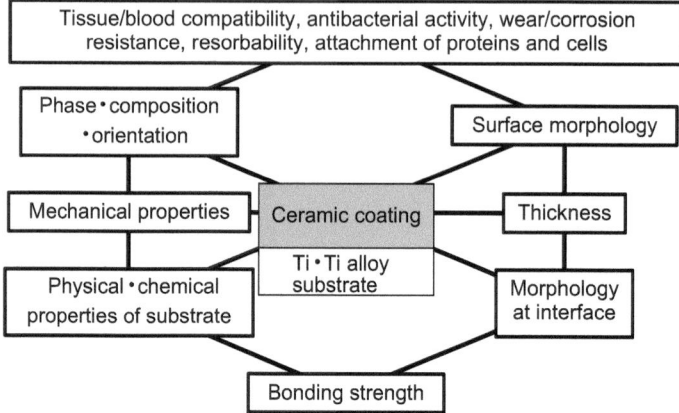

Fig. 2.9 Factors in biomedical ceramic coatings on Ti and its alloy substrates

Ti and its alloys are expected to see continued use as metallic biomaterials for implants. Figure 2.9 shows factors that must be considered when developing ceramic coatings on Ti and its alloy substrates for biomedical applications [5]. The interface between a coating film and the human body requires designed properties such as hard/soft tissue and blood compatibilities, antibacterial activity, wear and corrosion resistance, resorbability, and suppression/enhancement of protein and cell attachment. These properties are affected by surface morphology, phase, composition, and orientation of the coating films. High-strength and durable bonding between a coating film and substrates, which are affected by physical and chemical properties of the substrates and the morphology at the interface, are also crucial. Therefore, the mechanical properties and thickness of coating films, as well as the reaction between coating film and substrates, must be taken into consideration.

Acknowledgments This study was supported in part by a Grant-in-Aid for Scientific Research from the Ministry of Education, Culture, Sports, Science and Technology (MEXT), Japan (Nos. 25249094 and 26709049).

References

1. Narushima T. Titanium and its alloys as biomaterials. J Jpn Light Metals. 2005;55:561–5.
2. Niinomi M, Hanawa T, Narushima T. Japanese research and development on metallic biomedical, dental, and healthcare materials. JOM. 2005;57(4):18–24.
3. Narushima T. Surface modification for improving biocompatibility of titanium materials with bone. J Jpn Light Metals. 2008;58:577–82.
4. Hanawa T. Biofunctionalization of titanium for dental implant. Jpn Dent Sci Rev. 2010;46:93–101.
5. Goto T, Narushima T, Ueda K. Bio-ceramic coating on titanium by physical and chemical vapor deposition. In: Zhang S, editor. CRC handbook of biological and biomedical coatings. Boca Raton: CRC Press; 2011. p. 299–332.

6. de Jonge LT, Leeuwenburgh SCG, van den Beucken JJJP, te Riet J, Daamen WF, Wolke JGC, Scharnweber D, Jansen JA. The osteogenic effect of electrosprayed nanoscale collagen/calcium phosphate coatings on titanium. Biomaterials. 2010;31:2461–9.
7. Xie CM, Lu X, Wang KF, Meng FZ, Jiang O, Zhang HP, Zhi W, Fang LM. Silver nanoparticles and growth factors incorporated hydroxyapatite coatings on metallic implant surfaces for enhancement of osteoinductivity and antibacterial properties. ACS Appl Mater Interfaces. 2014;6:8580–9.
8. Ramazanoglu M, Lutz R, Ergun C, von Wilmowsky C, Nkenke E, Schlegel KA. The effect of combined delivery of recombinant human bone morphogenetic protein-2 and recombinant human vascular endothelial growth factor 165 from biomimetic calcium-phosphate-coated implants on osseointegration. Clin Oral Implants Res. 2011;22:1433–9.
9. Hickok NJ, Shapiro IM. Immobilized antibiotics to prevent orthopaedic implant infections. Adv Drug Deliv Rev. 2012;64:1165–76.
10. Haenle M, Fritsche A, Zietz C, Bader R, Heidenau F, Mittelmeier W, Gollwitzer H. An extended spectrum bactericidal titanium dioxide (TiO$_2$) coating for metallic implants: in vitro effectiveness against MRSA and mechanical properties. J Mater Sci Mater Med. 2011;22:381–7.
11. Goodman SB, Yao Z, Keeney M, Yang F. The future of biologic coatings for orthopaedic implants. Biomaterials. 2013;34:3174–83.
12. Ferraris S, Spriano S. Antibacterial titanium surfaces for medical implants. Mater Sci Eng C. 2016;61:965–78.
13. Du W-L, Niu S-S, Xu Y-L, Xu Z-R, Fan C-L. Antibacterial activity of chitosan tripolyphosphate nanoparticles loaded with various metal ions. Carbohydr Polym. 2009;75:385–9.
14. Monteiro DR, Gorup LF, Takamiya AS, Ruvollo-Filho AC, de Camargo ER, Barbosa DB. The growing importance of materials that prevent microbial adhesion: antimicrobial effect of medical devices containing silver. Int J Antimicrob Agents. 2009;34:103–10.
15. Narushima T, Ueda K, Goto T, Masumoto H, Katsube T, Kawamura H, Ouchi C, Iguchi Y. Preparation of calcium phosphate films by radiofrequency magnetron sputtering. Mater Trans. 2005;46:2246–52.
16. Ueda K, Narushima T, Goto T, Taira M, Katsube T. Fabrication of calcium phosphate films for coating on titanium substrates heated up to 773 K by RF magnetron sputtering and their evaluations. Biomed Mater. 2007;2:S160–6.
17. Yokota S, Nishiwaki N, Ueda K, Narushima T, Kawamura H, Takahashi T. Evaluation of thin amorphous calcium phosphate coatings on titanium dental implants deposited using magnetron sputtering. Implant Dent. 2014;23:343–50.
18. Chen W, Liu Y, Courtney HS, Bettenga M, Agrawal CM, Bumgardner JD, Ong JL. In vitro anti-bacterial and biological properties of magnetron co-sputtered silver-containing hydroxyapatite coating. Biomaterials. 2006;27:5512–7.
19. Bai X, More K, Rouleau CM, Rabiei A. Functionally graded hydroxyapatite coatings doped with antibacterial components. Acta Biomater. 2010;6:2264–73.
20. Trujillo NA, Oldinski RA, Ma H, Bryers JD, Williams JD, Popat KC. Antibacterial effects of silver-doped hydroxyapatite thin films sputter deposited on titanium. Mater Sci Eng C. 2012;32:2135–44.
21. Jelinek M, Kocourek T, Remsa J, Weiserová M, Jurek K, Mikšovský J, Strnad J, Galandáková A, Ulrichová J. Antibacterial, cytotoxicity and physical properties of laser – silver doped hydroxyapatite layers. Mater Sci Eng C. 2013;33:1242–6.
22. Ivanova AA, Surmeneva MA, Tyurin AI, Pirozhkova TS, Shuvarin IA, Prymak O, Epple M, Chaikina MV, Surmenev RA. Fabrication and physico-mechanical properties of thin magnetron sputter deposited silver-containing hydroxyapatite films. Appl Surf Sci. 2016;360:929–35.
23. Degidi M, Nardi D, Piattelli A. 10-year follow-up of immediately loaded implants with TiUnite porous anodized surface. Clin Implant Dent Relat Res. 2012;14:828–38.
24. Fujishima A, Zhang X, Tryk DA. TiO$_2$ photocatalysis and related surface phenomena. Surf Sci Rep. 2008;63:515–82.

25. Aita H, Hori N, Takeuchi M, Suzuki T, Yamada M, Anpo M, Ogawa T. The effect of ultraviolet functionalization of titanium on integration with bone. Biomaterials. 2009;30:1015–25.
26. Sawase T, Jimbo R, Baba K, Shibata Y, Ikeda T, Atsuta M. Photo-induced hydrophilicity enhances initial cell behavior and early bone apposition. Clin Oral Implants Res. 2008;19:491–6.
27. Sunada K, Watanabe T, Hashimoto K. Studies on photokilling of bacteria on TiO_2 thin film. J Photochem Photobiol A Chem. 2003;156:227–33.
28. Sousa VM, Manaia CM, Mendes A, Nunes OC. Photoinactivation of various antibiotic resistant strains of Escherichia coli using a paint coat. J Photochem Photobiol A Chem. 2013;251:148–53.
29. Xu M, Gao Y, Moreno EM, Kunst M, Muhler M, Wang Y, Idriss H, Wöll C. Photocatalytic activity of bulk TiO_2 anatase and rutile single crystals using infrared absorption spectroscopy. Phys Rev Lett. 2011;106:138302.
30. Okazumi T, Ueda K, Tajima K, Umetsu N, Narushima T. Anatase formation on titanium by two-step thermal oxidation. J Mater Sci. 2011;46:2998–3005.
31. Umetsu N, Sado S, Ueda K, Tajima K, Narushima T. Formation of anatase on commercially pure Ti by two-step thermal oxidation using N_2-CO gas. Mater Trans. 2013;54:1302–7.
32. Sado S, Ueda T, Ueda K, Narushima T. Formation of TiO_2 layers on CP Ti and Ti–Mo and Ti–Nb alloys by two-step thermal oxidation and their photocatalytic activity. Appl Surf Sci. 2015;357:2198–205.
33. Su R, Bechstein R, Sø L, Vang RT, Sillassen M, Esbjörnsson B, Palmqvist A, Besenbacher F. How the anatase-to-rutile ratio influences the photoreactivity of TiO_2. J Phys Chem C. 2011;115:24287–92.
34. Mangram AJ, Horan TC, Pearson ML, Silver LC, Jarvis WR. Guideline for prevention of surgical site infection, 1999. Infect Control Hosp Epidemiol. 1999;20:247–78.
35. Att W, Hori N, Iwasa F, Yamada M, Ueno T, Ogawa T. The effect of UV-photofunctionalization on the time-related bioactivity of titanium and chromium–cobalt alloys. Biomaterials. 2009;30:4268–76.
36. Li XZ, Li FB. Study of Au/Au^{3+}-TiO_2 photocatalysts toward visible photooxidation for water and wastewater treatment. Environ Sci Technol. 2001;35:2381–7.
37. Silva CG, Juárez R, Marino T, Molinari R, García H. Influence of excitation wavelength (UV or visible light) on the photocatalytic activity of titania containing gold nanoparticles for the generation of hydrogen or oxygen from water. J Am Chem Soc. 2011;133:595–602.
38. Ueda T, Sado S, Ueda K, Narushima T. Formation of TiO_2 layers on Ti-Au alloy using two-step thermal oxidation and their visible-light photocatalytic activity. Mater Lett. in press.

Open Access This chapter is distributed under the terms of the Creative Commons Attribution 4.0 International License (http://creativecommons.org/licenses/by/4.0/), which permits use, duplication, adaptation, distribution and reproduction in any medium or format, as long as you give appropriate credit to the original author(s) and the source, provide a link to the Creative Commons license and indicate if changes were made.

The images or other third party material in this chapter are included in the work's Creative Commons license, unless indicated otherwise in the credit line; if such material is not included in the work's Creative Commons license and the respective action is not permitted by statutory regulation, users will need to obtain permission from the license holder to duplicate, adapt or reproduce the material.

Chapter 3
Dealloying Toxic Ni from SUS316L Surface

Hidemi Kato, Takeshi Wada, and Sadeghilaridjani Maryam

Abstract SUS316L (18 %Cr-12 %Ni-2.5 %Mo-<0.03 %C) is one of the most pop-ular metallic materials for biomedical uses such as implants and surgical instru-ments owing to its lower cost, good corrosion resistance, and excellent workability compared with those of other types of biomedical metallic materials. Although Ni stabilizes the austenitic FCC structure to give the excellent workability of SUS316L, elution of toxic Ni ions into the human body can cause various health problems. In SUS316L, the elution of Ni ions is suppressed by the stable passive thin film arising from the Cr component. However, there remains some risk of Ni elution. In this paper, we remove Ni element from the surface of SUS316L using a novel dealloy-ing technique with a metallic melt and study its effect on the corrosion resistance and Ni ion elution of the steel in a simulated human body fluid.

Keywords SUS316L • Biomedical material • Ni • Dealloying • Surface improvement

3.1 Introduction

SUS316L (18 % Cr, 12 % Ni, 2.5 % Mo, <0.03 %C) is an austenitic stainless steel that contains 18 wt% Cr, 12 wt% Ni, 2.5 % Mo, and less than 0.03 % C (the "L" in 316L standing for low carbon), which tends to degrade the stability of the face-centered cubic (FCC) structure and segregate as carbides in grain boundaries, caus-ing intergranular cracking. Owing to these elements, SUS316L has the advantages of high corrosion resistance, excellent workability, low cost, and highly reliable mechanical properties and has thus been widely used as a biomedical material, such as in surgical instruments and bone fixtures. The Ni element in SUS316L stabilizes

H. Kato (✉) • T. Wada
Institute for Materials Research, Tohoku University,
2-1-1 Katahira, Aoba-ku, Sendai, Miyagi 980-8577, Japan
e-mail: hikato@imr.tohoku.ac.jp

S. Maryam
Department of Materials Science, Tohoku University, Sendai, Miyagi, Japan

© The Author(s) 2017
K. Sasaki et al. (eds.), *Interface Oral Health Science 2016*,
DOI 10.1007/978-981-10-1560-1_3

35

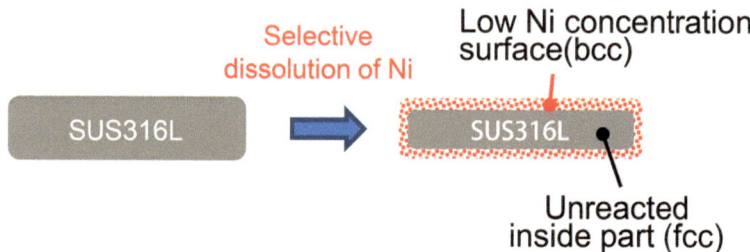

Fig. 3.1 Strategy of the present research for improving the biocompatibility of SUS316L

the FCC structure to produce excellent workability. However, Ni is a toxic element, meaning that using SUS316L potentially risks causing health problems. Therefore, Ni is necessary for workability but is unwanted for the biocompatibility of SUS316L. To solve this problem, we investigated the removal of Ni element from the surface of SUS316L using a novel dealloying technique, "dealloying in metallic melt," as shown in Fig. 3.1. This work intends to demonstrate the dealloying of Ni element from the SUS316L surface and investigate its effects on the corrosion resistance and Ni ion release of the SUS316L in Hank's balanced salt solution (HBSS).

3.1.1 Dealloying in a Metallic Melt

This method is first reported in Ref. [1], and its principle was mentioned in Ref. [2] as follows. When two metal elements are mixed, the free energy change is

$$\Delta G_{mix} = \Delta H_{mix} - T\Delta S_{mix}, \tag{3.1}$$

where ΔH_{mix} is the heat of mixing, ΔS_{mix} is the entropy of mixing, and T is the absolute temperature. Usually, the entropy increases after mixing. Therefore, if $\Delta H_{mix} < 0$, then $\Delta G_{mix} < 0$, and the mixing reaction can occur spontaneously from a thermodynamic point of view. Conversely, if $\Delta H_{mix} > 0$, the sign (positive or negative) of ΔG_{mix} depends on the temperature. If the temperature is adequately controlled to make the enthalpy term larger than the entropy term, then $\Delta G_{mix} > 0$, and the mixture of the two elements is prevented. Here, we dip an A–B binary alloy precursor into a metallic melt consisting of element C. If the heat of mixing between elements B and C is negative, i.e., $\Delta H_{mix,B-C} < 0$ and if the heat of mixing between elements A and C is positive, i.e., $\Delta H_{mix,A-C} > 0$, then by adequately controlling the temperature, only element B will dissolve from the precursor into the C melt. Because in this case element A is rejected from the C melt, it is expected to self-organize into a porous structure by surface diffusion in the same manner as that of the ordinary dealloying method in aqueous solution [3, 4]. Figure 3.2 shows a schematic of this novel dealloying method involving the selective dissolution of B atoms (orange) in the C atom melt (pink) and surface diffusion of the remaining A atoms

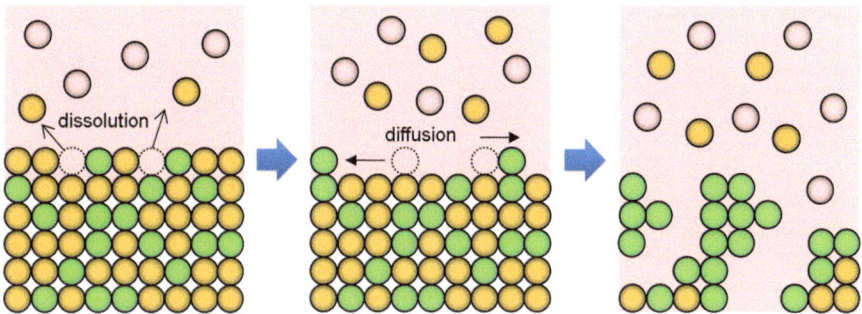

Fig. 3.2 Schematic of the dealloying method using a metallic melt, in which B atoms (*orange*) dissolve into a melt composed of C atoms (*pink*), and the remaining A atoms (*yellowish green*) self-organizes into a porous structure by surface diffusion [3]

Fig. 3.3 Triangle relationship of the enthalpies of mixing among elements A, B, and C for dealloying in a metallic melt [1]

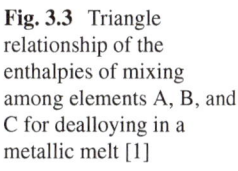
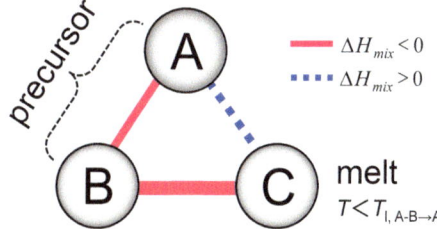

(yellowish green). Figure 3.3 summarizes this "triangle" relationship in terms of the heat of mixing among elements A, B, and C required for the dealloying reaction in a metallic melt. Because the heat of mixing is usually expressed using temperature and chemical composition, we must calculate a rigid value for the heat of mixing with these parameters to design the dealloying reaction. However, this can be complicated. The heat of mixing between the transition metals and between the transition metals and metalloids can be obtained from the table in Ref. [5], the values of which were approximately calculated by the Miedema model, while that of other metals can be obtained from the table constructed by Takeuchi et al. In this study, we first identified the candidates for elements A, B, and C from these values and then confirmed the relationships A–B, B–C (mixture), and A–C (separation) using the related binary phase diagrams if they were available in relevant databases.

3.1.2 Reaction Design for Dealloying Ni from SUS316L

SUS316L consists of Fe, Cr, Mo, and Ni. Mg is miscible with Ni but immiscible with the other elements, which can be confirmed from the corresponding mixing enthalpies: $\Delta H_{mix}= -4$ kJ/mol (Ni-Mg), +18 kJ/mol (Fe-Mg), +24 kJ/mol (Cr-Mg), and +36 kJ/mol (Mo-Mg) [5]. Thus, as shown in Fig. 3.4, Fe, Cr, and Mo are

Fig. 3.4 Reaction design for removing Ni element from the surface of SUS316L by dealloying in a metallic melt. Here, the carbon component is ignored owing to its low concentration. Based on the mixing enthalpy between each component, Fe, Cr, Mo=A, Ni=B and Mg=C group in Fig. 3.3

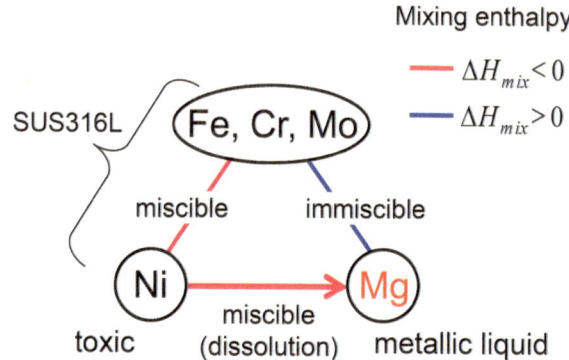

regarded as "A" group elements in the triangle relationship in Fig. 3.3, Ni is regarded as a "B" group element, and Mg is regarded as a "C" group element. Therefore, if the temperature is adequately controlled, we can expect that only Ni will dissolve from the surface when SUS316L is dipped into liquid Mg, resulting in the construction of a Fe-Cr-Mo porous alloy in the Mg melt.

3.2 Experimental Procedure

The method used for sample preparation is schematically illustrated in Fig. 3.5. The SUS316L used in this research was a commercial product purchased from Nilaco, Japan, in the form of rods of 10 mm in diameter. The rods were sliced into disks of 1 mm thickness, and these were then cold rolled to ~200 μm thickness. Mg was inductively melted in a chamber under inert argon gas using a graphite crucible.

The temperature of the Mg liquid bath was set to 700 or 900 °C, and the cold rolled SUS316L was dipped for 30 min. The resulting dealloyed sample was then dipped in an aqueous solution of HNO_3 to etch away the surrounding Mg-Ni phase, followed by washing and drying. The treated samples were evaluated using scanning electron microscopy-energy dispersive X-ray analysis (SEM-EDX).

Polarization tests were conducted in HBSS solution with a potentiostat/galvanostat (HZ5000, Hokuto Denko, Japan) within a voltage range of −800 to 500 mV vs the Ag/AgCl reference electrode. A pure platinum electrode was used as a counter electrode.

The ion release test was conducted in 1.7 g of HBSS solution at 310 K for 10 days, after which the amount of ions released was evaluated. The HBSS was then replaced with the same weight of fresh solution, and the test was continued for another 14 days. The ion release of Fe, Cr, and Ni was analyzed by inductively coupled plasma optical emission spectrometry (ICP-OES), the detection limit of which was ~0.1 μg.

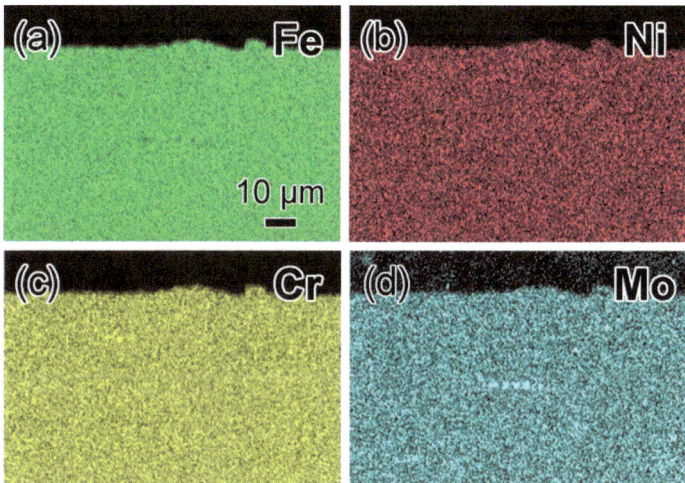

Fig. 3.9 Element mappings for Fe (**a**), Cr (**b**), Ni (**c**), and Mo (**d**) for the polished cross section of SUS316L dealloyed in Mg melt at 900 °C for 30 min

much lower than in the untreated part. Quantitative analysis of point A in the reacted porous region and point B in the solid inner revealed respective compositions of $Fe_{76.4}Cr_{20.0}Ni_{2.2}Mo_{1.4}$ (at. %) and $Fe_{68.9}Cr_{18.1}Ni_{11.8}Mo_{1.2}$ (at.%). The Ni concentration at point A point decreased to 2.2 at.%, which is less than 19 % of that at B point in the nonreacted inside. These results indicate that in the Mg melt at 700 °C, almost only the Ni element was removed from the SUS316L by a dealloying reaction, leaving a constricted porous structure.

3.3.2 Polarization Tests

Figure 3.12 shows the polarization curve of the SUS316L dealloyed at 700 °C for 30 min compared with that of the original SUS316L after cold rolling. After the dealloying, the corrosion potential of the SUS316L decreased from −240 to −400 mV, indicating that the dealloying treatment increased the corrodibility of the sample. However, after the corrosion potential, the passive region was maintained up until almost the same potential as that of the original sample. Therefore, it was concluded that the excellent corrosion resistance of the SUS316L was not degraded by the dealloying treatment.

Fig. 3.10 SEM image of surface (**a**), polished cross section of SUS316L dealloyed in a Mg melt at 700 °C for 30 min (**b**), and enlarged image of the reacted surface part (**c**). *Red dotted line* in (**b**) indicates the initial dimensions of the cross section before the dealloying treatment

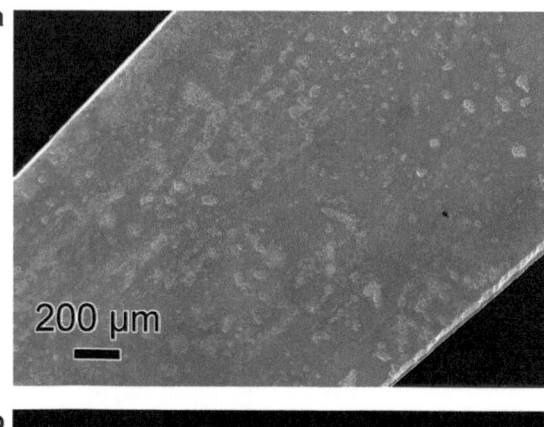

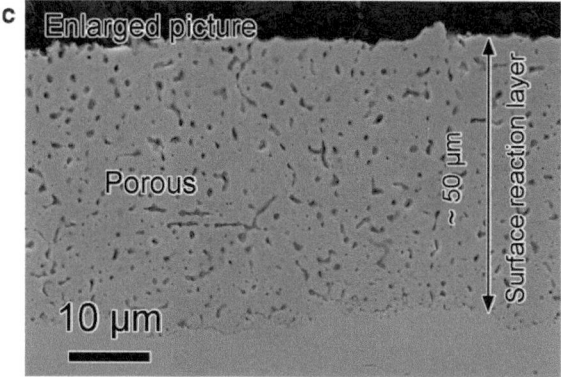

3.3.3   Ion Release Tests

For ion release tests, the original and dealloyed samples were dipped in 1.7 g of HBSS at 37 °C for 10 days, after which the HBSS was replaced with fresh solution and the immersion was continued for another 14 days. Herein, HBSS without samples was used as a reference. The observed ion release of Fe, Cr, and Ni is

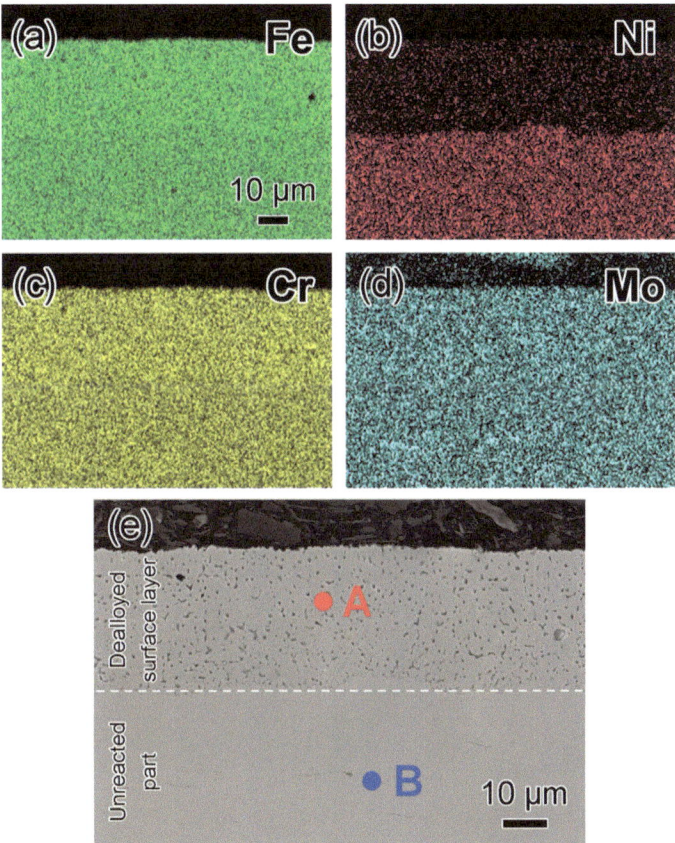

Fig. 3.11 Elemental mappings for Fe (**a**), Ni (**b**), Cr (**c**), and Mo (**d**) for the polished cross section of SUS316L dealloyed in Mg melt at 700 °C for 30 min, and SEM image taken near the boundary of the dealloyed and unreacted regions (**e**). *Red* and *blue dots* marked A and B in the reacted region and unreacted region, respectively, in (**e**) are the points at which quantitative EDX analysis was conducted

summarized in Fig. 3.13. During the first 10 days, the Fe ion release from both the original and dealloyed samples was comparatively large, but then decreased over the next 10–24 days. Little Cr release was observed, just within the detection limit, because surface Cr is considered to construct a passivated Cr hydroxide thin film on SUS316L. Ni ion release from the dealloyed sample, which is the most important for us to control, was found to be slightly increased compared with that of the original sample, especially during days 10–24. However, it then decreased after 24 days.

The main reasons for the increased Ni ion release were:

When the Ni in the SUS316L was dealloyed in the Mg melt, a Mg-Ni phase solidified on the surface. This Mg-Ni phase should have been etched away completely in the aqueous HNO_3 solution. However, if Mg-Ni remained after this etching

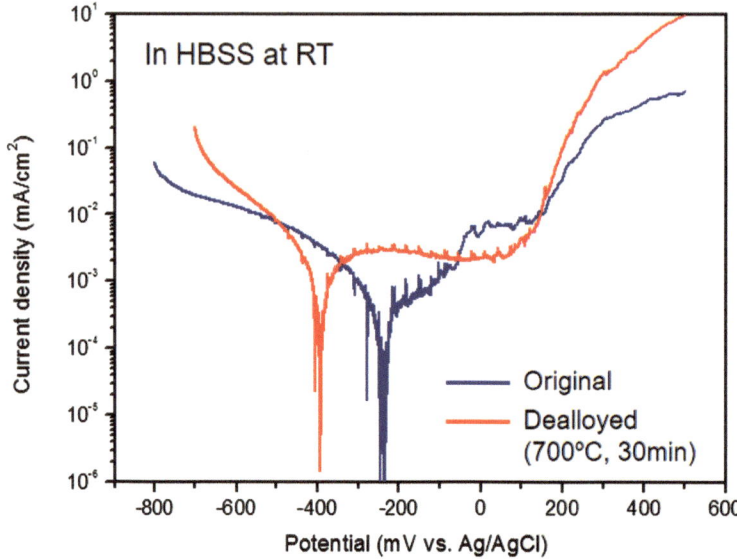

Fig. 3.12 Polarization curves for SUS316L dealloyed in Mg melt at 700 °C for 30 min (*red*) and original SUS316L after cold rolling (*blue*)

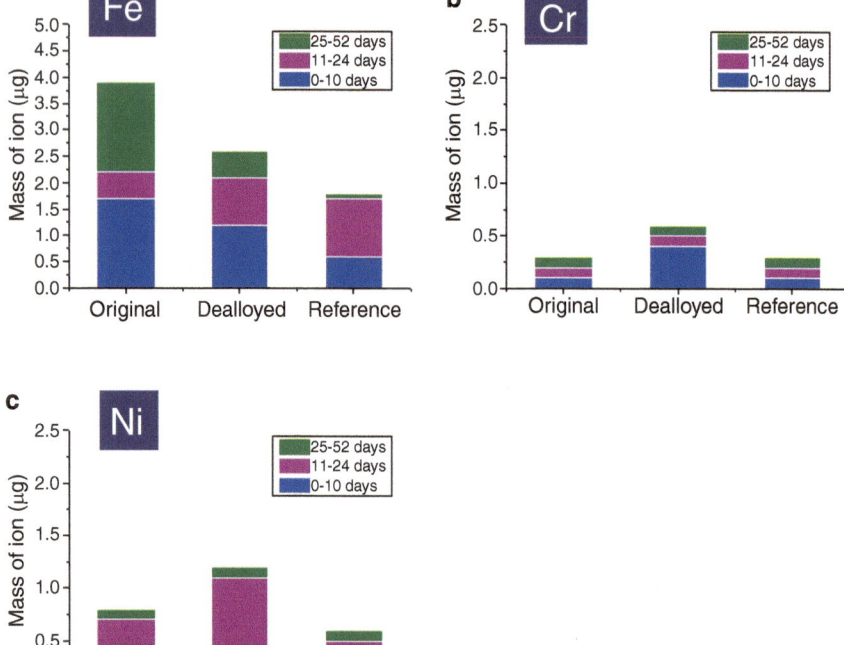

Fig. 3.13 Ion release of Fe (**a**), Cr (**b**), and Ni (**c**) from the surface of SUS316L dealloyed at 700 °C for 30 min compared with that of the original cold rolled SUS316L during the first 10 days (*blue*), 11–24 days (*pink*), and 25–52 days (*green*) immersion in HBSS. The HBSS was replaced with fresh solution after sampling at 10 and 24 days. Reference values are those measured with the same setup without the presence of a SUS316L sample

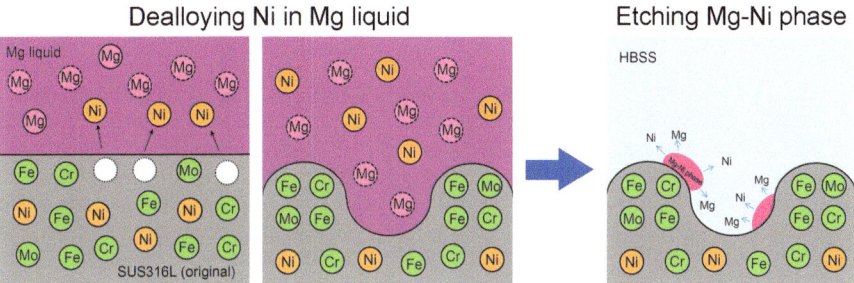

Fig. 3.14 Schematic illustration explaining origin of increased Ni ion release during 10–24 days immersion of the dealloyed SUS316L sample in HBSS. After dealloying treatment (*left*), Mg-Ni liquid solidified on the surface of the sample. Some Mg-Ni phase, perhaps located deep inside of the pores, remained even after the etching treatment in the aqueous HNO_3 solution (*right*). This phase likely dissolved slowly into the HBSS, resulting in a higher Ni ion release during this period

treatment, Ni-ion release could have been increased in the HBSS as shown in the following illustration (Fig. 3.14).

Dealloying tends to increase the surface area of the steel. This would increase the rate and amount of Ni ion release even after decreasing the concentration of Ni at the surface. If this is correct, then it should be possible to decrease the Ni ion release by decreasing the thickness of the reaction layer to decrease the surface area of treated steel.

3.4   Conclusions

1. The surface Ni concentration of SUS316L was successfully decreased from 11.8 to 2.2 at.% by dealloying the Ni in Mg liquid at 700 °C for 30 min.
2. Dealloyed SUS316L maintains excellent corrosion resistance in HBSS at room temperature.
3. Mass of Ni ion release from the Ni-dealloyed SUS316L surface was almost comparable or slightly increased compared with that of untreated SUS316L.
4. The main reasons for (3) are:

   - Increased surface area caused by the dealloying treatment, although the Ni concentration per unit area was decreased
   - Mg-Ni phase remaining on the sample surface after the etching in HNO_3 aqueous solution

Acknowledgment This research was financially supported by the interuniversity cooperative research program "Innovative Research for Biosis-Abiosis Intelligent Interface" from the Ministry of Sports, Culture, and Education of Japan.

References

1. Wada T, Yubuta K, Inoue A, Kato H. Dealloying by metallic melt. Mater Lett. 2011;65:1076–8. doi:10.1016/j.matlet.2011.01.054.
2. Fukuzumi Y, Wada T, Kato H. Surface improvement for biocompatibility of Ti-6Al-4V by dealloying in metallic melt. In: Sasaki K, Suzuki O, Takahashi N, editors. Interface oral health science 2014. Innovative research on biosis-abiosis intelligent interface. Springer Open; 2014. p. 93–101.
3. Forty AJ. Corrosion micromorphology of noble metal alloys and depletion gilding. Nature. 1979;282:597–8. doi:10.1038/282597a0.
4. Erlebacher J, Aziz MJ, Karma A, Dimitrov N, Sieradzki K. Evolution of nanoporosity in dealloying. Nature. 2001;410:450–3. doi:10.1038/35068529.
5. Takeuchi A, Inoue A. Classification of bulk metallic glasses by atomic size difference, heat of mixing and period of constituent elements and its application to characterization of the main alloying element. Mater Trans. 2005;46:2817–29. doi:10.2320/matertrans.46.2817.
6. Wada T, Setyawan AD, Yubuta K, Kato H. Nano- to submicro-porous β-Ti alloy prepared from dealloying in a metallic melt. Scripta Mater. 2011;65:532–5. doi:10.1016/j.scriptamat.2011.06.019.
7. Kim JW, Tsuda M, Wada T, Yubuta K, Kim SG, Kato H. Optimizing niobium dealloying with metallic melt to fabricate porous structure for electrolytic capacitors. Acta Mater. 2015;84:497–505. doi:10.1016/j.actamat.2014.11.002.
8. Wada T, Ichitsubo T, Yubuta K, Segawa H, Yoshida H, Kato H. Bulk-nanoporous-silicon negative electrode with extremely high cyclability for lithium-ion batteries prepared using a top-down process. Nano Lett. 2014;14:4505–10. doi:10.1021/nl501500g.
9. Yu SG, Yubuta K, Wada T, Kato H. Three-dimensional bicontinuous porous graphite generated in low temperature metallic liquid. Carbon. 2016;96:403–10. doi:10.1016/j.carbon.2015.09.093.

Open Access This chapter is distributed under the terms of the Creative Commons Attribution 4.0 International License (http://creativecommons.org/licenses/by/4.0/), which permits use, duplication, adaptation, distribution and reproduction in any medium or format, as long as you give appropriate credit to the original author(s) and the source, provide a link to the Creative Commons license and indicate if changes were made.

The images or other third party material in this chapter are included in the work's Creative Commons license, unless indicated otherwise in the credit line; if such material is not included in the work's Creative Commons license and the respective action is not permitted by statutory regulation, users will need to obtain permission from the license holder to duplicate, adapt or reproduce the material.

Chapter 4
Bio-ceramic Coating of Ca–Ti–O System Compound by Laser Chemical Vapor Deposition

Hirokazu Katsui and Takashi Goto

Abstract Bio-ceramic Ca–Ti–O system compound films were prepared by laser chemical vapor deposition (laser CVD). Laser CVD is a high-speed technique for coating films with versatile controllability of microstructures and crystal phases. Highly oriented $CaTiO_3$ films with specific textures and $Ca_{n+1}Ti_nO_{3n+1}$ films with the Ruddlesden–Popper-type structure were prepared at high deposition rates. The formation of calcium phosphate in simulated body fluid (SBF) was promoted by $Ca_{n+1}Ti_nO_{3n+1}$ films.

Keywords Calcium titanate • Laser CVD • Microstructure • Bioactive coating • High deposition rate

4.1 Introduction

Ti and Ti-based alloys are used as artificial bones and dental implants because of their acceptable mechanical properties, low weight, and adequate corrosion resistance in the human body. However, they suffer certain disadvantages, such as poor osteoinductive properties and a duration of several months for the reconstruction of the bone/implant interface with adequate adhesion. The osseointegration of an orthopedic implant involves a cascade of cellular and extracellular biological events that occur at the bone/implant interface [1]. The processes can be enhanced by the surface treatments and bio-ceramic coating on implants [2, 3]. The plasma-sprayed hydroxyapatite ($Ca_{10}(PO_4)_6(OH)_2$, HAp) coating on Ti is practically used for dental implants [3, 4]. However, the low interface bonding strength and coating toughness can cause a fracture in the interface between HAp and Ti implants. HAp films with low crystallinity coated on Ti implants dissolve rapidly when Ti is implanted into a

H. Katsui (✉) • T. Goto
Institute for Materials Research, Tohoku University,
2-1-1 Katahira, Aoba-ku, Sendai, Miyagi 980-8577, Japan
e-mail: katsui@imr.tohoku.ac.jp

© The Author(s) 2017
K. Sasaki et al. (eds.), *Interface Oral Health Science 2016*,
DOI 10.1007/978-981-10-1560-1_4

human body. The crystallinity and microstructure of coated films is an important factor for establishing a good interface between the bone and implants [4, 5].

Recently, calcium titanate ($CaTiO_3$) has gained considerable attention as a bio-material. $CaTiO_3$ coatings with controlled thickness and crystallinity are effective for bone formation because $CaTiO_3$ is chemically stable at low pH and can form HAp in SBF [6–8]. $CaTiO_3$ has also been proposed as an intermediate layer to improve the adhesion between HAp and Ti-based implants [9–13]. To date, a variety of techniques, such as sol–gel [13], hydrothermal reactions [14], ion implantation [15], sputtering [8], and anode oxidation techniques [7], were employed for $CaTiO_3$ coating. Chemical vapor deposition (CVD) is a versatile technique to prepare various ceramic films and is widely used in the industry. Sato et al. reported the synthesis of $CaTiO_3$ films by CVD using metal organic precursors followed by apatite formation on the film surface upon immersion in SBF [16]. Auxiliary energies such as plasmas and lasers could be employed to accelerate chemical reactions and prepare highly crystalline films with controlled morphology and crystal phases at high deposition rates [17–21]. In this study, we demonstrate the synthesis of Ca–Ti–O films by laser CVD, and the effects of deposition parameters on crystal phases, morphology, and deposition rate are investigated. Laser CVD can produce $Ca_{n+1}Ti_nO_{3n+1}$ films which exhibited a significant formability of calcium phosphate precipitates on the coating surface in the SBF immersion.

4.2   Laser Chemical Vapor Deposition

CVD is a gas-phase deposition process, comprising several chemical reactions between source gases (precursors). Dense films can be coated by CVD even on rough surfaces with high adherence and good conformal coverage. This is advantageous for bio-ceramic coatings on complex-shaped dental implants and artificial bones. Hence, bio-ceramic coatings of well-crystallized Ca–P–O system compounds, such as HAp, α- and β-$Ca_3P_2O_8$, $Ca_4P_2O_9$, and α- and β-$Ca_2P_2O_7$, have been performed using CVD [22–24]. Generally, the deposition rate of CVD is lower than that of plasma spray and electron beam physical vapor deposition. In conventional thermal CVD, the chemical reaction at the interface between the gas and substrate surface is driven by thermal energy. Laser irradiation can accelerate the chemical reactions and enable low-temperature deposition to avoid degradation and corrosion of the substrate materials. Figure 4.1 shows a schematic of the laser CVD apparatus for the coating of Ca–Ti–O compounds. The source materials (precursors) of Ca and Ti were evaporated, and the source vapors were introduced into a CVD reaction chamber. Oxygen was separately introduced into the chamber. A substrate was placed on a hot stage for preheating. The substrate surface was irradiated by an Nd:YAG laser (wavelength 1064 nm) through a quartz window. By controlling deposition parameters, such as laser power, deposition temperature, total pressure, and precursor supply conditions, various forms of deposits can be obtained, e.g., amorphous, fine crystals, columnar crystals, dendritic crystals, whiskers, plate-like

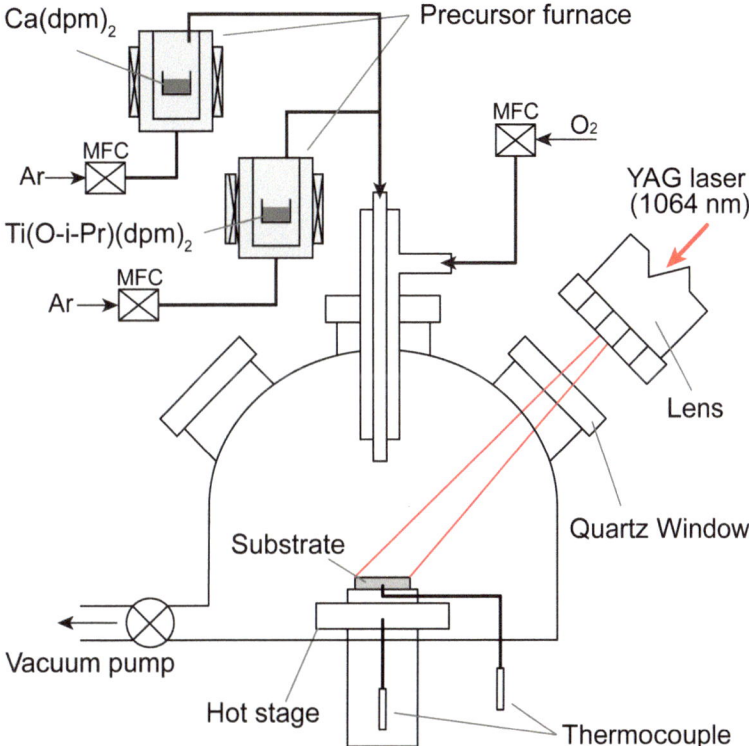

Fig. 4.1 Schematic of laser CVD apparatus

crystals, and epitaxial single-crystal films. In this study, aluminum nitride (AlN) was first used as the substrate, because it is thermochemically stable at high temperature, and its good workability enables us to investigate the effect of a wide range of CVD parameters on the Ca–Ti–O film characteristics. Based on the insight into the correlation between the CVD parameters and the film characteristics using the AlN substrates, bioactive Ca–Ti–O films were coated on metallic Ti substrates under optimum laser CVD conditions.

4.3   Bio-ceramic Coating of Ca–Ti–O by Laser CVD [25, 26]

The phase diagram of a CaO–TiO_2 pseudo-binary system is shown in Fig. 4.2 [27, 28]. At a Ca/Ti ratio of 1.0, the $CaTiO_3$ phase exists, which is the most common calcium titanate compound. No other phases are stable in the Ti-rich region between TiO_2 and $CaTiO_3$, whereas $Ca_{n+1}Ti_nO_{3n+1}$ phases exist in the Ca-rich region between $CaTiO_3$ and CaO. The crystal structures of $CaTiO_3$ and $Ca_{n+1}Ti_nO_{3n+1}$ are illustrated in Fig. 4.3. Further $CaTiO_3$ has a perovskite structure with a space group of *Pnma*,

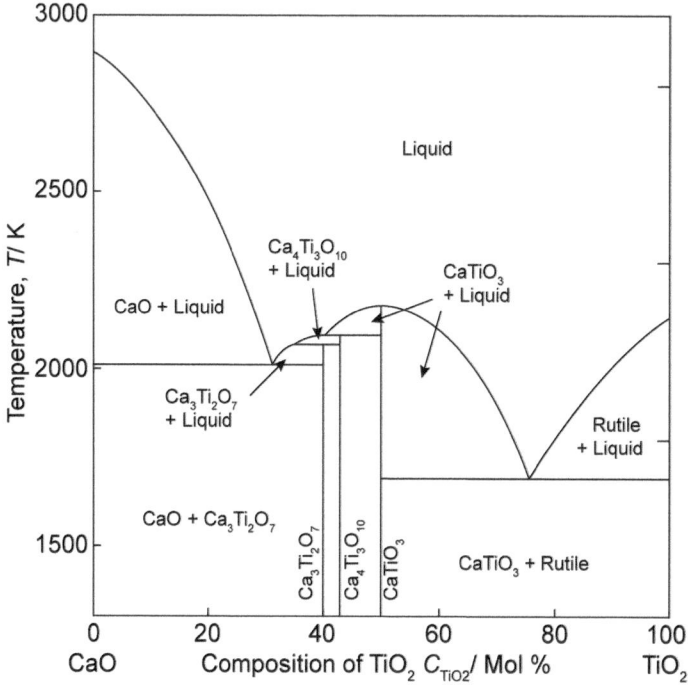

Fig. 4.2 Phase diagram of CaO–TiO$_2$ system [27, 28]

comprising the corner-sharing TiO$_6$ octahedra surrounded by Ca ions with a 12-fold coordination [29]. The Ca$_{n+1}$Ti$_n$O$_{3n+1}$ phases have perovskite-related structures, the so-called Ruddlesden–Popper structure, formed by alternate stacking of perovskite blocks and CaO layers, as shown in Fig. 4.3 [30]. The stacking sequence in a unit cell corresponds to the n value in Ca$_{n+1}$Ti$_n$O$_{3n+1}$. Two phases, Ca$_2$Ti$_3$O$_7$ ($n = 2$) and Ca$_3$Ti$_4$O$_{10}$ ($n = 3$), have been reported to exist in the TiO$_2$–CaO system. Although CaTiO$_3$ films fabricated by various methods and their bioactivities were investigated using in vivo and in vitro experiments [3, 6, 8, 10, 11, 13], there are few reports of the synthesis of Ca$_{n+1}$Ti$_n$O$_{3n+1}$ films as a biomaterial [31]. Laser CVD can be used to synthesize CaTiO$_3$ and Ca$_{n+1}$Ti$_n$O$_{3n+1}$ by controlling deposition parameters, such as the Ca/Ti supply ratio of the precursors and deposition temperature depending on the laser power. Figure 4.4 depicts the effects of deposition temperature and Ca/Ti supply ratio on the phase formation of Ca–Ti–O films by laser CVD. At a Ca/Ti supply ratio of approximately 1.0, single-phase CaTiO$_3$ films were formed at deposition temperatures below 1100 K. At deposition temperatures above 1100 K, CaTiO$_3$ films contained TiO$_2$, Ca–Al–O compounds (e.g., CaAl$_2$O$_4$ and CaAl$_4$O$_7$) and Al$_2$O$_3$, resulting in a reaction between the source gases and the AlN substrate at high temperatures. At Ca/Ti supply ratios <0.8, Ti-rich Ca–Ti–O compounds were formed; however, no phases were thermodynamically stable according to the phase diagrams [27, 28]. Under Ca-rich conditions (Ca/Ti supply ratio >1.0), Ca$_{n+1}$Ti$_n$O$_{3n+1}$

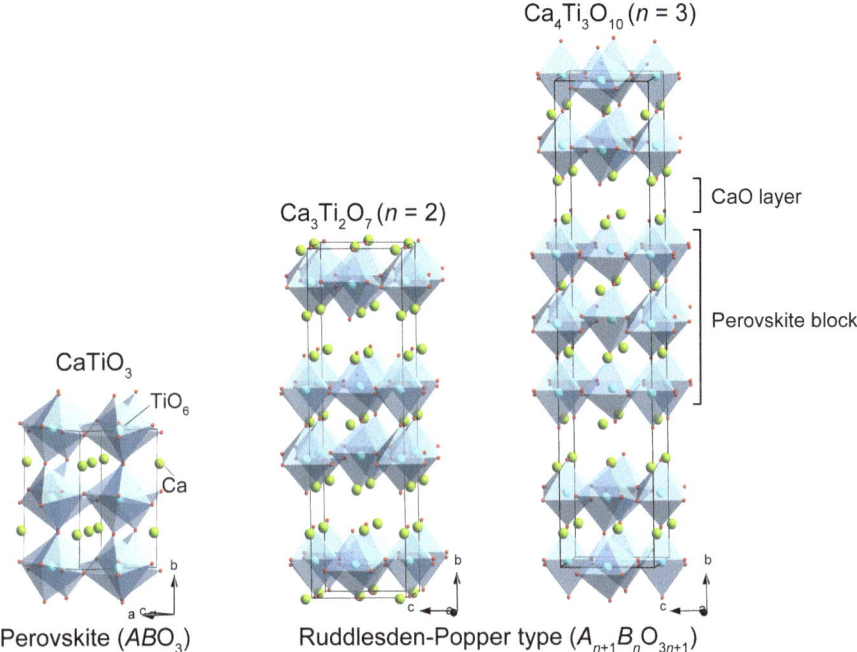

Fig. 4.3 Crystal structures of $CaTiO_3$, $Ca_3Ti_2O_7$, and $Ca_4Ti_3O_{10}$

films were deposited at relatively low deposition temperatures (<1000 K), whereas films prepared at deposition temperatures higher than 1000 K comprised CaO and $CaTiO_3$. In the Ti-rich compositional region between $CaTiO_3$ and TiO_2, several Ca–Ti–O compounds were reported. Bertaut and Blum [32] and Bright et al. [33] reported the synthesis of $CaTi_2O_4$ by electrolysis of TiO_2 and $CaTiO_3$ in a $CaCl_2$ melt. $CaTi_2O_4$ single crystals were synthesized by a flux method from $CaTiO_3$ in $CaCl_2$ and Ti metal [34]. The existence of $CaTi_4O_9$ and $CaTi_2O_5$ was reported in a wet chemical method and sol–gel method [35–38]. Ancora et al. published patents on the production of $CaTi_2O_5$ and $CaTi_5O_{11}$ [39], where the $CaTi_2O_5$ crystal structure differed from that produced by Limar and Kisel [35, 36]. Since these Ti-rich phases were considered to be metastable and decomposed into $CaTiO_3$ and TiO_2 at high temperatures and the synthesis process was limited, the detailed crystal structures and compositions remain unknown. In this study, the X-ray diffraction (XRD) patterns of Ti-rich Ca–Ti–O films by laser CVD in this study were similar to those of $CaTi_2O_5$ and $CaTi_5O_{11}$ reported by Ancora; however, the phase identification was difficult because the films may comprise a mixture of phases and have preferred orientations. The Ti-rich Ca–Ti–O films were transformed into TiO_2 and $CaTiO_3$ by heat treatment (post annealing) at 1273 K. Further investigation of the detailed chemical compositions and microstructure was required for the Ti-rich Ca–Ti–O compounds.

Fig. 4.4 Phase formation relation between deposition temperature and Ca/Ti precursor supply ratio for Ca–Ti–O films formed by laser CVD

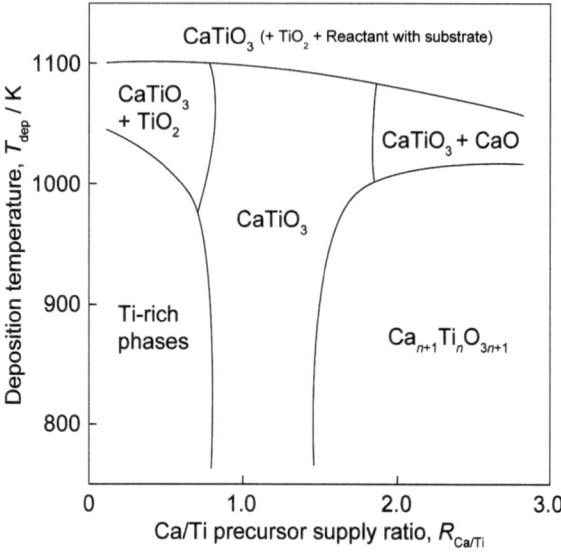

Figure 4.5 illustrates the effect of the deposition temperature on the crystal orientations of CaTiO$_3$ films prepared at a total pressure of 800 Pa, where the texture coefficient (*TC*) is the degree of crystal orientation. A *TC* value of 10 corresponds to perfect orientation, whereas a *TC* value of 1 corresponds to random orientation [40]. At temperatures below 800 K, CaTiO$_3$ films with (011) orientation were formed. With an increase in the deposition temperature, the preferred orientation changed from (011) to (101) at approximately 800 K. Further increases in the deposition temperature resulted in the formation of CaTiO$_3$ films having no preferred orientation. The preferred orientation during the growth of CaTiO$_3$ films can be controlled not only by the deposition temperature but also by the total pressure in the chamber. The (121)-oriented CaTiO$_3$ films were deposited in the total pressure range of 400–600 Pa at a deposition temperature of 825–855 K, whereas the preferred orientation was (101) at a total pressure of 800 Pa in the same deposition temperature range (Fig. 4.5). Figure 4.6 shows the typical surface and cross-sectional morphologies of CaTiO$_3$ films with the preferred orientations ((011), (101), and (121)) and without orientation. (011)-oriented CaTiO$_3$ films have a cone-like morphology with pyramidal facets, as shown in Fig. 4.6a. Square facets, which are several micrometers in size, were formed in (101)-oriented CaTiO$_3$ films, as shown in Fig. 4.6c. (121)-oriented CaTiO$_3$ films had a granular morphology with fine grains smaller than several micrometers in size (Fig. 4.6e). These CaTiO$_3$ films with strongly preferred orientations were grown in the columnar regime (Fig. 4.6b, d, f). CaTiO$_3$ films without preferred orientation prepared at a high deposition temperature composed randomly arranged faceted grains (several micrometers in size) with a dense and smooth cross section, as shown in Fig. 4.6c, f.

Figure 4.7 depicts the detailed microstructures with crystal structure models of the corresponding textures in the (011)- and (101)-oriented CaTiO$_3$ films. The

Fig. 4.5 Effect of deposition temperature on $TC(022)$ and $TC(101)$ of $CaTiO_3$ films

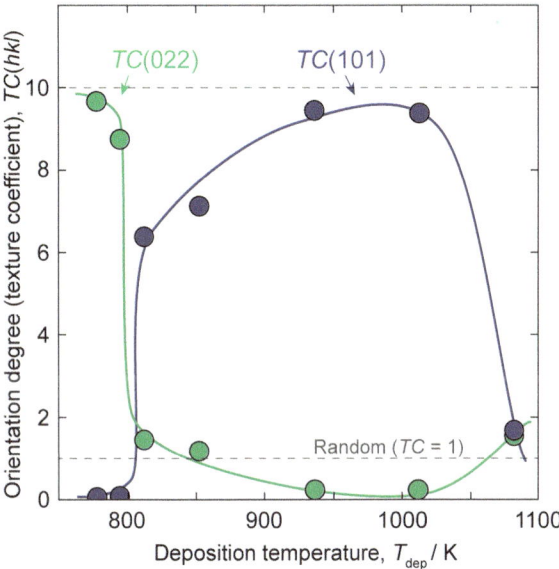

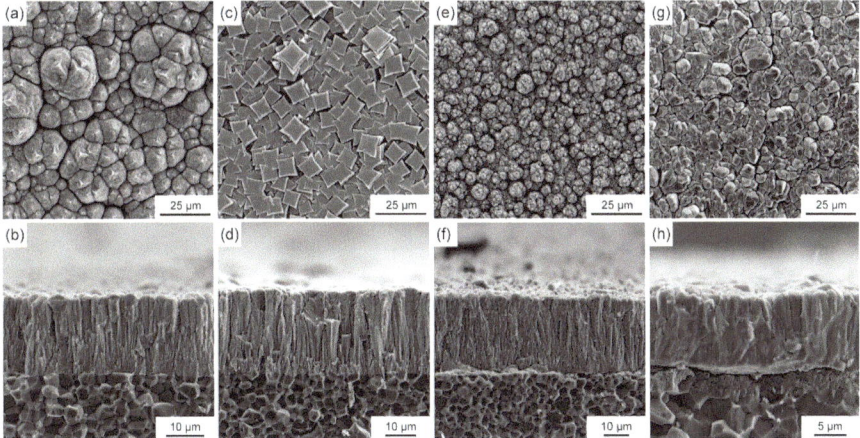

Fig. 4.6 SEM images of typical $CaTiO_3$ films deposited by laser CVD on AlN substrates. (**a, b**) (011)-oriented $CaTiO_3$ film at 795 K and 800 Pa, (**c, d**) (101)-oriented $CaTiO_3$ film at 935 K and 800 Pa, (**e, f**) (121)-oriented $CaTiO_3$ film at 855 K and 400 Pa, and (**g, h**) $CaTiO_3$ film with random orientation at 1080 K and 800 Pa

cone-like morphology of the (011)-oriented $CaTiO_3$ film comprised pyramidal facets, which are several tens nanometers in size. Considering the preferred (011) orientation and the shapes of the grains, the pyramidal texture could be associated with the $CaTiO_3$ crystal structure, and the faceted planes would be {010} and {110} as shown in Fig. 4.7b. The microstructure of the square facets in the (101)-oriented $CaTiO_3$ film is shown in Fig. 4.8a. Figure 4.8b shows the terrace on the top surface

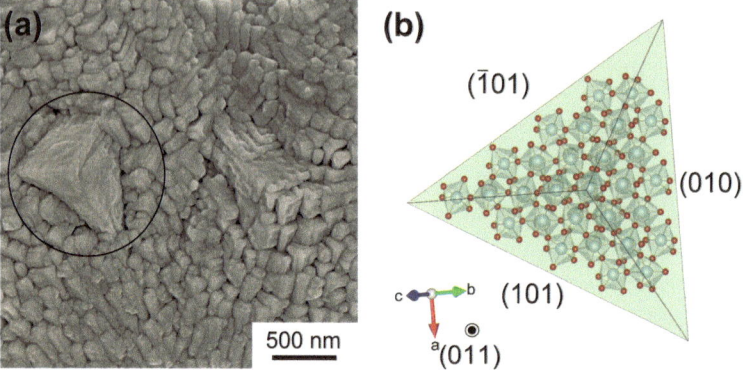

Fig. 4.7 Microstructure and oriented texture of (011)-oriented CaTiO₃ film. (**a**) Surface SEM image at high magnification and (**b**) crystallographic texture of the pyramidal facet

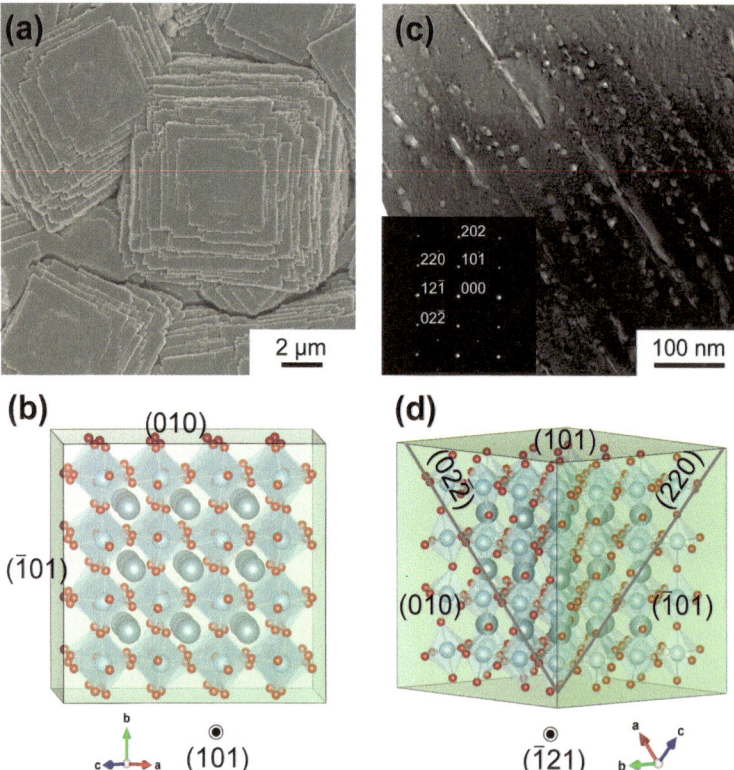

Fig. 4.8 Microstructure and oriented texture of (101)-oriented CaTiO₃ film. (**a**) Surface SEM image at high magnification, (**b**) crystallographic texture of the square facet, (**c**) cross-sectional TEM image of the square facet, and (**d**) relation between crystallographic texture and the formation of nanopores

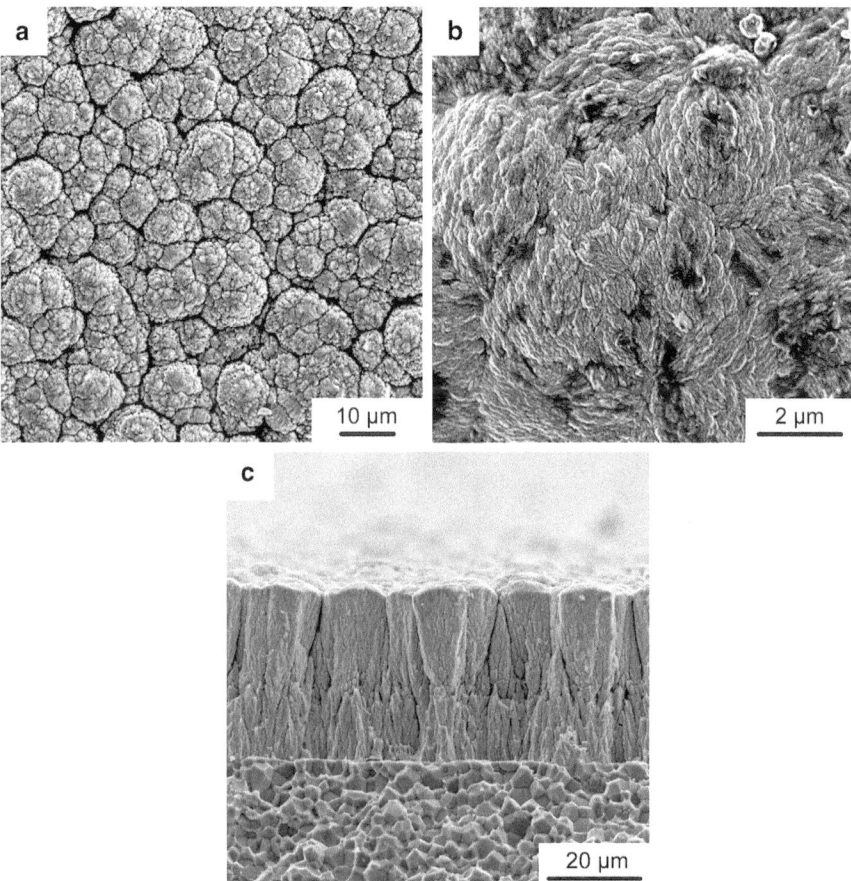

Fig. 4.9 SEM images of a $Ca_nTi_{n+1}O_{3n+1}$ film at 777 K and 800 Pa. (**a**) Surface morphology, (**b**) enlarged image of cauliflower-like grains, and (**c**) cross-sectional image

of the square facet corresponding to the (101) plane, along which the corner-sharing TiO_6 octahedra are aligned. Here, the lateral planes of the square facet were {101} and {010}. In Fig. 4.8c, the cross-sectional transmission electron microscopy (TEM) image of the square facet revealed that nanopores formed along the (110) and (011) planes, which are the close-packed planes of Ca–O atoms. These nanopores may relax the stress between the bio-ceramic films and metallic substrates [41–43].

Figure 4.9 shows the surface and cross-sectional morphologies of the $Ca_{n+1}Ti_nO_{3n+1}$ film prepared at a Ca/Al supply ratio of 1.6 and a deposition temperature of 777 K. The surface exhibited a cone-like morphology with a grain size of approximately 5–10 μm (Fig. 4.10a). Each cone-like grain comprised granules that were several tens nanometers in size. The cross section was cone-like, which is a typical morphology for CVD-deposited films [44]. The Ca/Ti composition of this film was

Fig. 4.10 Effect of
deposition temperature on
deposition rates of CaTiO₃
films formed by laser
CVD. For comparison,
plots of the deposition
rates of CaTiO₃ films
formed by conventional
thermal CVD are also
included

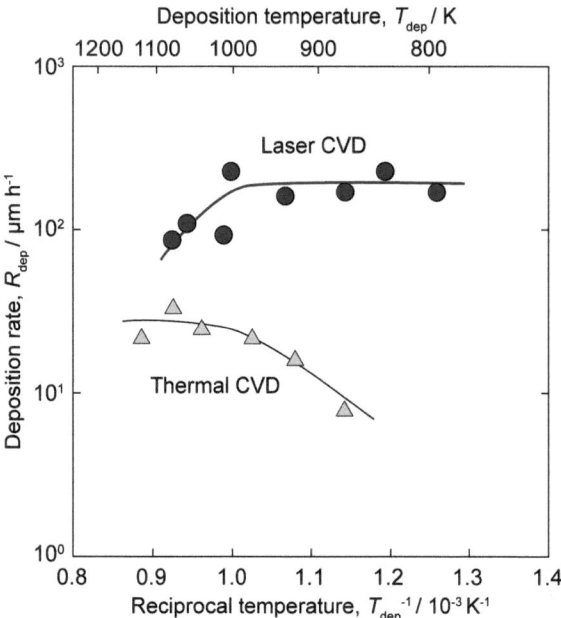

1.54 by EPMA, which was nearly the same as that of $Ca_3Ti_2O_7$. However, it was difficult to identify the detailed phases in the $Ca_{n+1}Ti_nO_{3n+1}$ films, because the XRD powder pattern of $Ca_3Ti_2O_7$ was similar to that of $Ca_4Ti_3O_{10}$ owing to the same type of long-range perovskite-related structure.

Figure 4.10 shows the temperature dependence of the deposition rate for $CaTiO_3$ films by laser CVD and conventional thermal CVD in an Arrhenius format. The deposition rates of the $CaTiO_3$ films by laser CVD reached 230 μm h⁻¹ in the temperature range of 800–1000 K. For the case of conventional thermal CVD, $CaTiO_3$ films without preferred crystal orientation were grown at the deposition rates in the range of 10–30 μm h⁻¹ and at deposition temperatures above 900 K. Laser CVD enables the preparation of $CaTiO_3$ with several types of oriented textures at lower deposition temperatures and considerably higher growth rates compared with those obtainable by thermal CVD.

Figure 4.11 depicts the surface morphologies of the $CaTiO_3$ and $Ca_{n+1}Ti_nO_{3n+1}$ films coated on the AlN substrates before and after immersion in SBF (Hanks' solution) for 3 days. Although no significant change in the randomly faceted grains of the $CaTiO_3$ films without preferred orientation occurred during immersion, the grain boundaries and the faceted edges became slightly obscured (Fig. 4.11a, b). On the other hand, for the as-deposited $CaTiO_3$ film comprising square-faceted grains with strong (101) orientation, the edges and corners of the facets became round and smooth after immersion in Hanks' solution (Fig. 4.11c, d). These changes in grain boundaries and facet edges could be caused by the dissolution of $CaTiO_3$ into Hanks' solution, indicating the biosolubility of the $CaTiO_3$ coating. The surface cone-like morphology with pyramidal facets of (011)-oriented $CaTiO_3$ films became

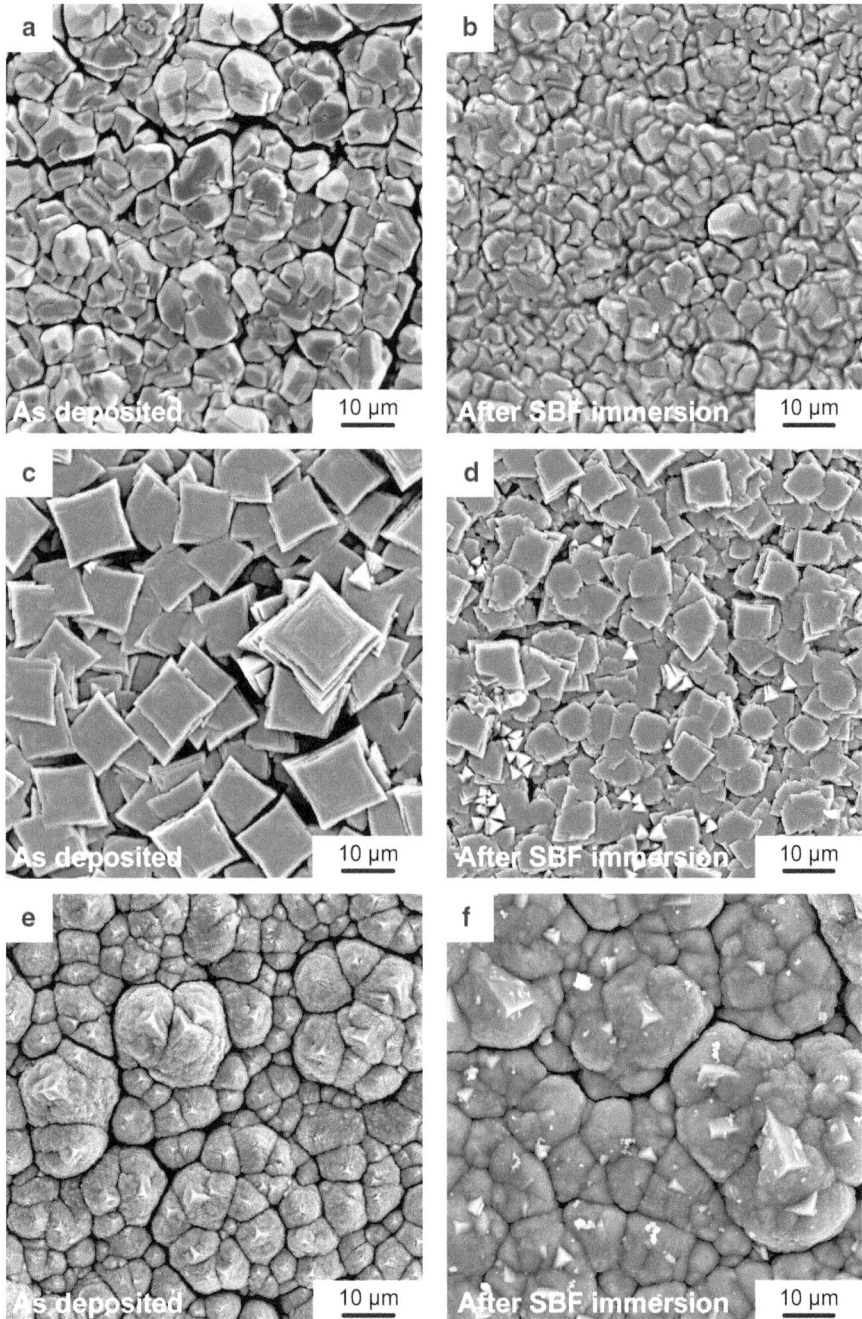

Fig. 4.11 Effect of immersion in Hanks' solution for 3 days on the surface morphologies of CaTiO$_3$ films coated on AlN substrates. (**a, b**) CaTiO$_3$ film with random orientation, (**c, d**) CaTiO$_3$ film with (101)-orientation, and (**e, f**) CaTiO$_3$ film with (011)-orientation. Images (**a, c**, and **e**) show as-deposited films, whereas (**b, d**, and **f**) show films after the immersion

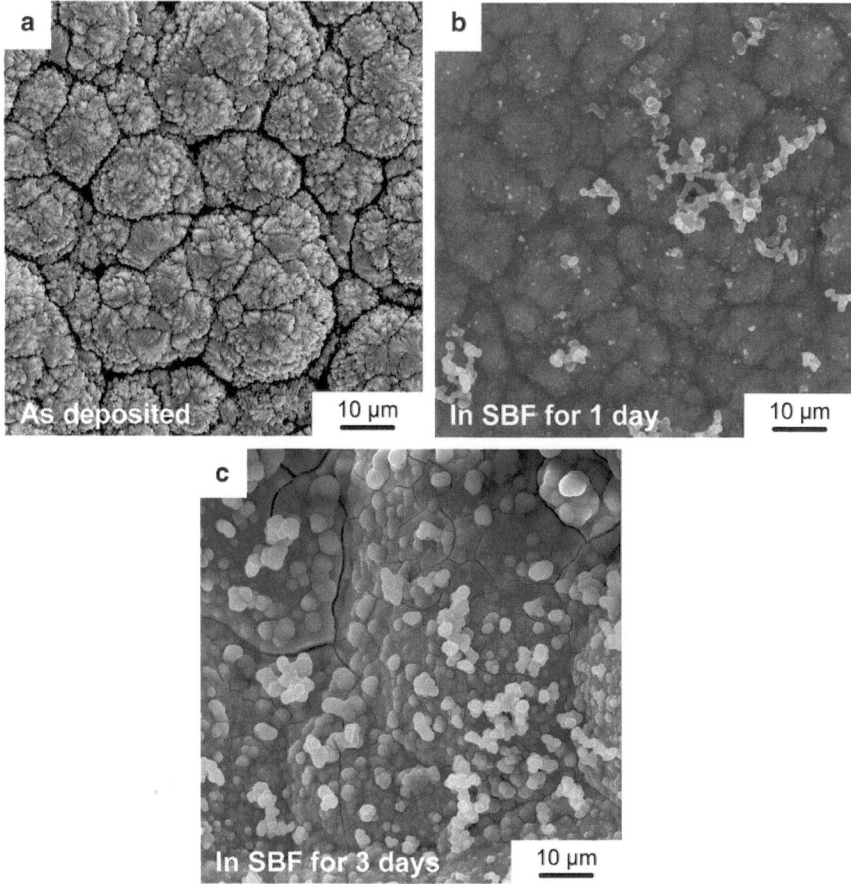

Fig. 4.12 The effect of immersion in Hanks' solution on the surface morphologies of $Ca_nTi_{n+1}O_{3n+1}$ films coated on AlN substrates; (**a**) as-deposited, (**b**) after immersion for 1 day, and (**c**) after immersion for 3 days

smooth, as shown in Fig. 4.11f. Figure 4.11f shows that a small amount of calcium phosphate precipitate (several hundred nanometers in size) with a bright contrast appeared on the film's surface after immersion in Hanks' solution. The biosolubility and calcium phosphate formation of $CaTiO_3$ films are affected by the morphology and preferred orientation. Figure 4.12 depicts the change in the surface morphology of $Ca_{n+1}Ti_nO_{3n+1}$ films caused by immersion in the Hanks' solution. The cauliflower-like grains of the as-deposited $Ca_{n+1}Ti_nO_{3n+1}$ film became smooth, and calcium phosphate precipitate was formed after immersion for 1 day (Fig. 4.12b). The entire surface of the film was covered by calcium phosphate precipitate after 3 days, as shown in Fig. 4.12c. Compared with the conventional perovskite $CaTiO_3$ films (Fig. 4.11), the Ruddlesden–Popper-type $Ca_{n+1}Ti_nO_{3n+1}$ films exhibited significant changes in the surface morphology and high calcium phosphate formation ability after the short-term immersion in SBF.

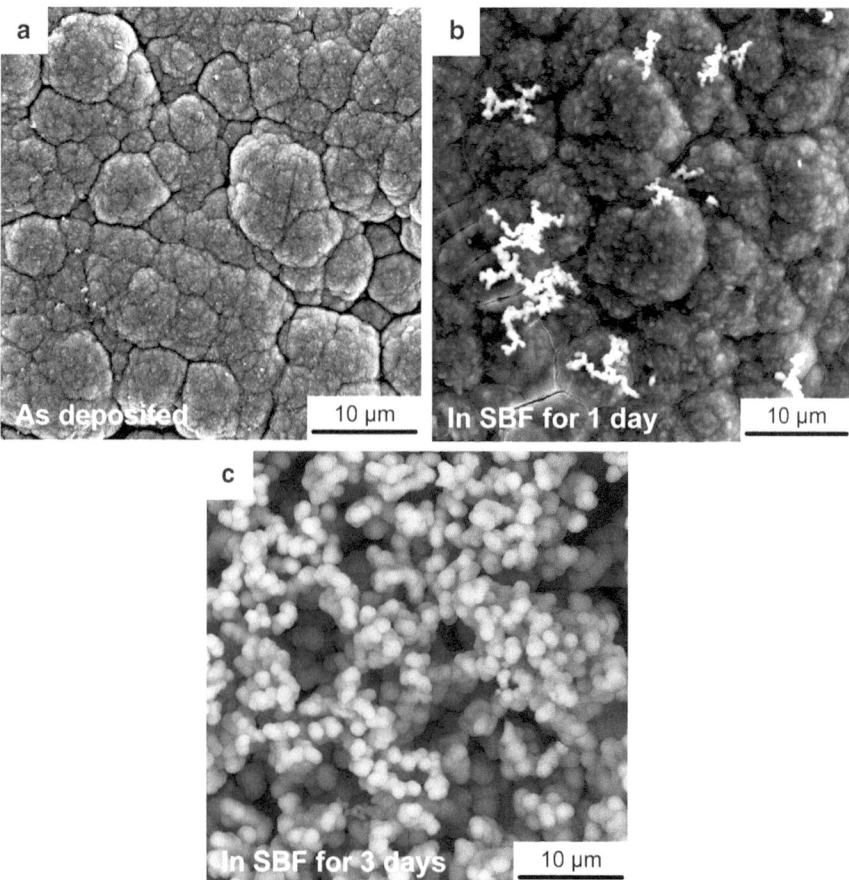

Fig. 4.13 The effect of immersion in Hanks' solution on the surface morphologies of $Ca_nTi_{n+1}O_{3n+1}$ films coated on Ti substrates; (**a**) as-deposited, (**b**) after immersion for 1 day, and (**c**) after immersion for 3 days

$Ca_{n+1}Ti_nO_{3n+1}$ film was coated on a CP-Ti substrate. Figure 4.13 shows the surface morphologies of the $Ca_{n+1}Ti_nO_{3n+1}$ film prepared at a deposition temperature of 620 K on a CP-Ti substrate. The as-deposited $Ca_{n+1}Ti_nO_{3n+1}$ film had a cauliflower-like morphology similar to that on an AlN substrate (Fig. 4.12a). After the SBF immersion for a day (Fig. 4.13b), the surface of the cauliflower-like grains became smooth, and the grain boundaries were obscured. The entire film surface was covered with calcium phosphate precipitate after immersion for 3 days, as shown in Fig. 4.13c. Therefore, laser CVD enables the bio-ceramic coating of $Ca_{n+1}Ti_nO_{3n+1}$ film on Ti substrates, and this coating is promising for enhancing the osteoinductivity of Ti-based implants.

4.4 Summary

Well-crystallized Ca–Ti–O films with various crystal phases and microstructures were produced at high deposition rates by laser CVD. Highly (011)-, (101)-, and (121)-oriented $CaTiO_3$ films were obtained, forming cauliflower-like, granular, and faceted morphologies. These various preferred orientations and morphologies affected the solubility, regeneration of calcium phosphate, and bio-inertness of $CaTiO_3$ films. For the Ca-rich compositions, $Ca_{n+1}Ti_nO_{3n+1}$ films with a Ruddlesden–Popper-type crystal structure were formed and exhibited promising bioactivity for calcium phosphate regeneration.

Acknowledgment This work was supported by the Ministry of Education, Culture, Sports, Science and Technology (MEXT), Scientific Research (A) No. 16H06121 and a cooperative program of the ARCMG-IMR No. 16G0405, Tohoku University.

References

1. Cao H, Liu X. Plasma-sprayed ceramic coatings for osseointegration. Int J Appl Ceram Technol. 2013;10:1–10. doi:10.1111/j.1744-7402.2012.02770.x.
2. Hanawa T. Biofunctionalization of metallic materials: creation of biosis–abiosis intelligent interface. In: Sasaki K, Suzuki O, Takahashi O, editors. Interface oral health science 2014. Tokyo: Springer; 2015. p. 53–64.
3. Hanawa T. Biofunctionalization of titanium for dental implant. Jpn J Dent Sci Rev. 2010;46:93–101. doi:10.1016/j.jdsr.2009.11.001.
4. Ohtsu N, Sato K, Yanagawa A, Saito K, Imai Y, Kohgo T, Yokoyama A, Asami K, Hanawa T. $CaTiO_3$ coating on titanium for biomaterial application—optimum thickness and tissue response. J Biomed Mater Res. 2007;82A:304–15. doi:10.1002/jbm.a.31136.
5. Narushima T. Surface modification for improving biocompatibility of titanium materials with bone. J Jpn Inst Light Metals. 2008;58:577–82. doi:10.2464/jilm.58.577.
6. Ohtsu N, Sato K, Saito K, Asami K, Hanawa T. Calcium phosphates formation on $CaTiO_3$ coated titanium. J Mater Sci Mater Med. 2007;18:1009–16. doi:10.1007/s10856-006-0114-x.
7. Iwasaki M. Fabrication of artificial bone by anodic oxidation of titanium. Surf Finish Soc Jpn. 2014;65:272–5. doi:10.4139/sfj.65.272.
8. Ohtsu N, Sato K, Saito K, Hanawa T, Asami K. Evaluation of degradability of $CaTiO_3$ thin films in simulated body fluids. Mater Trans. 2004;45:1778–81. doi:10.2320/matertrans.45.1778.
9. Rakngarm A, Miyashita Y, Mutoh Y. Formation of hydroxyapatite layer on bioactive Ti and Ti–6Al–4V by simple chemical technique. J Mater Sci Mater Med. 2007;19:1953–61. doi:10.1007/s10856-007-3285-1.
10. Wei D, Zhou Y, Jia D, Wang Y. Formation of $CaTiO_3/TiO_2$ composite coating on titanium alloy for biomedical applications. J Biomed Mater Res. 2008;84B:444–51. doi:10.1002/jbm.b.30890.
11. Ohba Y, Watanabe T, Sakai E, Daimon M. Coating of HAp/$CaTiO_3$ multilayer on titanium substrates by hydrothermal method. J Ceram Soc Jpn. 1999;107:907–12. doi:10.2109/jcersj.107.907.
12. Kačiulis S, Mattogno G, Pandolfi L, Cavalli M, Gnappi G, Montenero A. XPS study of apatite-based coatings prepared by sol-gel technique. Appl Surf Sci. 1999;151:1–5. doi:10.1016/S0169-4332(99)00267-6.

13. Holliday S, Stanishevsky A. Crystallization of CaTiO₃ by sol-gel synthesis and rapid thermal processing. Surf Coat Technol. 2004;188–189:741–4. doi:10.1016/j.surfcoat.2004.07.044.
14. Hamada K, Kon M, Hanawa T, Yokoyama K, Miyamoto Y, Asaoka K. Hydrothermal modification of titanium surface in calcium solutions. Biomaterials. 2002;23:2265–72. doi:10.1016/S0142-9612(01)00361-1.
15. Hanawa T, Ukai H, Murakami K. X-ray photoelectron spectroscopy of calcium ion- implanted titanium. J Electron Spectrosc. 1993;63:347–54. doi:10.1016/0368-2048(93)80032-H.
16. Sato M, Tu R, Goto T. Preparation conditions of CaTiO₃ film by metal-organic chemical vapor deposition. Mater Trans. 2006;47:1386–90. doi:10.2320/matertrans.47.1386.
17. Goto T. High-speed deposition of zirconia films by laser-induced plasma CVD. Solid State Ion. 2004;172:225–9. doi:10.1016/j.ssi.2004.02.034.
18. Chi C, Katsui H, Tu R, Goto T. Preparation of Na–Al–O films by laser chemical vapor deposition. Mater Chem Phys. 2015;160:456–60. doi:10.1016/j.matchemphys.2015.05.024.
19. Chi C, Katsui H, Goto T. Preparation of Na-beta-alumina films by laser chemical deposition. Surf Coat Technol. 2015;276:534–8. doi:10.1016/j.surfcoat.2015.06.019.
20. Chi C, Katsui H, Tu R, Goto T. Oriented growth and electrical property of LiAl₅O₈ film by laser. J Ceram Soc Jpn. 2016;124:111–5. doi:10.2109/jcersj2.15220.
21. Ito A, You Y, Katsui H, Goto T. Growth and microstructure of Ba β-alumina films by laser chemical vapor deposition. J Eur Ceram. 2013;33:2655–6. doi:10.1016/j.jeurceramsoc.2013.04.003.
22. Sato M, Tu R, Goto T. Preparation of hydroxyapatite and calcium phosphate films by MOCVD. Mater Trans. 2007;48:3149–53. doi:10.2320/matertrans.MRA2007145.
23. Sato M, Tu R, Goto T, Ueda K, Narushima T. Hydroxyapatite formation on Ca–P–O coating prepared by MOCVD. Mater Trans. 2008;49:1848–52. doi:10.2320/matertrans.MRA2008097.
24. Goto T, Katsui H. Chemical vapor deposition of Ca–P–O film coating. In: Sasaki K, Suzuki O, Takahashi O, editors. Interface oral health science 2014. Tokyo: Springer; 2015. p. 103–15.
25. Katsui H, Kumagai Y, Goto T. High-speed deposition of highly-oriented calcium titanate film by laser CVD. J Jpn Soc Powder Powder Metall. 2016;63:123–7. doi:10.2497/jjspm.63.123.
26. Katsui H, Kumagai Y, Goto T. High-speed deposition of highly-oriented calcium titanate film by laser CVD. J Jpn Soc Powder Metall. 2016;in press.
27. Kaufman L. Calculation of multicomponent ceramic phase diagrams. Physica B+C. 1988;150:99–114. doi:10.1016/0378-4363(88)90111-8.
28. DeVries RC, Roy R, Osborn EF. Phase equilibria in the system CaO–TiO₂. J Phys Chem. 1954;58:1069–73. doi:10.1021/j150522a005.
29. Sasaki S, Prewitt CT, Bass JD, Schulze WA. Orthorhombic perovskite CaTiO₃ and CdTiO₃: structure and space group. Acta Cryst. 1987;43:1668–74. doi:10.1107/S0108270187090620.
30. Beznosikov BV, Aleksandrov KS. Perovskite-like crystals of the Ruddlesden-Popper series. Crystallogr Rep. 2000;45:792–8. doi:10.1134/1.1312923.
31. Haenle M, Lindner T, Ellenrieder M, Willfahrt M, Schell H, Mittelmeier W, Bader R. Bony integration of titanium implants with a novel bioactive calcium titanate (Ca₄Ti₃O₁₀) surface treatment in a rabbit model. J Biomed Mater Res. 2012;100A:2710–6. doi:10.1002/jbm.a.34186.
32. Bertaut EF, Blum P. Détermination de la Structure de Ti₂CaO₄ par la Méthode Self-Consistante d'Approche Directe. Acta Crystallogr. 1956;9:121–6.
33. Bright NFH, Rowland JF, Wurm JG. The compound CaO.Ti2O3. Can J Chem. 1958;36:492–5. doi:10.1139/v58-070.
34. Rogge MP, Caldwell JH, Ingram DR, Green CE, Geselbracht MJ, Siegrist T. A new synthetic route to pseudo-brookite-type CaTi₂O₄. J Solid State Chem. 1998;141:338–42. doi:10.1006/jssc.1998.7932.
35. Limar' TF, Kisel' NG, Cherednichenko IF, Savos'kina AI. Calcium tetratitanate. Russ J Inorg Chem. 1972;17:292–4.
36. Kisel' NG, Limar' TF, Cherednichenko IF. Calcium dititanate. Inorg Mater Transl Izv Akad Nauk SSSR. 8 (1972):1568–70.

37. Pfaff G. Peroxide route to synthesize calcium titanate powders of different composition. J Eur Ceram Soc. 1992;9:293–9. doi:10.1016/0955-2219(92)90064-K.
38. Pfaff G. Synthesis of calcium titanate powders by the sol-gel process. Chem Mater. 1994;6:58–62. doi:10.1021/cm00037a013.
39. Ancora R, Borsa M, Marchi M. Photocatalytic composites containing titanium and limestone. US 2011/0239906 A1. 2011.
40. Rickerby DS, Jones AM, Bellamy BA. X-ray diffraction studies of physically vapour-deposited coatings. Surf Coat Technol. 1989;37:111–37. doi:10.1016/0257-8972(89)90124-2.
41. Kimura T, Goto T. Rapid synthesis of yttria-stabilized zirconia films by laser chemical vapor deposition. Mater Trans. 2003;44:421–4. doi:10.2320/matertrans.44.421.
42. Goto T. Thermal barrier coatings deposited by laser CVD. Surf Coat Technol. 2005;198:367–71. doi:10.1016/j.surfcoat.2004.10.084.
43. Lu TJ, Levi CG, Wadley HNG, Evans AG. Distributed porosity as a control parameter for oxide thermal barriers made by physical vapor deposition. J Am Ceram Soc. 2001;84:2937–46. doi:10.1111/j.1151-2916.2001.tb01118.x.
44. Weiss JR, Diefendorf RJ. Chemically vapor deposited SiC for high temperature and structural applications. Silicon Carbide. Proceedings of the Third International Conference on Silicon Carbide. 1973.

Open Access This chapter is distributed under the terms of the Creative Commons Attribution 4.0 International License (http://creativecommons.org/licenses/by/4.0/), which permits use, duplication, adaptation, distribution and reproduction in any medium or format, as long as you give appropriate credit to the original author(s) and the source, provide a link to the Creative Commons license and indicate if changes were made.

The images or other third party material in this chapter are included in the work's Creative Commons license, unless indicated otherwise in the credit line; if such material is not included in the work's Creative Commons license and the respective action is not permitted by statutory regulation, users will need to obtain permission from the license holder to duplicate, adapt or reproduce the material.

Part II
Symposium II: Innovation for Oral Science and Application

Chapter 5
Development of a Robot-Assisted Surgery System for Cranio-Maxillofacial Surgery

Chuanbin Guo, Jiang Deng, Xingguang Duan, Li Chen, Xiaojing Liu, Guangyan Yu, Chengtao Wang, and Guofang Shen

Abstract Medical robots have been developed rapidly in recent years. Clinical application of da Vinci system showed its advantages. Currently, there is no specialized robot system for cranio-maxillofcial surgery. We developed a cranio-maxillofacial surgical robot system focusing on the reconstruction of mandibular defects.

With the funding of the Chinese National High Technology Research and Development Program (863 Program), we developed a computer-aided design (CAD) system for four typical operations: reconstruction of mandibular defects, orbit reconstruction, skull base tumor resection, and orthognathic surgery, a navigation system and a robot with three arms for mandibular reconstruction. In the CAD system, the operation pattern was designed based on surgeons' habits and experiences. The software system was easy to be used with many functions for designing different surgical procedures. In surgical navigation system for guiding the robot, the hardware of the navigation system was assembled, and the software system of real-time registration was realized. The robot was designed and assembled. It had three arms and was able to finish the bone graft placement precisely and automatically under the navigation guidance according to the preoperative design. The whole system was assessed by model and animal experiment with good results.

Keywords Robot • Cranio-maxillofacial surgery • Navigation

C. Guo (✉) • J. Deng • X. Liu • G. Yu
School of Stomatology, Peking University, 22 Zhongguancun Nandajie,
Haidian District, Beijing 100081, People's Republic of China
e-mail: guodazuo@sina.com

X. Duan
Beijing Institute of Technology, Beijing, China

L. Chen
Tsinghua University, Beijing, China

C. Wang • G. Shen
Shanghai Jiao Tong University, Shanghai, China

© The Author(s) 2017   65
K. Sasaki et al. (eds.), *Interface Oral Health Science 2016*,
DOI 10.1007/978-981-10-1560-1_5

5.1 General Information of Surgical Robots

Surgical robots have been tried in many fields of surgery like in orthopedics, micro-surgery, neurosurgery, ENT surgery, catheterization procedure, etc. The first robot assisting orthopedic operations was made by Integrated Surgical System Company, called ROBODOC system, in 1992 [1, 2]. It was developed from industry robots and was able to perform operations like knee joint replacement. Since then the other robots assisting surgery were developed, but very few of them could be used in clinic. In August of 2015, a newly developed robot by a research team in Beijing was used to assist spine surgery in a real patient and got satisfactory result. Currently the commercially available surgical robot is da Vinci surgical robot which helped in finishing prostatectomy for the first time in the University of Frankfurt in 2000 and has become the main and much welcome surgical robot in the world. So far, as we know, there is no specially designed surgical robot for cranio-maxillofacial (CMF) surgery [3–6]. We began to develop our system in 2010 and have made two robots: one for mandibular reconstruction and the other for needle insertion to diagnose and treat lesions in the skull base. In this chapter we introduce the first robot system.

Our robot-assisted surgery (RAS) system for CMF surgery consisted of the following three parts: computer-aided design (CAD) system for CMF surgery, surgical robot for CMF surgery, and medical experiment for the CAD and robot. The outlines of RAS system for CMF surgery is showed in Fig. 5.1.

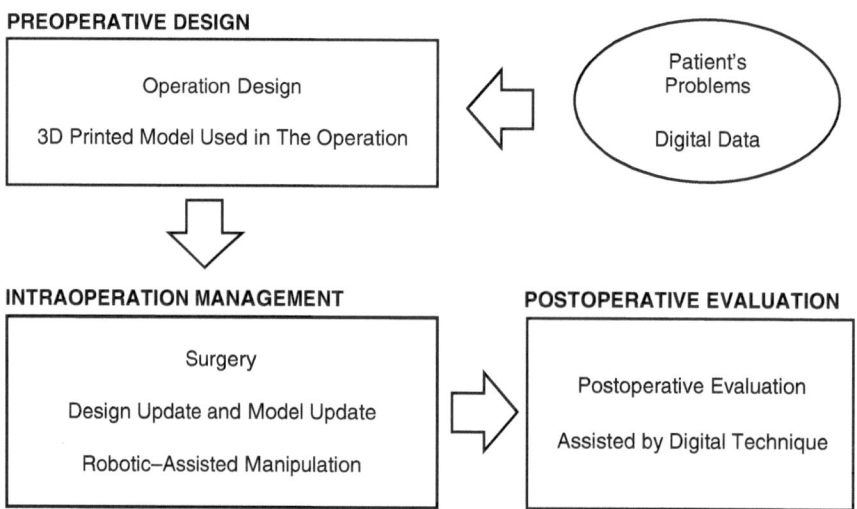

Fig. 5.1 The outlines of RAS system for CMF surgery

5.2 Computer-Aided Design (CAD) System for CMF Surgery

The CAD was designed for four typical and more or less difficult operations: orthognathic surgery, reconstruction of the orbital defects, resection of skull base tumors, and reconstruction of mandibular defects. It has the following five main functions: volume rendering of medical images, segmentation on medical images, CT/MRI image fusion, craniofacial database service, and virtual surgery designing.

The software is easy to use with good interactive regulation of transfer function. It can produce high-quality volume rendering of medical images and get quick and precise segmentation of soft tissue tumors based on graph-cut algorithm. Abstracted tumor from CT data can be visualized three dimensionally and measured for its volume with 99.5 % agreement of the true volume of the tumor. GPU marching cubes algorithm was used to build highly precise three-dimensional reconstruction of bony structures of the skeleton from CT images. The CT/MRI image fusion is another character of the CAD; it well fuses the bony structures abstracted from CT with the vessels abstracted from MRI and creates a new image with more information for more precise medical image analysis and treatment planning. It can also fuse other images obtained by the other image equipments, like fusion of CT data with 3D laser scan image data for the dentition.

We also built a CMF database with more than 20,000 CT data of CMF to assist preoperative design. Currently, one of its functions we already used in clinic is to search for "similar skull." We selected 71 landmarks on the CMF bones. Based on the landmarks, it can retrieve "similar skull" in a few seconds.

With the above mentioned basic functions, the CAD performs its most important task: carrying virtual surgery design for different operations, mainly for orthognathic surgical management, mandibular reconstruction, orbital reconstruction, and removal of the skull base tumors. The designing system is easy to be learned and used for virtual surgical planning step by step. All procedures can be stored and repeated, which helps young surgeons to learn how to design and do these operations. After the virtual planning is done, 3D stereolithographic model (STL model) can be manufactured from the virtual planning model by using rapid prototyping technique for model surgery. Or the virtual planning data is inputted into the navigation system to guide surgeons or robots to perform operations according to the virtual preoperative plan. Thus the precise real-time registration of the CAD system with navigation system is the key point for realization of accurate navigation.

5.3 Surgical Robot for CMF Surgery

The development of the robot had the following main contents: conformation of robotic system, control system, safety control of the robotic system, navigation and trail program, and test of the robot.

5.3.1 Conformation of the Robotic System

The robot had three arms and six degrees of freedom, its hold strength was 30–50 N, and its working square was $200 \times 200 \times 200$ mm. Each arm consisted of two segments and three joints. Figure 5.2 shows the mechanical structure of the robot and the appearance of the robot after assembling.

5.3.2 Safety Control of the Robotic System

Safety is the first consideration for clinical application of medical robots. It was reported that more than 60 % of adverse events in clinical use were caused by robots' malfunction. Our robot safety control involved the workflow from patients' diagnosis, virtual surgical planning, navigation to robot assisting surgery, robotic system itself like essential electro-circuit for safety control, interactive process, and

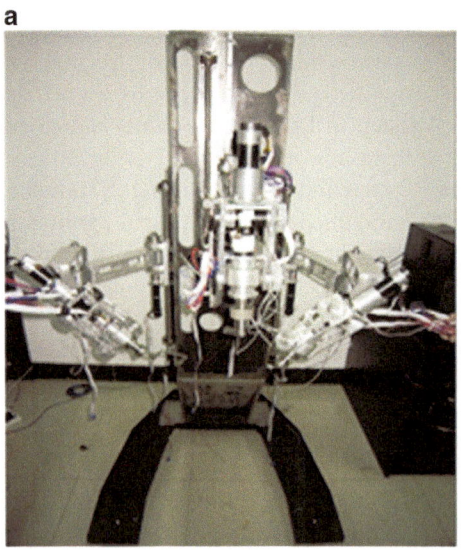
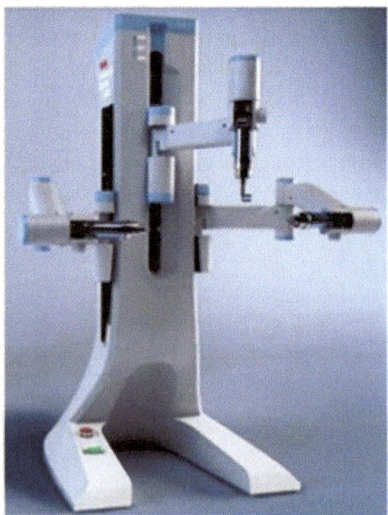

Fig. 5.2 (**a**) The mechanical structure of the robot. (**b**) The appearance of the robot

recovery mechanisms when unexpected faults occur. With these measures, patient safety and operator safety can be guaranteed.

5.3.3   Navigation and Trail Program

Navigation was realized by coordinate switch of four coordinates: 3D object space, image coordinate {V}; physical space, patient coordinate {P}; robotic space, robotic coordinate {R}; and optical space, optical coordinate {M}. The algorithms of trail program included cube polynomial and quintic polynomial.

After all these works were completed, the whole system of the robot was packaged. It consists of the robot, optical tracker, and workstation.

5.3.4   Test of the Robot

We assessed positioning accuracy of the robot and found its repeat positioning error was less than 0.10 mm and the systemic absolute positioning error <2.45 mm, which met the requirement of the design indicators for performance check. And the robot was ready for medical experiment.

5.4   Medical Experiment for the Robotic System

5.4.1   Evaluation of the CAD Software

Twenty-five young oral and maxillofacial surgeons were asked to use and evaluate the CAD software. The satisfaction degree was 91 %.

5.4.2   Accuracy of the Three-Dimensional Reconstruction of the CAD Software

The accuracy of three-dimensional reconstruction and the accuracy of the navigation system and the robot were all tested by model or animal study.

Thirty model skulls were used to test the accuracy of the three-dimensional reconstruction of the CAD software. Titanium screws of 0.1 mm were fixed on the models. The marked models were CT scanned. Reconstruction of the models was performed on the CT data. The comparative study between the reconstructed models and the real models showed that the error was 0.21 mm.

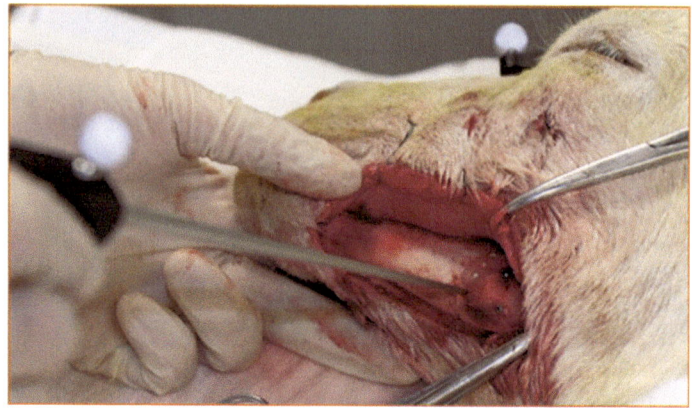

Fig. 5.3 Making osteotomy on the mandible under navigation

5.4.3   Accuracy of the Navigation System

For navigation assessment, the flowchart of the animal experiment includes the following: (1) use living goats as study animal, (2) fix marks on the cranio-maxillofacial region, (3) have CT scan, (4) have virtual surgical design, (5) do operation under navigation, (6) have postoperative CT scan, and (7) compare pre- and postoperative results. All of the four typical operations were tested for navigation accuracy. Comparison of postoperative results with preoperative design showed that the error of mandibular reconstruction was 2.9 mm, error of orthognathic surgery 2.7 mm, error of orbital reconstruction 1.2 mm, and error of skull base surgery 1.8 mm. There are many factors that can affect the accuracy of navigation. These factors are CT scan slice thickness, resolution and radiation dosage, shadow of titanium plates, accuracy of three-dimensional reconstruction, titanium plate bending, movements of remaining bone segments, etc. Current animal study results are acceptable for clinical application.

5.4.4   Animal Experiment of Robot-Assisted Operations

The experimental flow was similar to the abovementioned navigation study. After the animal CT data were collected, the preoperative design was planned using the CAD system. The robot and animal registration was performed, the surgery of making osteotomy on the mandible was done, and the fibular bone graft was fabricated under navigation according the preoperative design (Fig. 5.3). When the mandibular defect was made, the robots' two side arms held two remaining bone segments, the

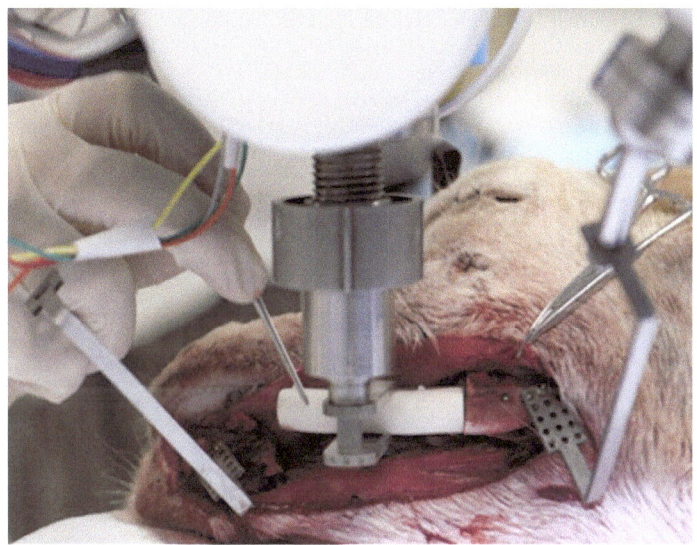

Fig. 5.4 The middle arm carried the bone graft automatically to the defect

middle arm held the bone graft, and under navigation the middle arm carried the bone graft automatically to the defect (Fig. 5.4). After the bone graft was positioned, it was fixed with titanium plates. Comparison of postoperative results with preoperative design showed that errors of left, right, and middle arms were 1.15 mm, 2.68 mm, 2.175 mm, respectively.

The robot animal study showed that the robot can move smoothly and do accurate placement of bone graft and aid fixation. Its three arms can coordinate well to aid bone graft placement for different types of mandibular defects. Its flexibility, bone holding method, miniaturization, etc. need improvement for higher stability and accuracy.

Conflict of Interest   There is no conflict of interest. This project was funded by the Chinese National High Technology Research and Development Program (863 Program), China.

References

1. Korb W, Marmulla R, Raczkowsky J, Muhling J, Hassfeld S. Robots in the operating theatre—chances and challenges. Int J Oral Maxillofac Surg. 2004;33(8):721–32. doi:10.1016/j.ijom.2004.03.015.
2. Taylor RH, Joskowicz L, Williamson B, Gueziec A, Kalvin A, Kazanzides P, et al. Computer-integrated revision total hip replacement surgery: concept and preliminary results. Med Image Anal. 1999;3(3):301–19.

3. Byeon HK, Holsinger FC, Kim DH, Kim JW, Park JH, Koh YW, et al. Feasibility of robot-assisted neck dissection followed by transoral robotic surgery. Br J Oral Maxillofac Surg. 2015;53(1):68–73. doi:10.1016/j.bjoms.2014.09.024.
4. Duan XG, Guo CB. Research and application of robot technology in surgical auxiliary operation. Zhonghua Kou Qiang Yi Xue Za Zhi. 2012;47(8):453–7.
5. Genden EM, Desai S, Sung CK. Transoral robotic surgery for the management of head and neck cancer: a preliminary experience. Head Neck. 2009;31(3):283–9. doi:10.1002/hed.20972.
6. Hassfeld S, Muhling J. Computer assisted oral and maxillofacial surgery—a review and an assessment of technology. Int J Oral Maxillofac Surg. 2001;30(1):2–13. doi:10.1054/ijom.2000.0024.

Open Access This chapter is distributed under the terms of the Creative Commons Attribution 4.0 International License (http://creativecommons.org/licenses/by/4.0/), which permits use, duplication, adaptation, distribution and reproduction in any medium or format, as long as you give appropriate credit to the original author(s) and the source, provide a link to the Creative Commons license and indicate if changes were made.

The images or other third party material in this chapter are included in the work's Creative Commons license, unless indicated otherwise in the credit line; if such material is not included in the work's Creative Commons license and the respective action is not permitted by statutory regulation, users will need to obtain permission from the license holder to duplicate, adapt or reproduce the material.

Chapter 6
Facilitating the Movement of Qualified Dental Graduates to Provide Dental Services Across ASEAN Member States

Suchit Poolthong and Supachai Chuenjitwongsa

Abstract The Association of Southeast Asian Nations (ASEAN) is a political and economic organization of ten countries located in Southeast Asia. Its one purpose is to promote a free trade area in services among member states. Dentistry is one sector subjected to the free movement policy in which dental practitioners can migrate freely across ASEAN to provide dental services. ASEAN policies and strategies to facilitate the free movement of dental professionals have been developed since 2009. To ensure that dental graduates from all dental schools across ASEAN possess comparable standards of practice, harmonization of undergraduate dental education across ASEAN was announced in 2015. Currently, ASEAN is focusing on developing common competencies for its dental graduates. To maximize the free movement, ASEAN dental education development will work toward developing curriculum guidelines and quality assurance for ASEAN undergraduate dental curricula as well as educational and academic development for ASEAN dental schools and staff.

Keywords Free movement of dental professionals • Harmonizing undergraduate dental education • ASEAN dental graduates

S. Poolthong
Faculty of Dentistry, Department of Operative Dentistry, Chulalongkorn University, 254 Phayathai Road, Bangkok 10100, Thailand

S. Chuenjitwongsa (✉)
Faculty of Dentistry, Department of Biochemistry, Chulalongkorn University, 254 Phayathai Road, Bangkok 10100, Thailand
e-mail: supachai.c@chula.ac.th

© The Author(s) 2017
K. Sasaki et al. (eds.), *Interface Oral Health Science 2016*,
DOI 10.1007/978-981-10-1560-1_6

6.1 Formation of ASEAN to AJCCD

The Association of Southeast Asian Nations (ASEAN) is a political and economic organization of ten countries located in Southeast Asia: Brunei Darussalam, Cambodia, Indonesia, Laos PDR, Malaysia, Myanmar, the Philippines, Singapore, Thailand, and Vietnam. Its aims include accelerating free trade area among its member states, protection of regional peace and stability, and opportunities for member countries to discuss differences peacefully. In 1995, ASEAN Framework Agreement on Services (AFAS) was established and signed by ASEAN Economic Ministers. Main purposes are to enhance cooperation in services among member states, eliminate substantially restrictions to trade in services, and liberalize trade in services with the aim to realizing a free trade area in services within ASEAN member states. Later on, Coordinating Committee on Services (CCS) was established on its purpose to undertake service integration initiatives within six service sectors, and one sector was healthcare services. Healthcare Services Sectoral Working Group (HSSWG) was then initiated to discuss matters pertaining to facilitation and cooperation within three healthcare professions, and dental profession was one of them. Finally, ASEAN Joint Coordinating Committee on Dental Practitioners (AJCCD), a committee under the HSSWG, was started to facilitate cooperation on Mutual Recognition Arrangement (MRA) for dental practitioners.

6.2 MRA for Free Flow of Services and Professionals

The MRA on dental practitioners was finally signed in February 2009 and planned for implementation in 2015. Its objectives are to facilitate mobility of dental practitioners within ASEAN, enhance exchange of information and expertise on standards and qualifications, promote adoption of best practices for professional dental services, and provide opportunities for capacity building and training of dental practitioners. According to the mobility of dental practitioners within ASEAN, there is a concern on the standard of dental services by dentists across the ASEAN. With this regard, harmonization of undergraduate dental education in order to create comparable dental curricula of all dental schools in ASEAN member states is necessary.

ASEAN MRA on dental practitioners was first established in Singapore in 2008. It has been the issue of discussion in the AJCCD meeting ever since. In the MRA, it was stated that this MRA would be discussed for improvement and be revised officially every 5 years. The detail of MRA consists of ten articles [1]. A summary of ten MRA articles is provided in Table 6.1.

For the content of MRA, an introduction provides some basic understanding and terminologies including objectives of AFAS, vision of ASEAN, and some notes regarding level of recognition on education and certification granted in other

Table 6.1 A summary of ten MRA articles

Article	Summary
1	Outlining the four objectives of the MRA: (1) to facilitate mobility of dental practitioners within ASEAN, (2) to exchange information and enhance cooperation in respect of mutual recognition of dental practitioners, (3) to promote adoption of best practices on standards and qualifications, and (4) to provide opportunities for capacity building and training of dental practitioners.
2	Providing definitions of professional terminologies used in dental services and the names of those authorized organizations, namely, Professional Dental Regulatory Authority (PDRA). The PDRA is responsible for dental regulations in their own countries. Those PDRAs vary from Medical Board, Medical Council, Dental Council, and Professional Regulation Commission to Ministry of Health. Majorities are for Dental Council and Ministry of Health.
3	Explaining recognition and eligibility of foreign dental practitioners in host countries. The foreign dental practitioners can apply for registration in the host country in accordance with its domestic regulations. With these regulations, they generate limitations which may not strongly facilitate free flow of dental practitioners such as the nationality issue and the examination using local languages. Though English is used as a common language in ASEAN, local languages play important role in the eligibility to apply for registration in many countries especially the countries outside the commonwealth.
4	Defining roles of PDRA and its duties including: to evaluate qualification, to impose other requirements where applicable, to grant recognition and register eligibility to practice in the host country, to monitor the compliance of foreign dental practitioners in order to conform with the domestic regulations, and to take action in case one fails to practice under domestic regulation.
5	An agreement from all member states indicating that they will not reduce, eliminate, or modify power and authority of PDRA.
6	Explaining about the establishment of AJCCD as a working group under HSSWG. The AJCCD committee comprises two representatives from each ASEAN country. The AJCCD's important role is to facilitate the implementation of MRA through better understanding of domestic regulation. AJCCD also encourages information exchange among all members and will review the MRA every 5 years if necessary.
7	Discussion about the possibility of mutual exemption where each PDRA has responsibility to protect health, safety, environment, and welfare of the community within its authority.
8	Containing the critical issue of dispute settlement. In order to avoid conflict, every attempt through communication, consultation, and cooperation is expected to mutually satisfy any matter that might affect the implementation of the MRA.
9	Stating about amendment that can be done through only mutual written agreement by the representatives of all member states. However, the PDRA can be done by its own country representative regardless of other members' agreements.
10	Representing the final provisions that was agreed for the deferment of the MRA, if desired, by any member to be informed within 6 months after MRA signing.

ASEAN countries as well as their expectation from the MRA to facilitate free flow of dental practitioners across the region. This is to form a highly competitive ASEAN economic region resulting in free flow of goods, services and investment, equitable economic development, reduced poverty and socioeconomic disparities, and enhanced political as well as economic and social stability. One part of free flow of goods, services, and investment is a free flow of dental professionals and dental services. The expected outcomes of the MRA are to strengthen professional capabilities by promoting flow of relevant information and exchanging expertise, experiences, and best practices suited to the specific needs of ASEAN member states.

6.3 Impact on Dental Education

To achieve high standard dental services from new dental graduates across the ASEAN member states, all dental schools ideally need an agreed upon common curriculum with similar learning outcomes and competencies of the graduates. The common curriculum must be of a high standard that is comparable to international standards such as the standards for undergraduate dental curricula in Europe [2]. Because there is currently no such curriculum, learning outcomes, or competencies in the ASEAN region, developing a standard curriculum is one of the keys to success for the free movement of dental professionals across the region and will provide new graduates with the common dental competencies required for all ASEAN dentists.

Competency is "a combination of knowledge, skills, professional attitude, personal attributes, an ability to work independently (without direct supervision), and context" [3]. Competency is a primary concern in competency-based dental curriculum [4] in which the societal needs for oral healthcare, institutional factors (e.g., structures of the dental school), and national factors (e.g., educational policy, politics) are taken into account [3]. In several parts of the world (e.g., USA, Europe), a list of competencies for new dental graduates has been already developed and implemented [5, 6]. However, in Southeast Asia, the common profile and competencies for ASEAN dentists have not yet been established. General competencies for dentists in any country could be similar; for example, a dentist should be able to perform a simple extraction. However, the community character, local sociocultural context, and oral health status can influence how the competencies are set up [3]. Thus, competencies for ASEAN dentists might be different from dentists in other areas outside Southeast Asia. Such competencies for ASEAN dentists, in particular, need to be identified and developed. New graduates from the ASEAN dental competencies would fulfill the purpose of facilitating the free movement of dental professionals and dentists across the Southeast Asia region.

Developing competencies for ASEAN dentists will also lead to further dental education development including standards for an undergraduate dental curriculum, quality assurance, teacher training, and collaborative research in dental education.

6.4 Facilitating Free Movement of Dental Professionals Across ASEAN

6.4.1 Current Work

In May 2015, the Faculty of Dentistry, Chulalongkorn University, hosted the ASEAN Dental Forum (ADF) where 26 of the AJCCD members from ASEAN member states and representatives from Thai dental schools have attended and signed the Bangkok Declaration of 22 May 2015. The declaration emphasizes that ASEAN dental schools will work together, on the basis of trust and willingness, toward primary goals to achieve the harmonization of dental education across the ASEAN region by (1) developing common competencies for the safe and independent practice of ASEAN general dental practitioners and (2) developing an ASEAN undergraduate dental curriculum. Currently the harmonization of ASEAN undergraduate dental education is one part of the AJCCD work plan 2016–2020. With an agreement of the AJCCD, the Faculty of Dentistry, Chulalongkorn University, is now the leading institution responsible for conducting an international meeting (ADF) to develop major competencies for ASEAN dentists. The next ADF meeting will be held in late August 2016.

In order to improve quality and gain acceptability of the common competencies developed by the ADF, information, opinions, and agreements from stakeholders within dental education are required. As a part of the ADF, representatives from ASEAN member states (educators and policy makers) are included in the meeting panel, and a wider audience (practitioners, academics, patients) will have an opportunity to get involved in a public hearing. ASEAN students – who are directly affected by the common competencies developed by the ADF – will be invited to attend a meeting in September 2016 (after the ADF meeting) to provide the "student voice" toward creating common dental competencies for ASEAN dentists. The student representatives will provide in-depth information and critique based on their perspectives (i.e., students as consumers of dental education). The consensual agreements from student representatives will be combined with the consensual agreements from the ADF (educators and policy makers) to develop the first draft of common competencies. When the first draft is written, a public hearing will be held, before being passed on to the main AJCCD body for final approval and implementation. Then the AJCCD member from each country will pass this message to the policy makers of the ASEAN member states and start to set the time line for the new curriculum of all dental schools across ASEAN to comply with the common competencies for ASEAN dentists. As a result, there will be less concern of the quality of dentists when they cross the border for free movement of dental services as required by the MRA.

6.4.2   Future Work

Only developing common competencies for ASEAN dental graduates may not fully support the free movement of dental professional across ASEAN. Future work that AJCCD and other political bodies need to focus includes four main areas:

1. The guidelines for an undergraduate dental curriculum in ASEAN emphasizing contemporary educational strategies, assessment, and credit transfer system in relation to the principles of competency-based dental education
2. Quality assurance for an undergraduate dental curriculum in ASEAN ensuring that the curriculum is developed and regularly improved in line with the ASEAN and international standards
3. Educational development policies and strategies embedding research-based education, enterprise education, public engagement, and modern educational methods (e.g., technology-enhanced learning, evidence-based practice) into the ASEAN undergraduate dental education
4. Academic training and continuing professional development for ASEAN dental educators/staff in the area of educational practice

6.5   Conclusion

Facilitating the free movement of dental professionals across ASEAN is not a straightforward process. This paper has summarized how the free movement of dental professionals across ASEAN was initiated and how harmonization of undergraduate dental education has been fundamental to the free movement process. The success of this process strongly relies on the mutual respect and collaboration among stakeholders especially ASEAN dental educators, students, practitioners, policy makers, and the whole ASEAN community.

References

1. ASEAN. ASEAN mutual recognition arrangement on dental practitioners. New York: The Association of Southeast Nations; 2009. http://investasean.asean.org/files/upload/MRA%20 Dental%20Practitioner%20(Feb09).pdf. Accessed on 26 Aug 2016.
2. Manogue M, McLoughlin J, Christersson C, Delap E, Lindh C, Schoonheim-Klein M, et al. Curriculum structure, content, learning and assessment in European undergraduate dental education – update 2010. Eur J Dent Educ. 2011;15:133–41.
3. Chuenjitwongsa S, Oliver R, Bullock A. Competence, competency-based education, and undergraduate dental education: a discussion paper. Eur J Dent Educ. 2016. doi:10.1111/eje.12213.

4. Chambers DW. Competency-based dental education in context. Eur J Dent Educ. 1998;2:8–13.
5. ADEA. ADEA competencies for the new general dentist. J Dent Educ. 2010;74:765–8.
6. Cowpe J, Plasschaert A, Harzer W, Vinkka-Puhakka H, Walmsley AD. Profile and competences for the graduating European dentist – update 2009. Eur J Dent Educ. 2010;14:193–202.

Open Access This chapter is distributed under the terms of the Creative Commons Attribution 4.0 International License (http://creativecommons.org/licenses/by/4.0/), which permits use, duplication, adaptation, distribution and reproduction in any medium or format, as long as you give appropriate credit to the original author(s) and the source, provide a link to the Creative Commons license and indicate if changes were made.

The images or other third party material in this chapter are included in the work's Creative Commons license, unless indicated otherwise in the credit line; if such material is not included in the work's Creative Commons license and the respective action is not permitted by statutory regulation, users will need to obtain permission from the license holder to duplicate, adapt or reproduce the material.

Chapter 7
Putting the Mouth into Health: The Importance of Oral Health for General Health

Christopher C. Peck

Abstract Good oral health is important to overall health and wellbeing. From the most fundamental perspectives of the oral–systemic health relationship, good oral health ensures an individual can eat, speak and socialise without pain or embarrassment. These capabilities improve an individual's wellbeing and ability to contribute to society. There are more complex oral–systemic health relationships in which poor oral health can contribute to systemic diseases including the links between orofacial disease and, for example, diabetes, cardiovascular disease, pneumonia or rheumatoid arthritis. Furthermore, the oral environment is a complex biological ecosystem which provides an opportunity to understand mechanisms underlying pathological processes and which can be a model for other body systems. Examples of these oral–systemic interactions are presented to demonstrate the importance of the orofacial environment and the need for it to be incorporated into health more broadly. By putting the mouth into health through integrated education, research and clinical care across the health disciplines, there will be advances in health that will improve individual and community health.

Keywords Health • Oral–systemic health • Pain • Disability

7.1 Introduvction

The orofacial region is unique because of its importance in social interactions, including communication and emotional expression, and in survival with mastication and lifesaving reflexes that protect other external sensory (e.g. retina, olfactory epithelium, taste receptors) and internal homeostatic (e.g. respiratory and digestive tracts) systems from damaging environmental changes. Consequently this region and good oral health are important components to overall health and wellbeing.

C.C. Peck (✉)
Faculty of Dentistry, The University of Sydney, Level 2, G12 – Services Building,
22 Codrington Street, Darlington, NSW 2006, Australia
e-mail: dentistry.dean@sydney.edu.au

© The Author(s) 2017
K. Sasaki et al. (eds.), *Interface Oral Health Science 2016*,
DOI 10.1007/978-981-10-1560-1_7

From the most fundamental perspectives of the oral–systemic health relationship, good oral health ensures an individual can eat, speak and socialise without pain or embarrassment. These capabilities improve an individual's wellbeing and ability to contribute to society. There are more complex oral–systemic health relationships in which poor oral health can contribute to systemic diseases including the links between orofacial disease and, for example, diabetes, cardiovascular disease, pneumonia or rheumatoid arthritis. Furthermore, the oral environment is a complex biological ecosystem which provides an opportunity to understand mechanisms underlying pathological processes, and which therefore can be a model for other body systems.

There have been significant advances in oral health improvement over the past 30 years, where there has been a change in focus from the acute management of the disease burden of dental caries and periodontal disease to an increased focus on prevention and more complex oral rehabilitation to improve function and aesthetics. Nevertheless, in Australia as in many parts of the world, there is still a worrying oral disease burden; oral cancer continues with significant morbidity and mortality [1], 30 % of adults have untreated tooth decay [2], 15 % of adults experience oral pain [3], 25 % adults are uncomfortable about their dental appearance [3] and oral problems account for the third highest level of acute preventable hospital admissions [4]. Furthermore they can affect self-esteem, social interaction, education, career achievement and emotional state [5] and lead to a deteriorating diet and compromised nutrition [6].

7.2   The Oral–Systemic Nexus

Whilst many countries, including Japan and Australia, have universal healthcare systems, frequently dental care is partly or completely excluded. This leads to societal inequality with groups, such as Aboriginal Australians, who are on low income, with limited education or living in remote areas suffering worse oral health. In a similar way that many governments exclude oral health from healthcare, many individuals, when prioritising household expenditure, consider oral health as a cosmetic issue rather than essential to overall health. In Australia over 45 % of adults over 25 years did not visit a dental practitioner in the past year [7].

There is much research indicating links between oral health and, for example, cardiovascular disease, mental health and diabetes. There are interesting findings showing that a person with fewer than ten of their own teeth is seven times more likely to die of coronary disease than someone with more than 25 of their own teeth [8] and that treating gum disease improves vascular health [9]. With Alzheimer's disease, a study of more than 4000 Japanese participants, aged 65 and older, found those with missing teeth were much more likely to have experienced memory loss or have early-stage Alzheimer's disease than those with an intact dentition [10]. The relationship between diabetes mellitus and periodontal disease appears bidirectional with diabetes increasing the risk for periodontitis and periodontal inflammation negatively affecting glycaemic control [11]. The relationship between the two diseases is strong and in fact periodontal disease has been coined the sixth complication of diabetes [12]. These and many other associations between oral and systemic health are important and the focus of much research, particularly exploring aetiological mechanisms underlying these associations.

As a result of the strong links between oral and systemic health, we have embarked on a strategic initiative to put the mouth into health, through education, research and clinical care. Our research endeavours are focussed on oral–systemic health interactions in exploring known associations, using the oral environment as a biomarker and as a model for disease processes or clinical care. Some examples of these endeavours follow.

7.3   Oral–Systemic Health Associations

There are strong associations between rheumatoid arthritis and periodontal disease, and it is important to determine if this is because they are two common chronic diseases or indeed if there is a causal link between the two diseases [13]. Interdisciplinary research suggests that the periodontal pathogen *Porphyromonas gingivalis* exacerbates collagen-induced arthritis (rheumatoid arthritis). The bacteria express an enzyme that causes human and bacterial protein citrullination (conversion of the amino acid arginine into citrulline) and autoantibodies to these citrullinated proteins. A pathogenic autoimmune response to these citrullinated proteins ensues causing earlier onset, accelerated progression and enhanced severity of the arthritic disease [14]. These findings may help with diagnosis and targeted treatments.

Chronic orofacial pain can be disabling affecting approximately 10 % of the adult population [15]. Pain is frequently not limited to the orofacial region; a local survey demonstrated up to 60 % of clinical patients reported pain elsewhere in their body (Fig. 7.1). Furthermore, subjects with temporomandibular disorders frequently had symptoms of widespread conditions including chronic fatigue syndrome and fibromyalgia [16]. We have demonstrated that temporomandibular disorders can be associated with catastrophic beliefs and with depression and that these psychological factors can impact jaw motor activity [17, 18].

Oral health appears to be related to physical health, specifically cardiorespiratory fitness. In a clinical study of 72 men, peak oxygen uptake during exercise testing was significantly lower, indicating lower levels of fitness, in those with moderate to severe periodontitis [19].

In an interesting study comparing genetic details of calcified dental plaque from 34 early European skeletons, it was shown that the transition from hunter-gatherer to farming shifted the oral microbial community to a disease-associated configuration. During the Industrial Revolution, cariogenic bacteria became dominant, and the modern oral microbial community is much less diverse than historic populations [20]. This may be contributing to contemporary chronic diseases, affecting both oral and systemic environments.

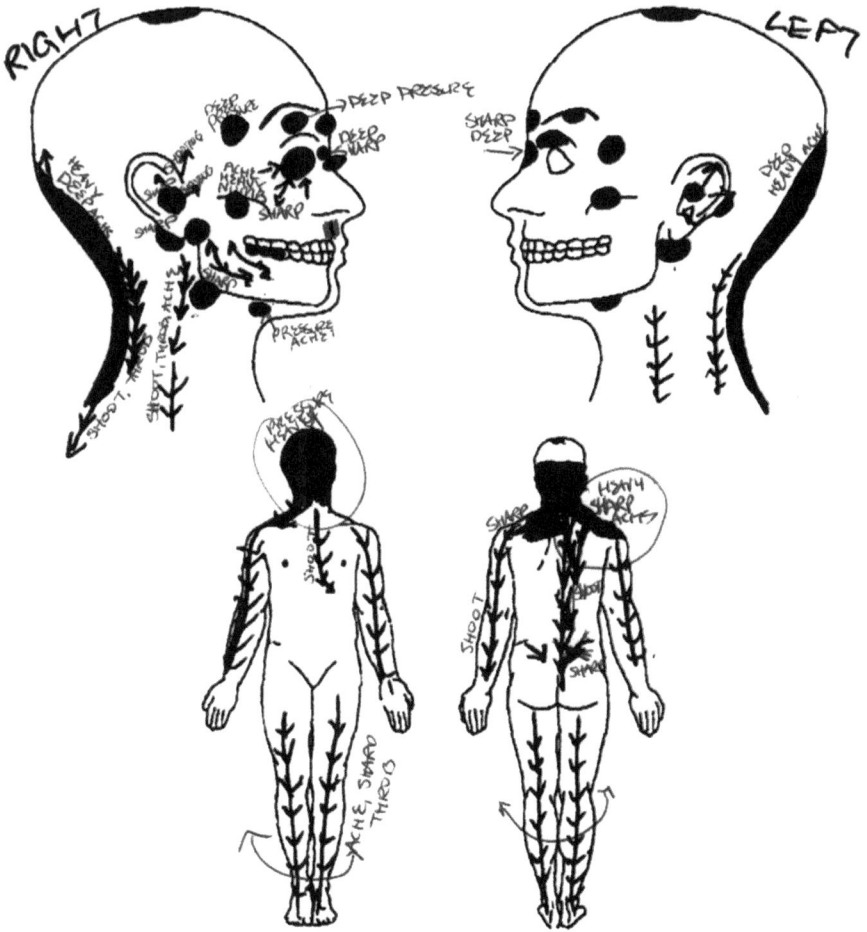

Fig. 7.1 An example of a pain map for an orofacial pain patient demonstrating pain extending beyond the oral and facial regions (*shaded* regions and *arrows* located painful regions and pain spread)

7.4   The Oral Environment: A Model for Other Body Systems

The oral environment is a complex and harsh biological setting connecting the internal body with the external environment. It consists of hard and soft tissues, biofilm and sensory and motor innervation and consequently can be a good model to monitor systemic health and demonstrate interactions between systems and the external environment. It is important for health professionals and especially dental practitioners to promote the importance of oral health to health overall.

The tooth can be a biomarker to determine, for example, past chemical exposure during development. One topical area is iron exposure during development in the form of iron-fortified supplements, and the possible relationship with age-related neurodegenerative disorders. Iron has been implicated in disorders such as Parkinson's disorder and it may be the iron levels during development that are implicated. Iron levels in tooth enamel and dentine can be directly mapped to the source and timing of infant nutrition during development of the tooth, thus providing an opportunity to correlate this with neurodegenerative disorders [21]. This retrospective method is attractive for exploring past chemical exposure during the development years. In turn, findings from this will be able to help inform on development of disorders and health policy for nutritional supplements.

Glial cells which make up the majority of cells in the brain were considered previously as scaffolds to support neurons. We now know that these cells' role is much more, and recent work has demonstrated an important role in the dental pulp. With dental caries, and associated hypoxic stress, there is angiogenic remodelling of the pulpal microvasculature through an interaction of endothelial cells, pericytes, microglia and telacytes in the pulp. This response provides vascular remodelling whilst maintaining a blood barrier. The directional expansion of the microvasculature is achieved through glial-assisted guided migration of endothelial cells [22]. This important function can be compared with other glial cell activities in the nervous system.

Temporomandibular disorders are the most common chronic pain disorders in the orofacial region and consist of a number of problems affecting the jaw's musculoskeletal system including the temporomandibular joints or masticatory muscles. The classification scheme for these disorders has been a model for other pain conditions and disorders as it includes both physical and psychosocial dimensions [23]. Recently it has been revised [24] with validated diagnostic criteria and an extended taxonomy has also been developed [25]. This research on temporomandibular disorders spanning 26 years provides a valuable record for developing diagnostic frameworks for other conditions throughout the body.

7.5 Summary

It is imperative to put the mouth into health to improve health and wellbeing of the individual and of society. There are many associations between oral and systemic health and research is investigating causal links. Embedding research, education and clinical care across the health disciplines provides opportunities to better understand the role of oral health in a much broader context.

References

1. Australian Institute of Health and Welfare. Head and neck cancers in Australia, Cancer series no. 83. Cat. no. CAN 80. Canberra: AIHW; 2014.
2. Roberts-Thomson K, Luzzi L, Brennan D. Social inequality in use of dental services: relief of pain and extractions. Aust N Z J Public Health. 2008;32(5):444–9.
3. Chrisopoulos S, Harford JE. Oral health and dental care in Australia: key facts and figures 2012, Cat. no. DEN 224. Canberra: AIHW; 2013.
4. Australian Institute of Health and Welfare. Australian hospital statistics 2012–13, Health services series no. 54. Cat. no. HSE 145. Canberra: Australian Institute of Health and Welfare; 2014.
5. US Department of Health and Human Services. Oral health in America: a report of the Surgeon General. Rockville: US Department of Health and Human Services, National Institute of Dental and Craniofacial Research, National Institutes of Health; 2000.
6. Locker D. The burden of oral health in a population of older adults. Commun Dent Health. 1992;9(2):109–24.
7. AIHW, Chrisopoulos S, Harford JE, Ellershaw A. Oral health and dental care in Australia: key facts and figures 2015, Cat. no. DEN 229. Canberra: AIHW; 2016.
8. Holmlund A, Holm G, Lind L. Number of teeth as a predictor of cardiovascular mortality in a cohort of 7,674 subjects followed for 12 years. J Periodontol. 2010;81(6):870–6.
9. Kapellas K, Maple-Brown LJ, Jamieson LM, Do LG, O'Dea K, Brown A, Cai TY, Anstey NM, Sullivan DR, Wang H, Celermajer DS, Slade GD, Skilton MR. Effect of periodontal therapy on arterial structure and function among aboriginal Australians: a randomized. Control Trial Hypertens. 2014;64:702–8.
10. Okamoto N, Morikawa M, Okamoto K, Habu N, Iwamoto J, Tomioka K, Saeki K, Yanagi M, Amano N, Kurumatani N. Relationship of tooth loss to mild memory impairment and cognitive impairment: findings from the fujiwara-kyo study. Behav Brain Funct. 2010;6:77.
11. Preshaw PM, Alba AL, Herrera D, Jepsen S, Konstantinidis A, Makrilakis K, Taylor R. Periodontitis and diabetes: a two-way relationship. Diabetologia. 2012;55(1):21–31.
12. Löe H. Periodontal disease: the sixth complication of diabetes mellitus. Diabetes Care. 1993;16:329–34.
13. de Smit MJ, Brouwer E, Vissink A, van Winkelhoff AJ. Rheumatoid arthritis and periodontitis; a possible link via citrullination. Anaerobe. 2011;17:196–200.
14. Maresz KJ, Hellvard A, Sroka A, Adamowicz K, Bielecka E, Koziel J, et al. Porphyromonas gingivalis facilitates the development and progression of destructive arthritis through its unique bacterial Peptidylarginine Deiminase (PAD). PLoS Pathog. 2013;9(9):e1003627.
15. Zakrzewska JM. Diagnosis and management of non-dental orofacial pain. Dent Update. 2007;34:134–6. 138–9.
16. Aaron LA, Burke MM, Buchwald D. Overlapping conditions among patients with chronic fatigue syndrome, fibromyalgia, and temporomandibular disorder. Arch Intern Med. 2000;160(2):221–7.
17. Henderson LA, Akhter R, Youssef AM, Reeves JM, Peck CC, Murray GM, Svensson P. The effects of catastrophizing on central motor activity. Eur J Pain. 2016;20:639–51.
18. Brandini D, Benson J, Nicholas M, Murray G, Peck C. Chewing in temporomandibular disorder patients: an exploratory study of an association with some psychological variables. J Orofac Pain. 2011;25(1):56–67.
19. Eberhard J, Stiesch M, Kerling A, Bara C, Eulert C, Hilfiker-Kleiner D, Hilfiker A, Budde E, Bauersachs J, et al. Moderate and severe periodontitis are independent risk factors associated with low cardiorespiratory fitness in sedentary non-smoking men aged between 45 and 65 years. J Clin Periodontol. 2014;41(1):31–7.
20. Adler C, Dobney K, Weyrich L, Kaidonis J, Walker A, Haak W, Bradshaw C, Townsend G, Soltysiak A, Alt K, et al. Sequencing ancient calcified dental plaque shows change in oral microbiota with dietary shifts of the Neolithic and Industrial revolutions. Nat Genet. 2013;45(4):450–5.

21. Hare DJ, et al. Is early-life iron exposure critical in neurodegeneration? Nat Rev Neurol. 2015;11(9):536–44. doi:10.1038/nrneurol.2015.100.
22. Farahani RM, Sarrafpour B, Simonian M, Li Q, Hunter N. Directed glia-assisted angiogenesis in a mature neurosensory structure: pericytes mediate an adaptive response in human dental pulp that maintains blood-barrier function. J Comp Neurol. 2012;520:3803–26.
23. Dworkin SF, LeResche L. Research diagnostic criteria for temporomandibular disorders: review, criteria, examinations and specifications, critique. J Craniomandib Disord. 1992;6(4):301–55.
24. Schiffman E, Ohrbach R, Truelove E, et al. Diagnostic Criteria for Temporomandibular Disorders (DC/TMD) for clinical and research applications: recommendations of the International RDC/TMD Consortium Network and Orofacial Pain Special Interest Group. J Oral Facial Pain Headache. 2014;28(1):6–27.
25. Peck CC, Goulet J-P, Lobbezoo F, et al. Expanding the taxonomy of the diagnostic criteria for temporomandibular disorders (DC/TMD). J Oral Rehabil. 2014;41(1):2–23.

Open Access This chapter is distributed under the terms of the Creative Commons Attribution 4.0 International License (http://creativecommons.org/licenses/by/4.0/), which permits use, duplication, adaptation, distribution and reproduction in any medium or format, as long as you give appropriate credit to the original author(s) and the source, provide a link to the Creative Commons license and indicate if changes were made.

The images or other third party material in this chapter are included in the work's Creative Commons license, unless indicated otherwise in the credit line; if such material is not included in the work's Creative Commons license and the respective action is not permitted by statutory regulation, users will need to obtain permission from the license holder to duplicate, adapt or reproduce the material.

Chapter 8
Orofacial Stem Cells for Cell-Based Therapies of Local and Systemic Diseases

Munira Xaymardan

Abstract Orofacial region and dental tissues harbour a wide range of stem/progenitor cells including mesenchymal stem cells and tissue-specific progenitors such as muscle satellite cells. The stem cells of the orofacial areas are readily available, highly proliferative and possess multi-differentiation abilities. These cells not only provide therapeutic and tissue engineering cell source for the defects of orofacial area and dental tissues but also provide additional cell source for the diseases of other organs. Understanding their differentiation pathways and mechanisms will be imperative in developing the most appropriate approaches for stem cell-based tissue engineering and therapeutic strategies.

Keywords Orofacial • Stem cells • Therapy

8.1 Overview

8.1.1 Need for Stem/Progenitor Cell-Based Therapy

Stem cells are defined as the cells that have two important abilities: (1) self-renewal and (2) differentiation. They have remarkable potential to develop into many different cell types in the body during early life and growth [1]. In adults, they serve as a cell source for continuous repair system, dividing essentially without limit to replenish other cells as long as the person or animal is still alive, as best exampled with the haematopoietic system, bone marrow (BM) stem cells providing blood cells throughout a person's life. Other adult organs are less regenerative in comparison; still, minor wears and tears are constantly repaired with stem cells by providing broader matrix turnover and cellular replenishment or by exerting more specialized

M. Xaymardan (✉)
The Faculty of Dentistry, The University of Sydney, Level 2, Westmead Hospital,
Hawkesbury Rd & Darcy Road, Westmead, NSW 2145, Australia
e-mail: Munira.xaymardan@sydney.edu.au

© The Author(s) 2017   89
K. Sasaki et al. (eds.), *Interface Oral Health Science 2016*,
DOI 10.1007/978-981-10-1560-1_8

function, such as differentiating into muscle cells, neurons, or, in the dental context, odontoblasts.

However, in adult mammalians, this reparative ability seems to be limited to minor injuries. In catastrophic events such as myocardial infarction (MI), spinal cord injury (SCI) or autoimmune and degenerative diseases, the endogenous stem cells do not seem to have the capacity to provide sufficient support for the parenchymal tissue regeneration. The lost tissue is often replaced by fibrous scar, compromising the integrity and function of the damaged organs. Therefore, developing approaches that can either activate the endogenous stem cells in vivo or expand them in vitro for transplantation still hold major research interest in the regenerative medicine.

Stem and progenitor cells exert their therapeutic abilities mainly through two mechanisms:

1. Functional cellular replacement of the damaged organs, for example, replacement of the neurons in SCI or replacement of the cardiomyocytes after MI. This aspect is the most desirable property of the stem cell therapeutics but currently is also difficult to achieve with adult stem cell sources. Embryonic stem (ES) cells and the induced pluripotent promise the pluripotency that can be used to differentiate various cell types for transplantation, however with caveats including uncontrolled cell proliferations and teratogenicity limiting the clinical use of these cells.
2. Modulation of the repair processes by providing extracellular matrix remodelling, paracrine trophic factors and immune modulation. More often than not, and in the mesenchymal stem cell (MSC) therapies in particular, the trophic effects seem to play a major role in the improvement of the organ function by modulating vascular supply and maintenance of tissue architectures rather than providing engraftment of the parenchymal cell types. The immunomodulatory effects of the MSCs have been exploited to treat various diseases including cardiac injuries and autoimmune diseases, with cells isolated and investigated from virtually all organs and tissues in the body.

Studies of the orofacial stem cells have been relatively newer events. Majority of the orofacial stem cell studies have been focused on the MSC cell types of the dental pulp, periodontal ligaments, facial BM, submucosal stromal and tongue. In the meantime, studies include generation of the parenchymal cell types of salivary glands, and recent speculation of the possible use of the tongue myogenic progenitors for cardiac repair also hold intriguing perspectives for future exploration of the stem cells in the orofacial region.

Orofacial region provides advantages over other cell sources:

1. Easy accessibility: Cells can be obtained from extracted third molars (pulp and periodontal cell sources), exfoliated deciduous teeth, oral mucosa biopsies and tissues that are extractable without major invasive procedures compared to other sources.

2. Higher potency: For example, MSCs isolated from the dental sources may possess high potencies compared to BM source as discussed below.
3. Complexity: Multi-lineage derivation of the orofacial tissues that contain all three germ layers of embryonic origins (endoderm, ectoderm and mesoderm, as discussed below) that may provide expanded versatility for stem cell therapies.

In this chapter, we discuss the developmental cell sources of the orofacial areas focusing on the discussion of the MSCs derived from orofacial region and their utility in the local and systemic diseases, as well as the potential promise of the orofacial muscle progenitor usage for cardiac regenerations.

8.1.2 Cell Sources During Orofacial Development

During vertebrate development, including humans, the oral and facial region develops very early in the embryos following formation of the cardiac tube and folding of the neural tube, approximately at the end of week 3 of prenatal life. The earliest orofacial progenitors originate posteriorly to the heart field and contain vascular (aortic arch) and muscular progenitors, which are joined by the neural crest mesenchymal (also known as ectomesenchyme) progenitor migration as the neural tube closes [2]. Expansion of these two populations facilitates the formation of the first and second pharyngeal arches that provide initial myogenic progenitors for orofacial musculatures, including masticatory, facial expressions and tongue [3, 4], and the multi-fated neural crest progenitors for formation of facial cartilage, flat bones, teeth and tooth support tissues such as periodontal ligaments [5]. Recent studies indicate that, albeit to a limited extent, adult tissues may retain some of these stem/progenitor cells at various stages of the development, become dormant and, when activated, are able to achieve minor repairs of the tissues during injury. Indeed, studies have shown the MSCs isolated from the orofacial area are neural crest in origin [6] and may have different differentiation potential compared to the MSCs isolated from other tissues such as the BM and heart which are thought to have been derived from the mesodermal origin [7].

At the later stages of the embryonic development, the mesodermal muscle progenitors (paraxial/occipital) and mesenchymal stromal cells aid the expansion of the muscle cell pools and the connective tissue (including bone) maturations. However, these cell sources provoke less interest from orofacial stem cell research point of view as the clonogenic stem cells isolated from this region seem to be more neural crest in origin [8].

The pharyngeal arches that give rise to orofacial tissues are internally lined by the endodermal epithelium and externally covered by the ectodermal epithelium; both form the oral mucosa lining as well as give rise to salivary glands [9]. Isolation of the myoepithelial progenitors from the salivary glands has shown to generate organoids in vitro and is able to engraft in the irradiated mice to form functional salivary glands [10].

8.2 Mesenchymal Stem Cell-Derived Oral Tissues

8.2.1 Markers and Sources of Orofacial MSCs

Majority of the orofacial stem cell research involves isolation of the so-called MSC population from various tissues. MSCs are first isolated by Friedenstain from BM and generally described as a group of plastic adherent cells that assume a spindle-shaped morphology and form so-called colony-forming unit fibroblasts (CFU-F), which can differentiate into connective tissue cells (adipocytes, osteocytes and chrondrocytes) *in vitro* [11]. In the orofacial region, the MSCs are isolated from adult human dental pulp, pulp of the exfoliated deciduous teeth, periodontal ligaments, apical follicles, mandibular BM and oral mucosal connective tissues that can be transplanted to various organs.

Similar to the MSCs derived from other tissues, orofacial MSCs do not display any tissue-specific markers in vivo. Some of the markers of in vitro derivatives of dental pulp stem cells (DPSCs) and periodontal ligament stem cells (PDLCs) include stromal markers such as CD90, CD44, CD13, CD19 and CD79 and pericyte markers such as CD146. In mouse, the MSCs are marked within a broader stromal population of the stem cell antigen 1 (Sca-1)-positive cohorts. In humans, Stro-1 and CD34 were described as MSC makers that can be used to isolate the in vitro CFU-F forming cells, albeit all these markers are expressed by other cell types and highly unspecific. Recently, platelet-derived growth factor receptor alpha (PDGFRa) has been described as a possible in vivo marker for the enrichment of MSC CFU-F forming ability in vitro and engraftment in vivo [12, 13]. Studies have shown that the PDGFRa expresses broadly in the neural crest cells, [14] and in the dental organs, clonogenic MSCs can be isolated from the dental pulp tissue using this marker [8]. However, no comprehensive enrichment studies have been conducted using PDGFRa . Similar to other MSCs, these cells generally lack haematopoietic cell markers such as CD45 and CD31 [1, 15].

Orofacial MSCs cells originate from the embryonic neural crest, generally described as possessing higher differentiation potencies than the MSCs isolated from other sites [16]. For example, MSCs isolated from the mandibular bone can engraft and regenerate tibial defect but the reverse is not possible. Dental pulp MSCs express embryonic markers that are not expressed in the MSCs isolated from other tissues as detailed below. This is partially due to a cordial form of Hox gene expression in these cells compared to the mesoderm-derived MSCs [17].

The presence of adult MSC populations within orofacial tissue, and their ability to adopt tissue-specific phenotypes, given the appropriate differentiation conditions, has led many investigators to suggest that the primary role of MSCs is to serve as cell replacement during the natural course of tissue turnover and homeostasis [18]. In addition, MSCs may serve important therapeutic roles because they appear to escape immune recognition and exert anti-inflammatory and immunomodulatory effects via the suppression of T, B, natural killer and antigen-presenting cells, both in vitro and in vivo. The immunomodulatory properties of these cells, and the ability to isolate and expand them in vitro without loss of their phenotypic or multi-lineage potential, have generated great interest in using MSCs as a therapeutic modality for immune-mediated diseases and tissue repair [19–21].

8.2.2   Mesenchymal Stem Cells of Dental Pulp Stem Cells (DPSCs)

Dental pulp tissue is derived from the embryonic dental papilla, which originates from neural crest mesenchymal cells during the morphogenesis of the first pharyngeal arches. DPSCs are presumably inherent descendants of this lineage that occupy a perivascular niche of the dental pulp and possess MSC-like properties [15]. Dental pulp tissue is isolated from coronal pulp of the routinely extracted human teeth or exfoliated deciduous teeth, therefore a non-invasive process. DPSCs have shown to generate bone when they are implanted in mice [22]; deciduous teeth MSCs were found to be able to induce bone formation, generate dentin and survive in mouse brain along with expression of neural markers that appear to be different from previously identified stem cells such as from BM [23].

In culture, DPSCs have been shown to express ES cell markers not currently identified in other adult MSCs; these markers include Oct-4, Nanog, SSEA-3, SSEA-4, TRA-1-60 and TRA-1-81 [24], and their presence may signify higher potency and enhanced stem cell activity for DSPCs compared to other adult tissue-derived stem cells. They can differentiate into dental structures, as well as smooth and skeletal muscles, neurons, cartilage and bone under defined in vitro culture conditions [15, 25], engraft to muscular dystrophic skeletal muscle and improve post-myocardial infarction cardiac function in animal models although without evidence of engraftment [26, 27].

Notably, recent reports further speculated that the DPSCs might have originated from a neuroectodermal lineage. Indeed, the dental pulp stem cells express neural markers, and they can differentiate into functionally active neurons [28–30] and, when stimulated, express neurogenic substance such as glutamate. Transplantation of human DPSCs into rat SCI model has shown marked recovery of hind limb locomotor functions via improved preservation of neuronal filaments and myelin sheaths by inhibition of apoptosis, promotion of regeneration of transected axons through paracrine mechanisms and replacement of lost cells by differentiating into mature oligodendrocytes [31].

8.2.3 MSCs of the Periodontal Ligament, Apical Papilla and Follicles

Periodontal tissue contains both hard and soft tissues. Periodontal tissue ligaments are and gingival tissues are derived neural crest mesenchyme reside in the dental follicles during tooth development. Similar to the dental pulp and BM MSCs, the periodontal ligament stem cells (PDLSCs) are also thought to be perivascular. Clonally selected Stro-1 and CD146 as markers have been isolated from extracted human third molars and shown to differentiate into collagen-producing fibrogenic and calcified cementogenic cells in vivo to repair attachment lost in the periodontium in immunosuppressed rats, showing the capacity of these cells to form collagen fibres, similar to Sharpey's fibres, connected to the cementum-like tissue suggesting the potential to regenerate PDL attachment [32].

Other studies, however, failed to reproduce the osteogenic potential of the PDLSCs. These variations in findings may support the notion that PDLSCs are heterogeneous MSC population that may contain subpopulations that have propensities for either collagen-producing fibrogenic subclones or cement/osteogenic cell types. PDLSC do not form dentin-pulp complexes and instead possess osteoblastic and cementoblastic lineages in order to regenerate periodontal tissue and maintain PDL integrity [33]. Currently, PDLSCs are not currently reported for use in nondental diseases; it is not clear whether this is due to a higher susceptibility of PDLSCs to periodontal disease states and age related changes.

CFU-F forming cells isolated from the dental apical follicle (DFS) and stem cells of apical papilla (SCAP) of the developing tooth display similar MSC characteristics in vitro, however with perhaps superior differentiation capacity of early/progenitor cells compared to other MSCs from PDL and BM. These cells rapidly attach to the culture dishes, proliferate at a higher rate than other MSCs and express putative stem cell markers such as Notch-1 and nestin, as well as ES cell markers Oct-3/Oct-4, Sox-2 and Nanog reported in SCAP [21, 33]. DFS and SCAP are shown to differentiate into dentine-like structures in in vitro, functional hepatocyte-like cells and neuronal tissue structures (Table 8.1).

8.3 Muscle Stem Cells

8.3.1 Potential of Orofacial Muscle Stem Cells in Cardiac Repair

Another intriguing potential source of stem cells may come from the progenitors of the orofacial muscle. The muscles of orofacial area including tongue, facial expression and mastication are classified as skeletal myocytes. These muscles are vital for

Table 8.1 Orofacial MSC sources and their applications

Location	Differentiation potential	Applications for systemic diseases
Dental pulp stem cells (DPSCs)	Dentine	Cementum, PLD regeneration [32, 33]
Human exfoliated deciduous tooth (SHED)	Osteocytes	Bone generation [22]
Periodontal ligament stem cells (PDLS)	Cementum	Myocardial infarction [27]
Stem cells of apical papilla (SCAP)	Sharpey's fibres	Muscular dystrophy [26]
	Cardiomyocytes	Limbal stem cell deficiency [34]
	Muscle cells	Hind limb ischemia [35]
	Neurons	Rat spinal cord injury [31]
	Hepatocytes	
	Corneal epithelium	
	Neurotrophic factor secretion	

mammalian survival dictated by their roles in the infant suckling and adult nutritional intake. As well, recent studies have shown that these muscles possess developmental kinship to the cardiac muscle cells that are marked by similar gene expressions such as cardiac-specific transcription factors *Isl1* and *Nkx2-5*. The similarities between these two groups also extend to the fatigue resistance of orofacial muscles [36], presumably due to similar glycolytic metabolic properties of the heart and presence of connexin molecules that are similar to cardiac configurations [37]. Unlike the cardiac muscles, the orofacial muscles have a remarkable ability to repair after injury, providing continuous myofibril remodelling throughout life, suggesting an active and proliferative myogenic properties of the satellite cells in these muscles. These attributes suggest that the orofacial muscle progenitors may emerge as an interesting and attractive source for regeneration of myocardium which only heals with non-contractile fibrosis after insults of MI. Thus far, limb muscle progenitors which have been isolated to transplant into myocardium failed to couple with the endogenous cardiomyocytes resulting in arrhythmia [38]. This may be presumably due to the developmental origin and characteristics of the skeletal muscles of the trunk and limbs being distinct from those of the cardiac and facial muscle. Other cell types such as MSCs from the BM and endogenous cardiac source have produced less than optimal results in cardiac stem cell treatments [7, 39], with former resulting in short-term benefits due to trophic effect rather than new cardiomyocyte generation [40] and the latter having limited therapeutic value as it is confined to neonates [41].

8.3.2 Developmental Similarities of Cardiac and Orofacial
Myogenic Progenitors

Recent lineage tracing and clonal analysis have shown that the orofacial and the heart myocytes originate from the multipotent pharyngeal mesoderm cells which ingress through the streak at the same stage [42, 43] and share a distinct lineage kinship in development. Two myogenic linkages have been identified [4]: (1) First, pharyngeal arch lineage gives rise to masticatory muscle and also contributes to myocardial cells in the right ventricle. (2) Second, pharyngeal arch lineage gives rise to muscles of facial expression and contributes myocardial cells to the arterial pole of the heart.

A key gene in early pharyngeal mesoderm genetic programme encodes the LIM homeodomain transcription factor *Islet1* (*Isl1*) which has been shown to identify the multipotent cardiac progenitor cells in the early embryo and the differentiating ES cells that can give rise to myocardial, endothelial and smooth muscle descendants. This gene is also expressed in pharyngeal myogenic progenitor cells and downregulated on differentiation to either cardiac or skeletal muscle fates [41]. Tzahor *et al.* have shown that pharyngeal myogenic muscle progenitors apparently have exactly the same subset of craniofacial muscles originating from *Isl1*-expressing progenitor cells [42]. Others have found that treatment of brachiometric satellite cells activates cardiac gene expression including *Isl1* and *Tbx20*. This finding suggests that satellite cells retain the environmental signals of their embryogenic progenitors, highlighting the potential use of these myoblasts for cardiac repair. Interestingly, while no equivalent population to satellite cells is found in cardiac muscle, a small number of residual ISl1-positive cells have been identified and proposed to be resident progenitor cells that may contribute to cardiac growth and repair in the foetal and early postnatal heart [44].

Additionally in our laboratory, initial findings of genetic lineage tracking and protein expression experiments using *Nkx2-5*, a cardinal cardiomyocyte marker, have shown remarkable similarity of the tongue and masticatory muscle throughout the embryonic development and in adults; as well, the adult tongue has a significant pool of satellite that is marked by the cardiac transcription gene (data not shown), supporting the notion that the activation and cardiogenic differentiation are possible using these cells.

8.3.3 An Interesting Example of Using Tongue Stem Cell
for Cardiac Repair

Shibuya et al., in 2010, characterized for the first time cardiomyocyte-like properties of cultured tongue muscle-derived stem cells. They showed that Sca-1-positive cells isolated from tongue muscles appear to differentiate into cardiomyocytes that electrically cooperate with adjacent cardiomyocytes. They presented the cardiomyocyte phenotype with beating and Nkx2-5 expression. Interestingly, these cells preserve their expression of connexin43 and appear to form gap junctions, as indicated by the transfer of dye and synchronization of calcium transients among adjacent cells. Collectively, these findings strongly suggest tongue progenitors may be ideal for cell therapy in heart disease [37]. Although their *in vitro* studies prove promising and their in vivo findings are limited, this may also be due to the markers that are used to isolate the progenitors which may not be targeted to the myogenic progenitors.

Myogenesis including cardiogenesis is complexly orchestrated. Understanding the developmental governance and the regulatory factors that determine the cell fate of the orofacial myocytes may shed light into both the function of the orofacial musculature and the cardiac repair mechanisms.

8.4   Closing Remark

The interest in organ regeneration using stem cells has increased in the last decade. In this context, orofacial stem cells are promising candidates, as they are readily available and highly proliferative and possess multi-differentiation abilities. These cells not only provide therapeutic and tissue engineering cell source for orofacial area and dental regeneration but also provide additional source for systemic cells.

References

 1. Xaymardan M, et al. Bone marrow stem cell: properties and pluripotency. In: Atala A, Lanza R, editors. Principles of regenerative medicine. San Diego: Elservier; 2008. p. 268–300.
 2. Gans C, Northcutt RG. Neural crest and the origin of vertebrates: a new head. Science. 1983;220(4594):268–73.
 3. Diogo R, et al. A new heart for a new head in vertebrate cardiopharyngeal evolution. Nature. 2015;520(7548):466–73.
 4. Lescroart F, et al. Clonal analysis reveals common lineage relationships between head muscles and second heart field derivatives in the mouse embryo. Development. 2010;137(19):3269–79.
 5. Hoch RV, Soriano P. Roles of PDGF in animal development. Development. 2003;130(20):4769–84.
 6. Kaltschmidt B, Kaltschmidt C, Widera D. Adult craniofacial stem cells: sources and relation to the neural crest. Stem Cell Rev. 2012;8(3):658–71.
 7. Chong JJ, et al. Adult cardiac-resident MSC-like stem cells with a proepicardial origin. Cell Stem Cell. 2011;9(6):527–40.
 8. Komada Y et al. Origins and properties of dental, thymic, and bone marrow mesenchymal cells and their stem cells. Plos One. 2012;7:e46436.
 9. Rothova M, et al. Lineage tracing of the endoderm during oral development. Dev Dyn. 2012;241(7):1183–91.
10. Ogawa M, et al. Functional salivary gland regeneration by transplantation of a bioengineered organ germ. Nat Commun. 2013;4:2498.

11. Friedenstein AJ, et al. Origin of bone marrow stromal mechanocytes in radiochimeras and heterotopic transplants. Exp Hematol. 1978;6(5):440–4.
12. Farahani RM, Xaymardan M. Platelet-derived growth factor receptor alpha as a marker of mesenchymal stem cells in development and stem cell biology. Stem Cells Int. 2015;2015:362753.
13. Houlihan DD, et al. Isolation of mouse mesenchymal stem cells on the basis of expression of Sca-1 and PDGFR-α. Nat Protoc. 2012;7:2103–11.
14. Tallquist MD, Soriano P. Cell autonomous requirement for PDGFRalpha in populations of cranial and cardiac neural crest cells. Development. 2003;130(3):507–18.
15. Shi S, Gronthos S. Perivascular niche of postnatal mesenchymal stem cells in human bone marrow and dental pulp. J Bone Miner Res. 2003;18(4):696–704.
16. Jensen J, et al. Dental pulp-derived stromal cells exhibit a higher osteogenic potency than bone marrow-derived stromal cells in vitro and in a porcine critical-size bone defect model. SICOT J. 2016;2:16.
17. Creuzet S, et al. Negative effect of Hox gene expression on the development of the neural crest-derived facial skeleton. Development. 2002;129(18):4301–13.
18. Caplan AI. Review: mesenchymal stem cells: cell-based reconstructive therapy in orthopedics. Tissue Eng. 2005;11(7–8):1198–211.
19. Ding G, et al. Effect of cryopreservation on biological and immunological properties of stem cells from apical papilla. J Cell Physiol. 2010;223(2):415–22.
20. Djouad F, et al. Mesenchymal stem cells: innovative therapeutic tools for rheumatic diseases. Nat Rev Rheumatol. 2009;5(7):392–9.
21. Li Z, et al. Immunomodulatory properties of dental tissue-derived mesenchymal stem cells. Oral Dis. 2014;20(1):25–34.
22. Yang KL, et al. A simple and efficient method for generating Nurr1-positive neuronal stem cells from human wisdom teeth (tNSC) and the potential of tNSC for stroke therapy. Cytotherapy. 2009;11(5):606–17.
23. Miura M, et al. SHED: stem cells from human exfoliated deciduous teeth. Proc Natl Acad Sci U S A. 2003;100(10):5807–12.
24. Kerkis I, et al. Isolation and characterization of a population of immature dental pulp stem cells expressing OCT-4 and other embryonic stem cell markers. Cells Tissues Organs. 2006;184(3–4):105–16.
25. Laino G, et al. In vitro bone production using stem cells derived from human dental pulp. J Craniofac Surg. 2006;17(3):511–5.
26. Kerkis I, et al. Early transplantation of human immature dental pulp stem cells from baby teeth to golden retriever muscular dystrophy (GRMD) dogs: local or systemic? J Transl Med. 2008;6:35.
27. Gandia C, et al. Human dental pulp stem cells improve left ventricular function, induce angiogenesis, and reduce infarct size in rats with acute myocardial infarction. Stem Cells. 2008;26(3):638–45.
28. Nor JE. Tooth regeneration in operative dentistry. Oper Dent. 2006;31(6):633–42.
29. Thesleff I, Aberg T. Molecular regulation of tooth development. Bone. 1999;25(1):123–5.
30. Tucker A, Sharpe P. The cutting-edge of mammalian development; how the embryo makes teeth. Nat Rev Genet. 2004;5(7):499–508.
31. Sakai K, et al. Human dental pulp-derived stem cells promote locomotor recovery after complete transection of the rat spinal cord by multiple neuro-regenerative mechanisms. J Clin Invest. 2012;122(1):80–90.
32. Zeng W, Wang C, Yang X. Identification of medicinal gualou (fruit-rind of Trichosanthes) and its mixed fruits in Sichuan. Zhongguo Zhong Yao Za Zhi. 1992;17(1):9–12. 62.
33. Huang GT, Gronthos S, Shi S. Mesenchymal stem cells derived from dental tissues vs. those from other sources: their biology and role in regenerative medicine. J Dent Res. 2009;88(9):792–806.

34. Gomes JA, et al. Corneal reconstruction with tissue-engineered cell sheets composed of human immature dental pulp stem cells. Invest Ophthalmol Vis Sci. 2010;51(3):1408–14.
35. Iohara K, et al. A novel stem cell source for vasculogenesis in ischemia: subfraction of side population cells from dental pulp. Stem Cells. 2008;26(9):2408–18.
36. van Eijden TMGJ, Turkawski SJJ. Morphology and physiology of masticatory muscle motor units. Crit Rev Oral Biol Med. 2001;12(1):76–91.
37. Shibuya M, et al. Tongue muscle-derived stem cells express connexin 43 and improve cardiac remodeling and survival after myocardial infarction in mice. Circ J. 2010;74(6):1219–26.
38. Abraham MR, et al. Antiarrhythmic engineering of skeletal myoblasts for cardiac transplantation. Circ Res. 2005;97(2):159–67.
39. Mazhari R, Hare JM. Mechanisms of action of mesenchymal stem cells in cardiac repair: potential influences on the cardiac stem cell niche. Nat Clin Pract Cardiovasc Med. 2007;4 Suppl 1:S21–6.
40. Fazel S, et al. Cardioprotective c-kit+cells are from the bone marrow and regulate the myocardial balance of angiogenic cytokines. J Clin Invest. 2006;116(7):1865–77.
41. Zhou B, et al. Nkx2-5- and Isl1-expressing cardiac progenitors contribute to proepicardium. Biochem Biophys Res Commun. 2008;375(3):450–3.
42. Tzahor E, Evans SM. Pharyngeal mesoderm development during embryogenesis: implications for both heart and head myogenesis. Cardiovasc Res. 2011;91(2):196–202.
43. Tirosh-Finkel L, et al. Mesoderm progenitor cells of common origin contribute to the head musculature and the cardiac outflow tract. Development. 2006;133(10):1943–53.
44. Kelly R. Core issues in craniofacial myogenesis. Exp Cell Res. 2010;316(18):3034–41.

Open Access This chapter is distributed under the terms of the Creative Commons Attribution 4.0 International License (http://creativecommons.org/licenses/by/4.0/), which permits use, duplication, adaptation, distribution and reproduction in any medium or format, as long as you give appropriate credit to the original author(s) and the source, provide a link to the Creative Commons license and indicate if changes were made.

The images or other third party material in this chapter are included in the work's Creative Commons license, unless indicated otherwise in the credit line; if such material is not included in the work's Creative Commons license and the respective action is not permitted by statutory regulation, users will need to obtain permission from the license holder to duplicate, adapt or reproduce the material.

Chapter 9
Biomaterials in Caries Prevention and Treatment

Lei Cheng, Yaling Jiang, Yao Hu, Jiyao Li, Hockin H.K. Xu, Libang He, Biao Ren, and Xuedong Zhou

Abstract Dental caries is one of the most widespread diseases in humans and has become a heavy economic burden. Tremendous progress aiming at improving the clinical performance of dental composite has been made in the past decades but secondary caries still remains a main problematic issue in composite restoration failure. Recently, some novel biomaterials have been developed in caries prevention and treatment and showed broad prospects of application. Nanoparticles of amorphous calcium phosphate (NACP) and CaF_2 were synthesized and could release calcium/phosphate and fluoride ions, contributing to remineralize tooth lesions and neutralize acids. Biomineralization agents, PAMAM, could mimic the natural mineralization process of dental hard tissues. Quaternary ammonium methacrylates (QAMs) were identified to be effective against dental biofilms. These biomaterials are promising for incorporation into dental composite in preventive and restorative dentistry, thus improving the efficacy and success rate in caries prevention and treatment.

Keywords Dental caries • Dental materials • Mineralization • Antibacterial • Quaternary ammonium methacrylates • Calcium phosphate nanoparticles

L. Cheng • Y. Jiang • Y. Hu • J. Li • L. He • X. Zhou (✉)
State Key Laboratory of Oral Diseases, West China Hospital of Stomatology, Sichuan University, No. 14, Section 3, Renmin South Road, Chengdu 610041, China

Department of Operative Dentistry and Endodontics, West China Hospital of Stomatology, Sichuan University, Chengdu, China
e-mail: zhouxd@scu.edu.cn

H.H.K. Xu
Biomaterials & Tissue Engineering Division, Department of Endodontics, Periodontics and Prosthodontics, University of Maryland Dental School, Baltimore, MD, USA

B. Ren
State Key Laboratory of Oral Diseases, West China Hospital of Stomatology, Sichuan University, No. 14, Section 3, Renmin South Road, Chengdu 610041, China

© The Author(s) 2017
K. Sasaki et al. (eds.), *Interface Oral Health Science 2016*,
DOI 10.1007/978-981-10-1560-1_9

Dental caries is a dietary carbohydrate-modified bacterial infectious disease caused by biofilm acids and is one of the most prevalent chronic diseases of people worldwide [1–4]. It's the main cause of oral pain and tooth loss, affecting the oral health of human beings seriously and also creating a heavy financial burden; meanwhile, it has been associated with some systematic diseases as well [5, 6]. Basically, dental caries is a multifactorial disease, but demineralization of susceptible dental hard tissues resulting from acidic by-products from bacterial fermentation of dietary carbohydrates is considered to be the fundamental mechanism. Once biofilm acids have caused tooth decay, the most effective treatment currently is to remove the carious tissues and fill the tooth cavity with a restorative material [7, 8].

Resin-based dental composites are increasingly used for tooth cavity restoration due to their excellent esthetics and improved performance [9–12]. Significant progress has been made to equip the esthetic composite restorations with less removal of tooth structures, enhanced load-bearing properties, and improved clinical performance [13–16]. However, as composites tend to accumulate more biofilms/plaques in vivo than other restorative materials, plaque adjacent to the restoration margins could lead to secondary caries [17, 18]. Indeed, previous studies demonstrated that secondary caries and bulk fracture are considered to be the most important factors leading to dental composite restoration failure, among which secondary caries are a main reason for replacing the existing restoration materials [19, 20]. As a result, half of all dental restorations fail within 10 years, and nearly 60 % of the average dentist's practice time is expended on replacing them [21, 22]. Replacement dentistry costs $5 billion annually in the USA alone [23]. Therefore, there is a great need to explore novel anticaries materials to combat secondary caries by incorporating bioactive agents possessing remineralization and antibacterial properties into the resin composites and bonding systems.

9.1 Synthesis of Novel Mineralization Materials in Preventive Dentistry

In the normal status, there is a physiological equilibrium between the remineralization and demineralization of dental hard tissues in the oral cavity [24–26]. The equilibrium will break toward demineralization when the organic acids produced by bacteria increase to a certain extent [7]. In consideration of this, it would be beneficial to employ mineralization materials in caries prevention. To date, two main strategies have been carried out to remineralize demineralized dental hard tissues. One is the ion-based strategy, and the other is the use of biomineralization agents.

Composites containing calcium phosphate (CaP) composites were previously synthesized [27, 28]; these composites could release supersaturating levels of calcium (Ca) and phosphate (PO_4) ions and achieve remineralization of tooth lesions in vitro. More recently, nanoparticles of amorphous calcium phosphate (NACP) of about 100 nm in size were firstly synthesized via a spray-drying technique [29].

Owing to the small size and high surface area effects, the "smart" NACP nanoparticles exhibit excellent characteristics, dramatically increasing the calcium and phosphate ions released at a cariogenic low pH while possessing mechanical properties nearly twofold of abovementioned CaP composite. In addition, NACP nanoparticles could neutralize a lactic acid solution of pH 4 by increasing the pH rapidly to nearly 6, which could avoid caries formation [30, 31]. Besides NACP, nanocomposites containing CaF_2 nanoparticles were also developed [32]. Studies have shown that these novel nanocomposites can release calcium and fluoride ions in a long-term stable manner. Both calcium and fluoride ions can inhibit demineralization and promote remineralization of dental hard tissues, and fluoride also has some antibacterial activity by interfering the formation and metabolism of dental plaque biofilm [33–35].

Although employed widely in the remineralization of carious dentin, such an ion-based strategy cannot be effective in locations where the crystallites are totally destroyed [36]. Another promising class of mineralization materials is the biomineralization agents. Inspired from the function of non-collagenous proteins (NCPs) in the biomineralization process of natural teeth, using biomimetic templates to remineralize the demineralized dentin is of great interest in recent years as NCPs, the natural nucleation templates, lose their abilities to induce in situ remineralization in the mature dentin [37, 38]. Poly(amino amine) (PAMAM)-type dendrimer is widely studied in dental biomineralization. It is a class of monodispersed polymeric nanomaterials with plenty of branches radiating from one central core and highly ordered architecture. It has been referred as "artificial protein" due to its biomimetic properties and well-defined/easily tailored structure, such as its functional group, generation, and spatial structure. Previous studies [39, 40] have clearly demonstrated that PAMAM and its derivatives could induce biomineralization of the demineralized dentin. Combined PAMAM with antibacterial agents also obtained double effects of mineralization, and antibacterial and needlelike crystals can precipitate both on the dentin surface and in the dentinal tubules, suggesting that PAMAM is a promising biomineralization material for dental use.

9.2 Quaternary Ammonium Methacrylates (QAMs)

Quaternary ammonium salts (QAMs),widely used in water treatment, food industry, textiles, and surface coating because of their low toxicity and broad spectrum of antimicrobial activity, are some ionic compounds which can be regarded as derived from ammonium compounds by replacing the hydrogen atoms with alkyl groups of different chain lengths [41]. They were first introduced to dental material industry in the 1970s as mouth rinses [42, 43]. Since then numerous researchers have devoted to exploring novel QAMs to meet the "immobilized bactericide" [44, 45] concept in dental materials. To date, a great many kinds of QAMs, such as 12-methacryloylox ydodecylpyridinium bromide (MDPB) [44, 46, 47], methacryloxylethylcetylammonium chloride (DMAE-CB) [48], quaternary ammonium dimethacrylate (QADM)

[49, 50], and quaternary ammonium polyethylenimine (QPEI) [51], have already been synthesized and incorporated into dental materials including glass ionomer cement (GIC) [52], etching-bonding systems [53, 54], and resin composites [55], attempting to achieve antibacterial effect. Recent researches have given us evidences that they make promising dental materials with properties which are suitable for treatments of dental caries and with much potential for preventing secondary caries.

9.2.1 Killing Bacteria and Inhibiting Biofilms

Many of the QAMs have been proven to be efficient in killing bacteria or inhibiting biofilms, including methacryloxylethylcetylammonium chloride (DMAE-CB), quaternary ammonium dimethacrylate (QADM), quaternary ammonium polyethylenimine, and quaternary ammonium dimethacrylate. An in vitro [49] study which intended to develop novel antibacterial dentin primers containing both quaternary ammonium dimethacrylate (QADM) and nanoparticles of silver (NAg) proved that uncured primer could kill the planktonic bacteria in caries cavity, as well as inhibit dental biofilms.

Recently, it has been revealed that glass ionomer cements (GIC) containing different concentrations of a novel material – dimethylaminododecyl methacrylate (DMADDM) – which also belongs to the QAM family, have antibacterial effects to different extents as well [56]. The results of this study indicated that with DMADDM, both the live bacteria and the EPS decreased in the biofilms of S. mutans. And the group with the higher concentration had more significant effect. Consistent with the reduced live bacteria and EPS production, the glucosyltransferase-encoding genes, gtfB, gtfC, and gtfD, of S. mutans were also significantly downregulated after adding DMADDM. Another research [57] found that with DMADDM added to dental adhesives, the ratio of S. mutans steadily dropped in the multispecies biofilms of Streptococcus mutans, Streptococcus gordonii, and Streptococcus sanguinis.

9.2.2 Mechanical Properties

The mechanical properties are critical indexes for the evaluation antimicrobial modified dental materials, such as dental adhesives and resin composites. QAMs have been proved that they did not change the adhesion and strength of the dental materials For example, the SEM examination of dentin-adhesive interfaces of a QADM modified dental adhesive revealed numerous resin tags "T" with no difference with the control [49]. Only at high mass fraction, QADM decreased the strength of the tetracalcium phosphate (TTCP) composite. Nevertheless, QADM can guarantee both the antibacterial and the mechanical properties of the material at suitable concentrations [58].

9.2.3 Durability

Resin aging is another reason for secondary caries or failure of restorations. Accordingly, tests for durability are needed if QAMs are making a difference in this aspect of composites.

A recent study [50] compared the flexural strength and elastic modulus of a novel nanocomposite containing nanoparticles of amorphous calcium phosphate (NACP) and quaternary ammonium dimethacrylate (QADM) with two different commercial composites in water immersion. Like the other two composites, the new nanocomposite displayed a moderate decrease during the first month of aging, with little decrease during 1–6 months. After a 180-day immersion, the strength and modulus of NACP-QADM nanocomposite were similar to those of commercial control composites. Also, NACP-QADM nanocomposite maintained its antibacterial properties during immersion. These statistics indicated that composites with QAMs maintain the same level of durability as common commercial composites.

9.2.4 Biological Safety

Researchers recently have provided us with the human gingival fibroblast cytotoxicity data of a certain kind of QAMs [59]. In this study, the control had fibroblast culture in fibroblast medium (FM) without resin eluents, and its cell viability was set at 100 %. For groups with resins, the fibroblasts were cultured in FM containing resin eluents. The fibroblast viability for all the bonding agent groups was nearly 100 %. There was generally little difference between 12-methacryloyloxydodecylp yridinium bromide (MDPB) and NAg groups and the non-antibacterial adhesive control or between bonding agent groups and the FM control without resin eluents ($p > 0.1$). Another measurement [52] of the DMADDM release showed that DMADDM concentrations decreased continuously and after 12 h values were found near to the zero level for both 1.1 wt.% and 2.2 wt.% DMADDM containing specimens in water. And in saliva, only one set presenting ten specimens of 2.2 wt.% DMADDM revealed release of the substance near the detection level.

With 12-methacryloyloxydodecylpyridinium bromide (MDPB) displaying no more fibroblast cytotoxicity than the routine medium and DMADDM's low release rate and time, we may draw the conclusion that QAMs are biologically safe with humans. But more researches are needed to further explain it.

9.2.5   Inhibiting MMPs

Proteolytic degradation of hybrid layer by host-derived matrix metalloproteinases (MMPs) leading to reduction of resin-dentin bond strength is believed to be among the major reasons for the failure of resin restorations [60, 61]. Fortunately, it has been found that the resin-dentin bonding systems containing DMADDM inhibited MMPs efficaciously.

Assay of the loss mass of demineralized dentin beams revealed that dentin incubated in DMADDM lost significantly less mass than the control group at 7 or 30 days. While with both DMADDM and chlorhexidine digluconate (CHX), dentin mass dropped even sharply. And spectrophotometrical measurement of hydroxyproline (HYP), used to determine the dissolution of collagen peptides, reflected that the combination of DMADDM and CHX dramatically lowered hydroxyproline dissolution weight [62].

9.2.6   Mechanism

The antibacterial mechanism of quaternary ammonium methacrylates (QAMs) is now widely thought to be "contact killing." That is, the positively charged (N^+) sites of the QAMs molecules would attract the negatively charged bacterial cells, and once contact is made, the electric balance of the cell membrane could be disturbed and the bacterium could explode under its own osmotic pressure. Previous studies [56, 63, 64] have confirmed that the surface charge density of the tested materials increased with the increase of the mass fraction of DMADDM, and higher mass fraction showed stronger antibacterial effects. A recent study [65] on the antibacterial effects of quaternary ammonium chain length revealed the increasing antibacterial potency with increasing CL from 3 to 16 while CL 18 may be a cutoff point. It was supposed that long-chained quaternary ammonium compounds could have additional antimicrobial activity by insertion into the bacterial membrane, resulting in physical disruption [66]. The evaluation of three-dimensional (3D) biofilms on antibacterial bonding agents containing QAMs showed that not only the bacteria contacting the resin surface but also through the bacteria in the 3D biofilm away from the resin surface were killed [67]. It was suggested that a stress condition in the bacteria could trigger a built-in suicide program in the biofilm, which was also termed the programmed cell death. However, the exact mechanism of QAMs still remained unclear and needed further research in the future.

9.2.7   Resistant/Persister Bacteria

Although the strong killing effects of QAMs are clinically beneficial, their frequent use may also have the possibility of leading to the development of bacterial drug resistance and tolerance as with other antibiotics, which are associated with the failure of antibacterial treatment and the relapse of many chronic infections. No reports have reported or investigated the possibility that oral bacteria can acquire resistance to QAMs so far. However, a new study recently has found that *Streptococcus mutans* could form a small portion of persisters which can survive lethal doses of DMADDM treatment. Persisters are nongrowing dormant cells that are produced in a clonal population of genetically identical cells and are generally believed to be responsible for drug tolerance [68, 69]. In contrast to the drug-resistant bacteria, persisters are not mutants but phenotypic variants of the wild-type population. The isolated persisters in the abovementioned study have the typical characteristic of multiple drug tolerance, and the experimental drugs of different mechanisms are proven to have little effects on the persisters. This finding will partially challenge the clinical use of QAMs, and further studies are needed to elucidate possible mechanisms behind this in order to tackle the problem.

9.2.8   Models

Different models were used to study the biomaterials in caries preventions and treatment. Most of the investigations applied in vitro models. Besides, an individual removable acrylic upper jaw splint in situ has already been built to test some properties of QAMs including antibacterial effect and biological safety [52]. Transparent custom-made acrylic splints (Thermoforming foils®, Erkodent, Pfalzgrafenweiler, Germany) were fabricated from alginate impressions as carrier of the GIC specimens. Six samples were fixed in the left and right buccal position in the molar and premolar regions with silicon impression material (President Light Body®, Coltène, Altstätten, Switzerland) onto the splints for intraoral exposure. Animal models are under construction and more figures are coming out soon.

References

1. Selwitz RH, Ismail AI, Pitts NB. Dental caries. Lancet. 2007;369(9555):51–9.
2. Takahashi N, Nyvad B. Caries ecology revisited: microbial dynamics and the caries process. Caries Res. 2008;42(6):409–18.
3. Takahashi N, Nyvad B. The role of bacteria in the caries process. JDR. 2011;90(3):294–303.
4. Paes Leme AF, Koo H, Bellato CM, et al. The role of sucrose in cariogenic dental biofilm formation – new insight. J Dent Res. 2006;85(10):878–87.
5. Hu DY, Hong X, Li X. Oral health in China-trends and challenges. Int J Oral Sci. 2011;3(1):7–12.

6. Fejerskov O, Kidd EAM. Dental caries: the disease and its clinical management. Commun Dent Oral Epidemiol. 2003;8(3):80–6.
7. Fejerskov O. Changing paradigms in concepts on dental caries: consequences for oral health care. Caries Res. 2004;38(3):182–91.
8. Banava S, Yazdi MS, Heshmat H. Restorative: bio-treatment of caries. Br Dent J. 2014;216(6):267–7
9. Ferracane JL. Resin composite – state of the art. Dent Mater Off Publ Acad Dent Mater. 2011;27(1):29–38.
10. Chen MH. Update of dental nanocomposites. J Dent Res. 2010;89(6):549–60.
11. Ilie N, Hickel R. Investigations on mechanical behaviour of dental composites. Clin Oral Investig. 2009;13(4):427–38.
12. Pereira-Cenci T, Cenci M, Fedorowicz Z, et al. Antibacterial agents in composite restorations for the prevention of dental caries. Cochrane Database Syst Rev. 2009;12(3):1559–9
13. Ruddell DE, Maloney MM, Thompson JY. Effect of novel filler particles on the mechanical and wear properties of dental composites. Dent Mater Off Publ Acad Dent Mater. 2002;18(1):72–80.
14. Pashley DH, Tay FR, Imazato S. How to increase the durability of resin-dentin bonds. Compend Contin Educ Dent. 2011;32(7):60–4. 66.
15. Satterthwaite JD, Maisuria A, Vogel K, et al. Effect of resin-composite filler particle size and shape on shrinkage-stress. Dent Mater Off Publ Acad Dent Mater. 2012;28(6):609–14.
16. Tyas MJ, Anusavice KJ, Frencken JE, et al. Minimal intervention dentistry—a review. Int Dent J. 2000;50(1):1–12.
17. Svanberg M, Mjör IA, Ørstavik D. Mutans streptococci in plaque from margins of amalgam, composite, and glass-ionomer restorations. J Dent Res. 1990;69(3):861–4.
18. Beyth N, Domb AJ, Weiss EI. An in vitro quantitative antibacterial analysis of amalgam and composite resins. J Dent. 2007;35(3):201–6.
19. Sakaguchi RL. Review of the current status and challenges for dental posterior restorative composites: clinical, chemistry, and physical behavior considerations. Dent Mater. 2005;21:3–6.
20. Sarrett DC. Clinical challenges and the relevance of materials testing for posterior composite restorations. Dent Mater. 2005;21(1):9–20.
21. Deligeorgi V, Mjör IA, Wilson NH. An overview of reasons for the placement and replacement of restorations. Prim Dent Care J Fac Gen Dent Pract. 2001;8(1):5–11.
22. Frost PM. An audit on the placement and replacement of restorations in a general dental practice. Prim Dent Care J Fac Gen Dent Pract. 2002;9(9):31–6.
23. Jokstad A, Bayne S, Blunck U, et al. Quality of dental restorations FDI Commission Project 2–95. Int Dent J. 2001;51(3):117–58.
24. Featherstone J. Dental caries: a dynamic disease process. Aust Dent J. 2008;53(3):286–91.
25. Moreno EC, Zahradnik RT. Demineralization and remineralization of dental enamel. J Dent Res. 1979;58(Spec Issue B):896–903.
26. Featherstone JD. The caries balance: the basis for caries management by risk assessment. Oral Health Prev Dent. 2004;2 suppl 1:259–64.
27. Skrtic D, Antonucci JM, Eanes ED, et al. Physiological evaluation of bioactive polymeric composites based on hybrid amorphous calcium phosphates. J Biomed Mater Res. 2000;53(4):381–91.
28. Langhorst SE, O'Donnell JNR, Skrtic D. In vitro remineralization of enamel by polymeric amorphous calcium phosphate composite: quantitative microradiographic study. Dent Mater. 2009;25(7):884–91.
29. Xu HHK, Moreau JL, Sun L, et al. Nanocomposite containing amorphous calcium phosphate nanoparticles for caries inhibition. Dent Mater. 2011;27(8):762–9.
30. Moreau JL, Sun L, Chow LC, et al. Mechanical and acid neutralizing properties and bacteria inhibition of amorphous calcium phosphate dental nanocomposite. J Biomed Mater Res B Appl Biomater. 2011;98B(98):80–8.

31. Weir MD, Chow LC, Xu HH. Remineralization of demineralized enamel via calcium phosphate nanocomposite. J Dent Res. 2012;91(10):979–84.
32. Xu HHK, Moreau JL, Sun L, et al. Strength and fluoride release characteristics of a calcium fluoride based dental nanocomposite. Biomaterials. 2008;29(32):4261–7.
33. Singh ML, Papas AS. Long-term clinical observation of dental caries in salivary hypofunction patients using a supersaturated calcium-phosphate remmeralizing rinse. J Clin Dent. 2009;20(3):87–92.
34. Kulshrestha S, Khan S, Hasan S, et al. Calcium fluoride nanoparticles induced suppression of Streptococcus mutans, biofilm: an in vitro and in vivo approach. Appl Microbiol Biotechnol. 2015;100(4):1–14.
35. Xu HH, Mdsun W. Strong nanocomposites with Ca, PO(4), and F release for caries inhibition. J Dent Res. 2010;89(1):19–28.
36. Frencken JE, Peters MC, Manton DJ, et al. Minimal intervention dentistry for managing dental caries – a review. Int Dent J. 2012;62(5):223–43.
37. Chen L, Liang K, Li J, et al. Regeneration of biomimetic hydroxyapatite on etched human enamel by anionic PAMAM template in vitro. Arch Oral Biol. 2013;58(8):975–80.
38. Tay FR, Pashley DH. Guided tissue remineralisation of partially demineralised human dentine. Biomaterials. 2008;29(8):1127–37.
39. Liang K, Yuan H, Li J, et al. Remineralization of demineralized dentin induced by amine-terminated PAMAM dendrimer. Macromol Mater Eng. 2014;300(1):107–17.
40. Wu D, Yang J, Li J, et al. Hydroxyapatite-anchored dendrimer for in situ remineralization of human tooth enamel. Biomaterials. 2013;34(21):5036–47.
41. Kourai H, Yabuhara T, Shirai A, et al. Syntheses and antimicrobial activities of a series of new bis-quaternary ammonium compounds. Eur J Med Chem. 2006;41:437–44.
42. Rosa M, Sturzenberger OP. Clinical reduction of gingivitis through the use of a mouthwash containing two quaternary ammonium compounds. Periodontology. 1976;47:535–7.
43. Bonesvoll P, Gjermo PA. Comparison between chlorhexidine and some quaternary ammonium compounds with regard to retention, salivary concentration and plaque-inhibiting effect in the human mouth after mouth rinses. Arch Oral Biol. 1978;23:289–94.
44. Imazato S, Ebi N, Takahashi Y, Kaneko T, Ebisu S, Russell RR. Antibacterial activity of bactericide-immobilized filler for resin-based restoratives. Biomaterials. 2003;24:3605–9.
45. Imazato S, Imai T, Russell RR, Torii M, Ebisu S. Antibacterial activity of cured dental resin incorporating the antibacterial monomer MDPB and an adhesion-promoting monomer. Biomed Mater Res. 1998;39:511–5.
46. Imazato S, Kuramoto A, Takahashi Y, Ebisu S, Peters MC. In vitro antibacterial effects of the dentin primer of Clearfil Protect Bond. Dent Mater. 2006;22:527–32.
47. Imazato S, Ohmori K, Russell RR, McCabe JF, Momoi Y, Maeda N. Determination of bactericidal activity of antibacterial monomer MDPB by a viability staining method. Dent Mater J. 2008;27:145–8.
48. Li F, Chai ZG, Sun MN, et al. Anti-biofilm effect of dental adhesive with cationic Monomer. J Dent Res. 2009;88(4):372–6.
49. Cheng L, Zhang K, Melo MAS, et al. Anti-biofilm dentin primer with quaternary ammonium and silver nanoparticles. J Dent Res. 2012;91(6):598–604.
50. Cheng L, Weir MD, Zhang K, et al. Antibacterial nanocomposite with calcium phosphate and quaternary ammonium. J Dent Res. 2012;91(5):460–6.
51. Beyth N, Yudovin-Farber I, Bahir R, et al. Antibacterial activity of dental composites containing quaternary ammonium polyethylenimine nanoparticles against Streptococcus mutans. Biomaterials. 2006;27(21):3995–4002.
52. Feng J, Cheng L, et al. In situ antibiofilm effect of glass-ionomer cement containing dimethylaminododecyl methacrylate. Dent Mater. 2015;31(8):992–1002.
53. Zhang K, Cheng L, Wu EJ, et al. Effect of water-ageing on dentine bond strength and anti-biofilm activity of bonding agent containing new monomer dimethylaminododecyl methacrylate. J Dent. 2013;41(6):504–13.

54. Cheng L, Weir MD, Zhang K, et al. Dental primer and adhesive containing a new quaternary ammonium monomer and nano-silver with anti-biofilm properties. J Dent. 2013;41(4):345–55.
55. Zhou C, Weir MD, Zhang K, Cheng L, et al. Synthesis of novel antibacterial quaternary ammonium monomer for incorporation into calcium phosphate nanocomposite to inhibit caries. Dent Mater. 2013;29(8):859–70.
56. Wang S, Cheng L, et al. Int J Oral Sci. 2015, accepted.
57. Zhang K, Wang S, Cheng L, et al. Effect of antibacterial dental adhesive on multispecies biofilms formation. J Dent Res. 2015;94(4):622–9.
58. Cheng L, Weir MD, et al. Tetracalcium phosphate composite containing quaternary ammonium dimethacrylate with antibacterial properties. J Biomed Mater Res B Appl Biomater. 2012;100(3):726–34.
59. Zhang K, Li F, Imazato S, Cheng L, et al. Dual antibacterial agents of nano-silver and 12-met hacryloyloxydodecylpyridinium bromide in dental adhesive to inhibit caries. J Biomed Mater Res B Appl Biomater. 2013;101(6):929–38.
60. Osorio R, Yamauti M, Osorio E, et al. Effect of dentin etching and chlorhexidine application on metalloproteinase-mediated collagen degradation. Eur J Oral Sci. 2011;119:79–85.
61. Tjaderhane L, Nascimento FD, Breschi L, et al. Optimizing dentin bond durability: control of collagen degradation by matrix metalloproteinases and cysteine cathepsins. Dent Mater. 2013;29:116–35.
62. Zhou W, Cheng L et al. unpublished.
63. Wang SP, Zhang KK, Cheng L, et al. Antibacterial effect of dental adhesive containing dimethylaminododecyl methacrylate on the development of streptococcus mutans biofilm. Int J Mol Sci. 2014;15(7):12791–806.
64. Imazato S, Ma S, Chen JH, et al. Therapeutic polymers for dental adhesives: loading resins with bio-active components. Dent Mater Off Publ Acad Dent Mater. 2014;30(1):97–104.
65. Li F, Weir MD, Xu HHK. Effects of quaternary ammonium chain length on antibacterial bonding agents. J Dent Res. 2013;92(10):932–8.
66. Simoncic B, Tomsic B. Structures of novel antimicrobial agents for textiles – a review. Text Res J. 2010;80(14):1721–37.
67. Zhou H, Weir MD, Xu HHK, et al. Evaluation of three-dimensional biofilms on antibacterial bonding agents containing novel quaternary ammonium methacrylates. Int J Oral Sci. 2014;6(2):77–86.
68. Fauvart M, Groote VND, Michiels J. Role of persister cells in chronic infections: clinical relevance and perspectives on anti-persister therapies. J Med Microbiol. 2011;60(6):699–709.
69. Jermy A. Bacterial physiology: no rest for the persisters. Nat Rev Microbiol. 2013;11(3):148–9.

Open Access This chapter is distributed under the terms of the Creative Commons Attribution 4.0 International License (http://creativecommons.org/licenses/by/4.0/), which permits use, duplication, adaptation, distribution and reproduction in any medium or format, as long as you give appropriate credit to the original author(s) and the source, provide a link to the Creative Commons license and indicate if changes were made.

The images or other third party material in this chapter are included in the work's Creative Commons license, unless indicated otherwise in the credit line; if such material is not included in the work's Creative Commons license and the respective action is not permitted by statutory regulation, users will need to obtain permission from the license holder to duplicate, adapt or reproduce the material.

Part III
Symposium III: Regenerative Oral Science

Chapter 10
Efficacy of Calcium Phosphate-Based Scaffold Materials on Mineralized and Non-mineralized Tissue Regeneration

Osamu Suzuki, Takahisa Anada, and Yukari Shiwaku

Abstract Calcium phosphate materials have been advocated as useful implantable scaffolds for bone tissue engineering. We have reported that synthetic octacalcium phosphate (OCP) is capable of enhancing differentiation of hard tissue-forming cells including osteoblastic cells, osteoclast precursor cells, and odontoblastic cells. The differentiation of mesenchymal stem cells is promoted to form new bone in the presence of OCP with atelo-collagen in vivo condition. The stimulatory capacity of OCP to conduct new bone increases with the copresence of amorphous calcium phosphate (ACP). Physical and chemical analyses of the materials suggested that the bioactivity of such hydroxyapatite (HA) precursor phases is induced as a result of the progressive change of chemical property of these materials during the hydrolysis into HA under physiological environment. The composite materials composed of OCP and natural polymers are capable of repairing not only mineralized tissues but also non-mineralized tissue. The form of OCP-based materials and the tissue responses to the materials will be summarized.

Keywords Calcium phosphate • Octacalcium phosphate • Scaffold • Tissue regeneration

O. Suzuki (✉)
Division of Craniofacial Function Engineering, Tohoku University Graduate School of Dentistry, 4-1 Seiryo-machi, Aoba-ku, Sendai, Japan
e-mail: suzuki-o@m.tohoku.ac.jp

T. Anada
Division of Craniofacial Function Engineering, Tohoku University Graduate School of Dentistry, 4-1 Seiryo-machi, Aoba-ku, Sendai, Miyagi 980-8577, Japan

Y. Shiwaku
Division of Craniofacial Function Engineering, Tohoku University Graduate School of Dentistry, 4-1 Seiryo-machi, Aoba-ku, Sendai, Miyagi 980-8577, Japan

Liaison Center for Innovative Dentistry, Tohoku University Graduate School of Dentistry, Sendai, Miyagi, Japan

© The Author(s) 2017
K. Sasaki et al. (eds.), *Interface Oral Health Science 2016*,
DOI 10.1007/978-981-10-1560-1_10

10.1 Introduction

Calcium phosphate ceramics, such as sintered hydroxyapatite (HA, $Ca_{10}(PO_4)_6(OH)_2$) and β-tricalcium phosphate (β-TCP, β-$Ca_3(PO_4)_2$), have widely been used as filling materials for various bone defects due to the superior osteoconductivity which is capable of bonding to bone tissue without intervention of connective tissues [1]. HA and β-TCP have been used as implantable scaffold materials with mesenchymal stem cells (MSCs) for bone regeneration [2]. These calcium phosphate materials provide suitable site where osteoblastic cells can attach, proliferate, and differentiate [2]. The physicochemical property of calcium phosphate materials affects the cellular responses to some extent [3, 4]. Octacalcium phosphate (OCP, $Ca_8H_2(PO_4)_6 \cdot 5H_2O$) is a non-sintered calcium phosphate material which was first proven to directly bond to bone tissue by onlaying it on mouse calvaria [5]. OCP has been regarded as a precursor phase in HA formation from supersaturated calcium and phosphate solutions with respect to HA and is therefore suggested to be a precursor to bone apatite crystals [6, 7] as well as amorphous calcium phosphate (ACP, $Ca_3(PO_4)_2 \cdot nH_2O$) [8]. OCP shows diversity regarding the stoichiometry [9] and the microstructure [10] depending on the preparation condition [11], which controls its bioactivity in vitro and in vivo [11]. This article summarizes the bioactivity of OCP and the composite materials with natural polymers as scaffold materials we have reported previously.

10.2 Mineralized and Non-mineralized Tissue Responses

OCP displays a unique osteoconductive property which tends to biodegrade and be followed by new bone formation if implanted in various bone defects, including the defects in intramembranous bone and long bones in various animal models [11, 12]. The efficacy of OCP has recently been reported by implanting its composite with collagen in human maxilla after cystectomy [13]. The stimulatory capacity of OCP could be induced by the environmental changes around the crystals, including calcium ion concentration change, due to the progressive hydrolysis from OCP to HA under physiological conditions [5, 14, 15]. The hydrolysis of OCP to HA is enhanced if ACP coexists with OCP, and the bone regenerative capacity of OCP is augmented by mechanically mixing OCP with ACP in rat critical-sized calvaria defect [16]. A composite of OCP with gelatin matrix, prepared through a wet synthesis, showed a greater biodegradable property coupled with a greater bone regenerative property if the material was implanted in rat critical-sized calvaria defect [17, 18]. The in vivo studies indicated that OCP/gelatin composite repairs not only rabbit tibia defect model [19] but also reforms infraspinatus tendon insertion using rabbit rotator cuff tear model [20].

10.3 Matrix Materials for Calcium Phosphate

Various calcium phosphate composites with natural polymers have been developed for bone and other tissue engineering (Table 10.1). Collagen (Col) is the major component of extracellular matrix proteins in bone, and its reconstituted Col has been utilized as matrix materials for calcium phosphates. HA/Col composites have been prepared by different processes, such as direct physical mixing [21, 22], chemical deposition of HA [23], or biomimetic mineralization [24]. Biodegradable calcium

Table 10.1 Calcium phosphate composites with various natural polymers designed for bone tissue engineering

Natural polymers	Calcium phosphates	Composites	Notes (references)
Collagen	HA	Sponge	Chemical deposition [23]
	Nano HA	Sponge	Physical mixture [22]
		Nanofiber	Physical mixture [21]
		Fiber	Biomimetic mineralization [24]
	OCP	Sponge	Chemical deposition [26]
	β-TCP	Sponge	Physical mixture [25]
Gelatin	HA	Particles	Chemical deposition [29]
		Sponge	Chemical deposition [30]
			Porogen leaching [31]
		Film	Physical mixture [28]
	OCP	Sponge	Chemical deposition [17]
			Chemical deposition and physical mixture [18]
	β-TCP	Sponge	Physical mixture [49]
Chitosan	HA	Sponge	Biomimetic mineralization [36]
	Nano HA	Sponge	Physical mixture [35]
			Chemical deposition [33]
	β-TCP	Sponge	Physical mixture [37]
Silk fibroin	HA	Sponge	Physical mixture [34]
		Film	Biomimetic mineralization [50]
	Ca-deficient HA	Particles	Chemical deposition [32]
Alginate	HA	Sponge	Physical mixture [40]
			Biomimetic mineralization [39]
		Beads	Chemical deposition [41]
	OCP	Sponge	Chemical deposition [42]
			Physical mixture [43]
		Beads	Physical mixture [45]
	β-TCP	Beads	Physical mixture [38]
Hyaluronic acid	HA	Hydrogel	Physical mixture [46]
	OCP	Hydrogel	Physical mixture [48]
	β-TCP	Hydrogel	Physical mixture [47]

phosphates/Col composites, such as OCP/Col and β-TCP/Col, have also been obtained using coprecipitation methods [25, 26] and by physical mixing [27]. Gelatin (Gel) is a denatured form of collagen and has the advantage of higher biodegradability. HA/Gel composites have been obtained in various forms, not only in porous form but also in film and particle forms [28–31]. OCP/Gel composites have been prepared by physical mixing and coprecipitation [17, 18]. Other natural polymers, such as chitosan and silk fibroin, have also been combined with calcium phosphates [32–37].

Alginate (Alg) is a natural polysaccharide obtained from brown algae and applied for wound healing and drug delivery due to its biocompatibility. HA/Alg and β-TCP/Alg have been prepared as porous scaffolds or beads [38–41]. Our group has developed OCP/Alg porous scaffolds [42, 43]. Although it is known that Alg is not able to interact with mammalian cells [44], OCP/Alg was capable of enhancing osteoblastic cell attachment and bone formation depending on the pore size and the porosity of the composites [42]. OCP/Alg beads including osteoblastic cells have a potential to activate and deliver osteoblastic cells from the beads to the local sites of bone defects [45]. Hyaluronic acid (HyA) is a major component of the extracellular matrix in the connective tissue. HyA hydrogel can be used as injectable materials filling bone defects directly and has been reported to promote bone regeneration [46, 47]. OCP/HyA is a bone substitute material which was proven to show the injectable and bioactive properties [48].

10.4   Cell Responses to Calcium Phosphate Materials

We have reported that OCP induces osteoblastic differentiation of mouse stromal cells [15, 51]. When mouse ST2 cells were cultured on the OCP or HA coatings, expression of osteogenic markers, such as type I collagen, alkaline phosphatase, and osterix, was enhanced on the OCP coating plates in an OCP dose-dependent manner [51]. We also reported that OCP enhanced alkaline phosphatase (ALP) activity of mouse mesenchymal stem cell line D1 cells in a three-dimensional cell culture system compared to HA and β-TCP [52]. When rat dental pulp cells were cultured on the OCP or HA coatings, OCP promoted their odontoblastic differentiation more than HA, as confirmed by ALP activity, mineralization, and enhancement of dentine sialophosphoprotein expression [53]. Furthermore, OCP is capable of inducing osteoclast formation from bone marrow cells in vitro in the presence of osteoblasts without vitamin D_3 [54].

OCP can gradually convert into the crystal structure of apatite in the physiological conditions [11]. The conversion induces the release of phosphate ions to the periphery of the crystals and the uptake of calcium ion into the crystals [14]. We demonstrated that these changes in ion concentration have effects on osteoblastic differentiation through the phosphorylation of p38 MAP kinase [55], migration of macrophage-like cells [56], and osteoclastic differentiation by the increase of the expression of RANKL in osteoblasts [54].

10.5 Conclusion

We have designed OCP-based materials and reported their tissue regenerative properties not only in the calcified tissues but also in the noncalcified tissue. The physicochemical properties of OCP induced in the conversion process from OCP to HA could be involved in promoting osteoblastic and osteoclastic differentiation and bone formation and reformation of infraspinatus tendon insertion. The composite materials of OCP with natural polymers, described in this article, could be used as scaffold materials for mineralized tissue regeneration and potentially for non-mineralized tissue regeneration.

Acknowledgments The studies reported in the present article were supported in part by Grants-in-aid (23106010, 26293417, and 15K15720) from the Ministry of Education, Science, Sports, and Culture of Japan.

References

1. LeGeros RZ. Calcium phosphate-based osteoinductive materials. Chem Rev. 2008;108:4742–53.
2. Matsushima A, Kotobuki N, Tadokoro M, Kawate K, Yajima H, Takakura Y, et al. In vivo osteogenic capability of human mesenchymal cells cultured on hydroxyapatite and on beta-tricalcium phosphate. Artif Organs. 2009;33:474–81.
3. Liu Y, Cooper PR, Barralet JE, Shelton RM. Influence of calcium phosphate crystal assemblies on the proliferation and osteogenic gene expression of rat bone marrow stromal cells. Biomaterials. 2007;28:1393–403.
4. Sun JS, Tsuang YH, Chang WH, Li J, Liu HC, Lin FH. Effect of hydroxyapatite particle size on myoblasts and fibroblasts. Biomaterials. 1997;18:683–90.
5. Suzuki O, Nakamura M, Miyasaka Y, Kagayama M, Sakurai M. Bone formation on synthetic precursors of hydroxyapatite. Tohoku J Exp Med. 1991;164:37–50.
6. Brown W. Crystal growth of bone mineral. Clin Orthop Relat Res. 1966;44:205–20.
7. Brown WE, Smith JP, Lehr JR, Frazier AW. Crystallographic and chemical relations between octacalcium phosphate and hydroxyapatite. Nature. 1962;196:1050–5.
8. Eanes ED, Gillessen IH, Posner AS. Intermediate states in the precipitation of hydroxyapatite. Nature. 1965;208:365–7.
9. Miyatake N, Kishimoto KN, Anada T, Imaizumi H, Itoi E, Suzuki O. Effect of partial hydrolysis of octacalcium phosphate on its osteoconductive characteristics. Biomaterials. 2009;30:1005–14.
10. Honda Y, Anada T, Kamakura S, Morimoto S, Kuriyagawa T, Suzuki O. The effect of microstructure of octacalcium phosphate on the bone regenerative property. Tissue Eng Part A. 2009;15:1965–73.
11. Suzuki O. Octacalcium phosphate: osteoconductivity and crystal chemistry. Acta Biomater. 2010;6:3379–87.
12. Suzuki O. Octacalcium phosphate (OCP)-based bone substitute materials. Jpn Dent Sci Rev. 2013;49:58–71.
13. Kawai T, Echigo S, Matsui K, Tanuma Y, Takahashi T, Suzuki O, et al. First clinical application of octacalcium phosphate collagen composite in human bone defect. Tissue Eng Part A. 2014;20:1336–41.

14. Suzuki O, Kamakura S, Katagiri T. Surface chemistry and biological responses to synthetic octacalcium phosphate. J Biomed Mater Res B Appl Biomater. 2006;77:201–12.
15. Suzuki O, Kamakura S, Katagiri T, Nakamura M, Zhao B, Honda Y, et al. Bone formation enhanced by implanted octacalcium phosphate involving conversion into Ca-deficient hydroxyapatite. Biomaterials. 2006;27:2671–81.
16. Kobayashi K, Anada T, Handa T, Kanda N, Yoshinari M, Takahashi T, et al. Osteoconductive property of a mechanical mixture of octacalcium phosphate and amorphous calcium phosphate. ACS Appl Mater Interfaces. 2014;6:22602–11.
17. Handa T, Anada T, Honda Y, Yamazaki H, Kobayashi K, Kanda N, et al. The effect of an octacalcium phosphate co-precipitated gelatin composite on the repair of critical-sized rat calvarial defects. Acta Biomater. 2012;8:1190–200.
18. Ishiko-Uzuka R, Anada T, Kobayashi K, Kawai T, Tanuma Y, Sasaki K, et al. Oriented bone regenerative capacity of octacalcium phosphate/gelatin composites obtained through two-step crystal preparation method. J Biomed Mater Res B Appl Biomater 2016 in press.
19. Suzuki K, Honda Y, Anada T, Handa T, Miyatake N, Takahashi A, et al. Stimulatory capacity of an octacalcium phosphate/gelatin composite on bone regeneration. Phosphorus Res Bull. 2012;26:53–8.
20. Itoigawa Y, Suzuki O, Sano H, Anada T, Handa T, Hatta T, et al. The role of an octacalcium phosphate in the re-formation of infraspinatus tendon insertion. J Shoulder Elbow Surg. 2015;24:e175–84.
21. Asran AS, Henning S, Michler GH. Polyvinyl alcohol–collagen–hydroxyapatite biocomposite nanofibrous scaffold: Mimicking the key features of natural bone at the nanoscale level. Polymer. 2010;51:868–76.
22. Curtin CM, Cunniffe GM, Lyons FG, Bessho K, Dickson GR, Duffy GP, et al. Innovative collagen nano-hydroxyapatite scaffolds offer a highly efficient non-viral gene delivery platform for stem cell-mediated bone formation. Adv Mater. 2012;24:749–54.
23. Kikuchi M, Itoh S, Ichinose S, Shinomiya K, Tanaka J. Self-organization mechanism in a bone-like hydroxyapatite/collagen nanocomposite synthesized in vitro and its biological reaction in vivo. Biomaterials. 2001;22:1705–11.
24. Liu Y, Li N, Qi YP, Dai L, Bryan TE, Mao J, Pashley DH, Tay FR. Intrafibrillar collagen mineralization produced by biomimetic hierarchical nanoapatite assembly. Adv Mater. 2011;23:975–80.
25. Gotterbarm T, Richter W, Jung M, Berardi Vilei S, Mainil-Varlet P, Yamashita T, et al. An in vivo study of a growth-factor enhanced, cell free, two-layered collagen-tricalcium phosphate in deep osteochondral defects. Biomaterials. 2006;27:3387–95.
26. Honda Y, Kamakura S, Sasaki K, Suzuki O. Formation of bone-like apatite enhanced by hydrolysis of octacalcium phosphate crystals deposited in collagen matrix. J Biomed Mater Res B Appl Biomater. 2007;80:281–9.
27. Kamakura S, Sasaki K, Honda Y, Anada T, Suzuki O. Octacalcium phosphate combined with collagen orthotopically enhances bone regeneration. J Biomed Mater Res B Appl Biomater. 2006;79:210–17.
28. Bigi A, Panzavolta S, Roveri N. Hydroxyapatite-gelatin films: a structural and mechanical characterization. Biomaterials. 1998;19:739–44.
29. Chang MC, Ko CC, Douglas WH. Preparation of hydroxyapatite-gelatin nanocomposite. Biomaterials. 2003;24:2853–62.
30. Kim HW, Kim HE, Salih V. Stimulation of osteoblast responses to biomimetic nanocomposites of gelatin-hydroxyapatite for tissue engineering scaffolds. Biomaterials. 2005;26:5221–30.
31. Liu X, Smith L, Hu J, Ma PX. Biomimetic nanofibrous gelatin/apatite composite scaffolds for bone tissue engineering. Biomaterials. 2009;30:2252–8.
32. Choi Y, Cho S, Park DJ, Park HH, Heo S, Jin HJ. Silk fibroin particles as templates for mineralization of calcium-deficient hydroxyapatite. J Biomed Mater Res Part B. 2012;100B:2029–34.

33. Manjubala I, Shelor S, Bössert J, Jandt KD. Mineralisation of chitosan scaffolds with nano-apatite formation by double diffusion technique. Acta Biomater. 2006;2:75–84.
34. Shi P, Teh T, Toh SL, Goh JC. Variation of the effect of calcium phosphate enhancement of implanted silk fibroin ligament bone integration. Biomaterials. 2013;34:5947–57.
35. Thein-Han WW, Misra RD. Biomimetic chitosan-nanohydroxyapatite composite scaffolds for bone tissue engineering. Acta Biomater. 2009;5:1182–97.
36. Wang G, Zheng L, Zhao H, Miao J, Sun C, Liu H, Huang Z, Yu X, Wang J, Tao X. Construction of a fluorescent nanostructured chitosan-hydroxyapatite scaffold by nanocrystallon induced biomimetic mineralization and its cell biocompatibility. ACS Appl Mater Interfaces. 2011;3:1692–701.
37. Zhang Y, Zhang M. Synthesis and characterization of macroporous chitosan/calcium phosphate composite scaffolds for tissue engineering. J Biomed Mater Res. 2001;55:304–12.
38. Matsuno T, Hashimoto Y, Adachi S, Omata K, Yoshitaka Y, Ozeki Y, Umezu Y, Tabata Y, Nakamura M, Satoh T. Preparation of injectable 3D-formed beta-tricalcium phosphate bead/alginate composite for bone tissue engineering. Dent Mater J. 2008;27:827–34.
39. Suárez-González D, Barnhart K, Saito E, Vanderby Jr R, Hollister SJ, Murphy WL. Controlled nucleation of hydroxyapatite on alginate scaffolds for stem cell-based bone tissue engineering. J Biomed Mater Res A. 2010;95:222–34.
40. Turco G, Marsich E, Bellomo F, Semeraro S, Donati I, Brun F, Grandolfo M, Accardo A, Paoletti S. Alginate/Hydroxyapatite biocomposite for bone ingrowth: a trabecular structure with high and isotropic connectivity. Biomacromolecules. 2009;10:575–1583.
41. Xie M, Olderoy M, Andreassen JP, Selbach SM, Strand BL, Sikorski P. Alginate-controlled formation of nanoscale calcium carbonate and hydroxyapatite mineral phase within hydrogel networks. Acta Biomater. 2010;6:3665–75.
42. Fuji T, Anada T, Honda Y, Shiwaku Y, Koike H, Kamakura S, et al. Octacalcium phosphate-precipitated alginate scaffold for bone regeneration. Tissue Eng Part A. 2009;15:3525–35.
43. Shiraishi N, Anada T, Honda Y, Masuda T, Sasaki K, Suzuki O. Preparation and characterization of porous alginate scaffolds containing various amounts of octacalcium phosphate (OCP) crystals. J Mater Sci Mater Med 2010 in press.
44. Smetana Jr K. Cell biology of hydrogels. Biomaterials. 1993;14:1046–50.
45. Endo K, Anada T, Yamada M, Seki M, Sasaki K, Suzuki O. Enhancement of osteoblastic differentiation in alginate gel beads with bioactive octacalcium phosphate particles. Biomed Mater. 2015;10:065019.
46. Jeong SH, Fan Y, Baek JU, Song J, Choi TH, Kim SW, Kim HE. Long-lasting and bioactive hyaluronic acid-hydroxyapatite composite hydrogels for injectable dermal fillers: physical properties and in vivo durability. J Biomater Appl 2016 in press.
47. Lee JH, Kim J, Baek HR, Lee KM, Seo JH, Lee HK, Lee AY, Zheng GB, Chang BS, Lee CK. Fabrication of an rhBMP-2 loaded porous β-TCP microsphere-hyaluronic acid-based powder gel composite and evaluation of implant osseointegration. J Mater Sci Mater Med. 2014;25:2141–51.
48. Suzuki K, Anada T, Miyazaki T, Miyatake N, Honda Y, Kishimoto KN, et al. Effect of addition of hyaluronic acids on the osteoconductivity and biodegradability of synthetic octacalcium phosphate. Acta Biomater. 2014;10:531–43.
49. Takahashi Y, Yamamoto M, Tabata Y. Osteogenic differentiation of mesenchymal stem cells in biodegradable sponges composed of gelatin and beta-tricalcium phosphate. Biomaterials. 2005;26:3587–96.
50. Kino R, Ikoma T, Monkawa A, Yunoki S, Munekata M, Tanaka J, Asakura T. Deposition of bone-like apatite on modified silk fibroin films from simulated body fluid. J Appl Polym Sci. 2006;99:2822–30.
51. Anada T, Kumagai T, Honda Y, Masuda T, Kamijo R, Kamakura S, et al. Dose-dependent osteogenic effect of octacalcium phosphate on mouse bone marrow stromal cells. Tissue Eng Part A. 2008;14:965–78.

52. Anada T, Sato T, Kamoya T, Shiwaku Y, Tsuchiya K, Takano-Yamamoto T, Sasaki K, Suzuki O. Evaluation of bioactivity of octacalcium phosphate using osteoblastic cell aggregates on a spheroid culture device. Regen Ther. 2016;3:58–62.
53. Wang X, Suzawa T, Miyauchi T, Zhao B, Yasuhara R, Anada T, et al. Synthetic octacalcium phosphate-enhanced reparative dentine formation via induction of odontoblast differentiation. J Tissue Eng Regen Med. 2015;9:1310–20.
54. Takami M, Mochizuki A, Yamada A, Tachi K, Zhao B, Miyamoto Y, et al. Osteoclast differentiation induced by synthetic octacalcium phosphate through RANKL expression in osteoblasts. Tissue Eng Part A. 2009;15:3991–4000.
55. Nishikawa R, Anada T, Ishiko-Uzuka R, Suzuki O. Osteoblastic differentiation of stromal ST-2 cells from octacalcium phosphate exposure via p38 signaling pathway. Dent Mater J. 2014;33:242–51.
56. Hirayama B, Aanda T, Shiwaku Y, Miyatake N, Tsuchiya K, Nakamura M, Takahashi T, Suzuki O. Immune cell response and subsequent bone formation induced by implantation of octacalcium phosphate in a rat tibia defect. RSC Adv. 2016;6:57475–84.

Open Access This chapter is distributed under the terms of the Creative Commons Attribution 4.0 International License (http://creativecommons.org/licenses/by/4.0/), which permits use, duplication, adaptation, distribution and reproduction in any medium or format, as long as you give appropriate credit to the original author(s) and the source, provide a link to the Creative Commons license and indicate if changes were made.

The images or other third party material in this chapter are included in the work's Creative Commons license, unless indicated otherwise in the credit line; if such material is not included in the work's Creative Commons license and the respective action is not permitted by statutory regulation, users will need to obtain permission from the license holder to duplicate, adapt or reproduce the material.

Chapter 11
Gene Delivery and Expression Systems in Induced Pluripotent Stem Cells

Maolin Zhang, Kunimichi Niibe, Takeru Kondo, Yuya Kamano, Makio Saeki, and Hiroshi Egusa

Abstract Induced pluripotent stem (iPS) cells, which can be generated from somatic cells by genetic manipulation, are invaluable experimental and therapeutic tools for development of tissue regeneration technologies. Many studies have demonstrated that gene delivery to pluripotent stem cells is useful for basic studies in developmental biology and for driving differentiation toward a specific cell lineage for regenerative applications. Several gene delivery systems using viral and nonviral vectors have been used for stem cell research. These gene delivery systems are designed to accommodate specific research purposes; thus, each of them possesses its own advantages and disadvantages according to the experimental design. In addition, the type of constitutive promoter in the expression vector greatly affects the transcriptional activity of transgenes in pluripotent stem cells. Therefore, it is necessary to consider the characteristics of the vectors and their promoters when selecting a gene delivery system to transfer the target gene into iPS cells. In this mini-review, characteristics of commonly used viral (adenoviral, adeno-associated viral, retroviral, and lentiviral) vectors and a nonviral *piggyBac* transposon DNA vector with constitutive promoters are outlined to support the selection of an appropriate gene delivery and expression system for iPS cell research.

M. Zhang • K. Niibe • T. Kondo • Y. Kamano
Division of Molecular and Regenerative Prosthodontics, Tohoku University Graduate School of Dentistry, 4-1 Seiryo-machi, Aoba-ku, Sendai, Miyagi 980-8575, Japan

M. Saeki
Division of Dental Pharmacology, Niigata University Graduate School of Medical and Dental Sciences, Niigata, Japan

H. Egusa (✉)
Division of Molecular and Regenerative Prosthodontics, Tohoku University Graduate School of Dentistry, 4-1 Seiryo-machi, Aoba-ku, Sendai, Miyagi 980-8575, Japan

Center for Advanced Stem Cell and Regenerative Research, Tohoku University Graduate School of Dentistry, Sendai, Miyagi, Japan
e-mail: egu@dent.tohoku.ac.jp

© The Author(s) 2017
K. Sasaki et al. (eds.), *Interface Oral Health Science 2016*,
DOI 10.1007/978-981-10-1560-1_11

Keywords Constitutive promoter • Induced pluripotent stem (iPS) cells • *PiggyBac* transposon-based gene delivery • Tetracycline-controlled transcriptional regulation • Viral vector

11.1 Introduction

Stem cells, which are characterized as immature, self-renewal, and undifferentiated cells that can give rise to many different cell lineages, are expected to open new needed therapeutic avenues [1]. Induced pluripotent stem (iPS) cells are cells reprogrammed from somatic cells via genetic modification to obtain embryonic stem (ES) cell characteristics [2, 3]. Although multipotent adult stem cells such as mesenchymal stem cells (MSCs) have been well investigated for clinical application [4], basic research on pluripotent stem cells, such as ES cells and iPS cells, may lead to further understanding of in vitro tissue/organ development, which in turn could be applied to next-generation therapeutic approaches for whole-tissue/organ regeneration.

Gene delivery to pluripotent stem cells provides a powerful experimental system to investigate the early stages of tissue/organ development. In addition, genetic modification of patient-specific iPS cells, particularly disease-model iPS cells, could facilitate the study of pathological mechanisms and provide new therapeutic approaches in personalized medicine [5]. To obtain efficient and stable transgene expression, various gene delivery methods ranging from viral vectors to plasmid-based transient gene expression have been applied to pluripotent stem cells [6]. In these methods, many types of constitutive promoters have been utilized in the expression vectors, such as the cytomegalovirus (CMV), human elongation factor 1α (EF1α), CMV enhancer/β-actin promoter with β-actin intron (CA), Rous sarcoma virus (RSV) [7], human β-actin (ACTB), phosphoglycerate kinase (PGK) [8], and simian virus 40 early (SV40) [9] promoters. The transcriptional activity of transgenes considerably varies among these promoters depending on the cell type [10]. Notably, the CMV promoter, which is one of the most popular choices for gene delivery vectors because of its strong activity in most cell lines, shows considerably weak transgene activation in stem cells [7, 8, 11–14]; therefore, it is important to select an optimal promoter to obtain the expected transgene expression in iPS cell experiments.

Beyond the constitutive promoter systems described above, regulated control of gene expression has great significance for stem cell research because of the ability to avoid undesirable effects of constitutive transgene expression after cellular differentiation [15]. Several studies have employed tetracycline (tet)-regulated systems to control transgene expression in both ES cells and iPS cells, thereby enabling the transgene function to be explored in a spatiotemporal manner [16–18]. Each gene delivery and expression system has particular advantages and disadvantages depending on the desired outcome of the experimental design. Here, we briefly review representative gene delivery and expression systems from the perspective of their application to iPS cell research.

11.2 Viral-Based Gene Delivery Systems

Gene delivery systems are classified into two major classes: virus-based vectors and nonviral vectors (Table 11.1). Viruses are suitable for efficient gene delivery experiments because of their ability to penetrate into the cell nucleus and replicate [19]. Viral vectors have been widely used to deliver foreign genes into the cell nucleus because of their high transduction efficiency and capacity for long-term transgene expression [20]. Ideally, the vector should be nontoxic, minimally immunogenic, and capable of highly efficient penetration and delivery to numerous cell types [21]. The principal viral vectors currently in use include adenovirus, adeno-associated virus (AAV), retrovirus, and lentivirus [22].

11.2.1 Adenovirus Vectors

Adenovirus, a 70–100-nm, non-enveloped, double-stranded DNA virus [23], belongs to the family *Adenoviridae* and is well known to cause respiratory tract infections. Adenoviruses enter mammalian cells via attachment to the *Coxsackievirus* and adenovirus receptor (CAR) [24]. Adenoviruses rarely integrate with the host genome because their genome is maintained episomally in the cell nucleus. Because of their ability to transduce many cell types, including both dividing and nondividing cells, without genomic integration, adenoviral vectors have been considered as

Table 11.1 Comparison of commonly used viral vectors and *piggyBac* transposon system

Vector	Adenovirus	AAV	Retrovirus	Lentivirus	*piggyBac*
Vector	Viral (non-enveloped)	Viral (non-enveloped)	Viral (enveloped)	Viral (enveloped)	Nonviral (transposon)
Delivered molecule	DNA	DNA	RNA	RNA	DNA
Packaging capacity [32]	4–5 kb	5 kb	9–12 kb	8 kb	9–14 kb
Genome integration	No	No	Yes	Yes	Yes (removable: cut and paste)
Gene expression	Transient	Transient/stable	Stable	Transient/stable	Stable
Applicable cell types	Broad	Broad	Dividing cells only	Broad	Broad

promising delivery systems for gene transduction experiments. However, the transient nature of their transgene expression limits their utility in in vitro research designs, and the toxicity and associated immune responses may hamper their clinical application.

Adenovirus vectors can easily introduce exogenous genes into mouse ES cells [7] and iPS cells [13], and they are also used as effective gene delivery tools for human ES cells and iPS cells [25]. The transient expression mediated by adenovirus vectors is actually an advantage for stem cell research, in that undesirable effects of constitutive transgene expression after cell differentiation can be avoided. Indeed, adenovirally mediated transient expression of *Runx2* or *PPARγ* was shown to efficiently guide mouse iPS cells to differentiate into osteoblasts or adipocytes, respectively [13].

It should be noted that selection of an appropriate constitutive promoter, such as EF1α and CA promoters, is important for effective adenoviral transgene expression in pluripotent stem cells [26]. When used in adenovirus vectors, the CMV and RSV promoters show weak activity in mouse pluripotent stem cells [7, 13] because they are silenced by DNA methylation [27, 28]; therefore these promoters may not be the best choice for adenoviral transduction experiments in iPS cell research.

11.2.2   AAV Vectors

AAV is a nonpathogenic, nonautonomous single-stranded DNA parvovirus that requires a helper virus such as adenovirus or herpes virus for replication. AAV has many serotypes, and among them, AAV2 is well studied and widely used as a gene delivery vector. Without the helper virus, the AAV genome remains episomal in target cells [29, 30]. Genome integration is observed at a low frequency and at a specific site on chromosome 19 [31]. AAV vectors derived from AAV lack viral coding sequences and rarely cause toxic and immune reactions, and they are thus considered as a promising gene delivery system for clinical use. However, the limited packaging capacity of AAV (approximately ~5 kb) is a major limitation of this vector system [32].

One interesting property of AAV is that the inverted terminal repeats (ITRs) of the AAV genome permit AAV vectors to efficiently introduce gene-targeting constructs into homologous chromosomal loci in a cellular genome [33]. This unique property permits gene editing, and efficient gene targeting by AAV vectors has been achieved in human ES cells and iPS cells [34–36]. Damdindorj et al. [37] reported that when used in an AAV vector, the CMV promoter provided stable and robust gene expression in cancer cell lines; however, this promoter does not seem to be preferable for noncancerous cell lines and for the purpose of AAV-based gene targeting. Further studies are needed to identify the most suitable constitutive promoter for the application of AAV vectors to precise genetic manipulation of iPS cells, which could have great scientific and therapeutic potential.

11.2.3   Retrovirus Vectors

Retroviral vectors are among the most commonly used gene delivery systems for target gene transduction. They possess several advantages compared with other viral vectors, such as high-level transgene expression activity in long-term culture of most dividing somatic cells and their large DNA capacity (9–12 kb) [32]. Retroviral vectors reverse-transcribe their single-stranded RNA genome into DNA that is then integrated into target cell genome. According to their genome organization, retroviruses are broadly divided into two categories: simple onco-retroviruses, such as Moloney murine leukemia virus (MLV), and complex retroviruses including lentiviruses, such as human immunodeficiency virus [23]. Onco-retrovirus-based vectors are not capable of gene transfer to nondividing cells because they rely on cell division for transduction.

Importantly, transgene expression by MLV-based vectors is restricted in undifferentiated pluripotent stem cells by de novo DNA methylation [38, 39] and other mechanisms, such as TRIM28-mediated silencing of the promoter element within the MLV long-terminal repeat (LTR) [40, 41]. A retroviral mutant vector, the murine ES-cell virus (MESV), was developed to facilitate target gene expression in ES cells through its ability to escape immediate silencing and initiate proviral expression [38, 42]; however, the MESV vector is still prone to inactivation during long periods of culture [38]. The self-silencing property of retroviral vectors is advantageous during reprogramming for iPS cell generation because it is necessary for the forced expression of exogenous reprogramming factors to cease once the cell reaches the ES-cell-like state. Indeed, Yamanaka and colleagues first generated iPS cells from mouse and human fibroblasts using MLA-based vectors [2, 3].

However, the self-silencing property of retroviral transduction in pluripotent stem cells would be disadvantageous for molecular studies that require sustained expression of exogenous gene products. In addition, retrovirus vectors have the potential to induce insertional mutagenesis in iPS cells through their random integration into the host genome [43]. These aspects should be carefully considered when designing iPS cell experiments using retroviral vector systems.

11.2.4   Lentivirus Vectors

In contrast to onco-retroviruses, lentiviruses are transported to the nucleus of the target cells by active transport but do not require cell division for transduction, which allows them to transduce quiescent and nondividing cells [44]. Lentivirus vectors have an ~8-kb packaging capacity that limits the introduction of many genomic DNA sequences, but most cDNA sequences can be accommodated. Given their advantages, lentiviral vectors have become the predominant vectors for gene transduction in many types of cells and in transgenic animals [45].

Similar to other retroviral vectors, lentiviral vectors possess the ability to integrate into the host genome and stably express the delivered target gene. Although insertional mutagenesis by lentiviral vectors is still their major disadvantage, it occurs less frequently for lentiviruses than for onco-retroviruses [46, 47]. To reduce the risk of insertional mutagenesis, non-integrating lentiviral (NIL) vectors, which carry either mutant integrase or mutations in the integrase binding sits, have been developed [48–50]. In this vector system, the lentiviral genome remains episomal in the nucleus, with sustained transgene expression that does not require genome integration [48, 49]; therefore, it is expected to provide a safe and promising gene delivery system for laboratory and clinical use [51].

One important characteristic of lentivirus vectors is their resistance to silencing during propagation and differentiation of ES cells [52]; as a result, they have become widely used for gene transduction in ES cells [53–56]. Hong et al. [12] demonstrated that when used in a lentiviral vector, the EF1α promoter drove robust transgene expression in mouse ES cells from undifferentiated status to fully differentiated status during neuronal differentiation, whereas the CMV promoter activated transgene expression only in late stages of differentiation. Norrman et al. [8] reported that the use of ACTB, EF1α, and PGK promoters in lentiviral vectors permitted stable transgene expression in human ES cells, whereas the CMV promoter was less effective and expression was rapidly downregulated within 7 days.

11.3 Transposon-Based Gene Delivery Systems

As a nonviral gene delivery method, electroporation has been widely used for exogenous gene expression in ES cells [57]; however, a major drawback of this method is low transfection efficiency. Transposon DNA vectors have been recently used for nonviral gene delivery [58]. The *piggyBac* transposon, a transposon DNA vector identified from the cabbage looper moth *Trichoplusia ni* [59], has been reported as a highly efficient tool to insert exogenous genes into mammalian cells [60–62]. The *piggyBac* transposon system is constituted by two basic elements: a 2,472-bp transposon with 13-bp inverted terminal repeats (ITRs) and a 594 amino acid transposase [59, 63].

The *piggyBac* transposon mediates gene transfer through a cut-and-paste mechanism (Fig. 11.1) [58], where it is inserted into the genomic DNA at TTAA tetranucleotide sites and then integrates with the chromosomal DNA through the activity of the transposase [63]. This system can efficiently deliver DNA fragments sized 9 kb, even up to 14 kb, without significant decreases in transposition efficiency [60]. Another advantage is that the *piggyBac* transposon can be excised from the original insertion site without leaving any remnant sequence [64]; thus, *piggyBac*-mediated genetic insertions are reversible. Given its capacity for efficient and reversible gene transfer, the *piggyBac* transposon system is a promising vector for gene delivery.

The *piggyBac* transposon system can be used to generate transgene-free iPS cells [65–68], which may thus have increased therapeutic utility. Additionally, the

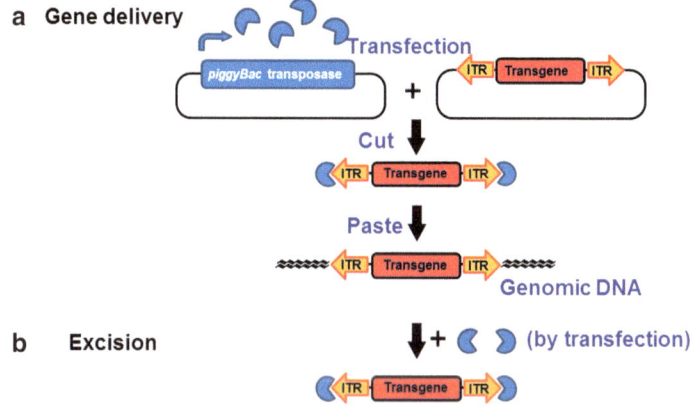

Fig. 11.1 Mechanism of *piggyBac* transposition. (**a**) In the transfected cell, transposase is expressed from the *piggyBac* transposase expression vector, and it then recognizes and binds to the specific inverted terminal repeats (ITRs) of the transgene vector plasmid and cuts the DNA sequence of the transgene from the original sites. Then, the transgene DNA sequence integrates into the genomic DNA of the target cell. (**b**) For excision, re-expression of transposase by transfection of the *piggyBac* transposase expression vector leads to cutting of the transgene at the ITRs in the genomic DNA, which results in removal of the inserted transgene from the genomic DNA

piggyBac transposon system has been used for efficient gene delivery to human ES cells, where the insertion can be removed from the ES genome without leaving any insertional mutation as described above [69, 70]. This system has also been used for gene delivery to human iPS cells [71–73]. In dental research, a *piggyBac* transposon-based gene expression system has been applied to human deciduous tooth dental pulp cell-derived iPS cells to express EGFP and tdTomato transgenes [74].

11.4 tet-Controlled Transcriptional Regulation System

Control of transgene expression is important for preventing potential adverse effects of the continued overexpression of the transgene. tet-regulated gene expression systems are among the most widely used gene regulation systems [75] and consist of two variants: the tet-off and the tet-on systems (Fig. 11.2) [76, 77]. In the tet-off system, when doxycycline, an analog of tet, is absent, the tet transactivators (tTAs) bind to their target element, a tet-operator sequence (tet response element, TRE) that is upstream of a promoter, to drive transgene expression. Conversely, in the presence of doxycycline, the tTAs cannot bind to the TRE; therefore, transgene expression is hindered. The tet-on system was derived from the tet-off system by inducing random mutations in the tTAs [77]. These mutant reverse transactivators (rtTAs) bind to the TRE in the presence of doxycycline to drive transgene expression, and transgene expression does not occur without doxycycline.

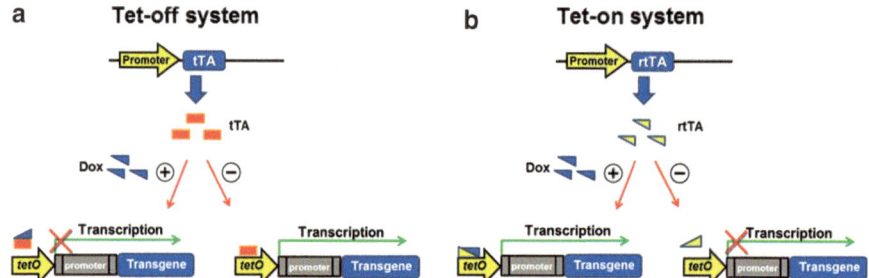

Fig. 11.2 Tetracycline (tet)-controlled transcriptional regulation systems. (**a**) tet-off system: in the presence of doxycycline (Dox), tet transactivators (tTAs) cannot bind to the tet-operator (tetO) sequence to induce target gene (transgene) expression. In the absence of Dox, tTAs bind to the tetO sequence to drive target gene expression. (**b**) tet-on system: reverse tet transactivators (rtTAs) bind to the tetO in the presence of Dox to induce transgene expression. In contrast, in the absence of Dox, rtTAs cannot bind to the tetO; thus, transgene expression does not occur

Controlled regulation of transgene expression has great significance for investigating molecular mechanisms of pluripotency and cellular differentiation in stem cells [16], and tet-controlled transcriptional activation systems have thus been applied to pluripotent stem cells [17, 18, 78]. Dox-inducible lentiviral and *piggyBac* vectors have also been used to direct reprogramming of somatic cells to iPS cells [66, 79]. This system also permits the regulation of transgene expression in iPS cells to drive their differentiation toward specific cell lineages such as myocytes [71–73].

11.5   Conclusions

Because each gene delivery system possesses its own characteristics, researchers should consider the suitability of the system, rather than technical convenience, for a particular iPS cell experiment. The choice of promoter is also important, especially for experiments in iPS cells. Although little systematic information is available regarding the activity of constitutive promoters in undifferentiated iPS cells, the EF1α and CA promoters, but not the CMV promoter, are expected to be suitable for high levels of stable transgene expression. The *piggyBac* transposon-based gene delivery system provides several benefits over classic viral and nonviral gene delivery systems. In addition, it can be combined with tet-controlled transcriptional regulation to achieve spatiotemporal control of transgene expression during iPS cell differentiation, which may provide a great impact on iPS cell research.

References

1. Egusa H, Sonoyama W, Nishimura M, Atsuta I, Akiyama K. Stem cells in dentistry – part I: stem cell sources. J Prosthodont Res. 2012;56:151–65. doi:10.1016/j.jpor.2012.06.001.
2. Takahashi K, Yamanaka S. Induction of pluripotent stem cells from mouse embryonic and adult fibroblast cultures by defined factors. Cell. 2006;126:663–76. doi:10.1016/j.cell.2006.07.024.
3. Takahashi K, Tanabe K, Ohnuki M, Narita M, Ichisaka T, Tomoda K, et al. Induction of pluripotent stem cells from adult human fibroblasts by defined factors. Cell. 2007;131:861–72. doi:10.1016/j.cell.2007.11.019.
4. Egusa H, Sonoyama W, Nishimura M, Atsuta I, Akiyama K. Stem cells in dentistry – part II: clinical applications. J Prosthodont Res. 2012;56:229–48. doi:10.1016/j.jpor.2012.10.001.
5. Avior Y, Sagi I, Benvenisty N. Pluripotent stem cells in disease modelling and drug discovery. Nat Rev Mol Cell Biol. 2016;17:170–82. doi:10.1038/nrm.2015.27.
6. Kobayashi N, Rivas-Carrillo JD, Soto-Gutierrez A, Fukazawa T, Chen Y, Navarro-Alvarez N, et al. Gene delivery to embryonic stem cells. Birth Defects Res C Embryol Today. 2005;75:10–8. doi:10.1002/bdrc.20031.
7. Kawabata K, Sakurai F, Yamaguchi T, Hayakawa T, Mizuguchi H. Efficient gene transfer into mouse embryonic stem cells with adenovirus vectors. Mol Ther. 2005;12:547–54. doi:10.1016/j.ymthe.2005.04.015.
8. Norrman K, Fischer Y, Bonnamy B, Sand FW, Ravassard P, Semb H. Quantitative comparison of constitutive promoters in human ES cells. PLoS ONE. 2010;5:e12413. doi:10.1371/journal.pone.0012413.
9. Eiges R, Schuldiner M, Drukker M, Yanuka O, Itskovitz-Eldor J, Benvenisty N. Establishment of human embryonic stem cell-transfected clones carrying a marker for undifferentiated cells. Curr Biol. 2001;11:514–8. doi:10.1016/S0960-9822(01)00144-0.
10. Qin JY, Zhang L, Clift KL, Hulur I, Xiang AP, Ren BZ, et al. Systematic comparison of constitutive promoters and the doxycycline-inducible promoter. PLoS One. 2010;5:e10611. doi:10.1371/journal.pone.0010611.
11. Chung SM, Andersson T, Sonntag KC, Bjorklund L, Isacson O, Kim KS. Analysis of different promoter systems for efficient transgene expression in mouse embryonic stem cell lines. Stem Cells. 2002;20:139–45. doi:10.1634/stemcells.20-2-139.
12. Hong S, Hwang DY, Yoon S, Isacson O, Ramezani A, Hawley RG, et al. Functional analysis of various promoters in lentiviral vectors at different stages of in vitro differentiation of mouse embryonic stem cells. Mol Ther. 2007;15:1630–9. doi:10.1038/sj.mt.6300251.
13. Tashiro K, Inamura M, Kawabata K, Sakurai F, Yamanishi K, Hayakawa T, et al. Efficient adipocyte and osteoblast differentiation from mouse induced pluripotent stem cells by adenoviral transduction. Stem Cells. 2009;27:1802–11. doi:10.1002/stem.108.
14. Wen S, Zhang H, Li Y, Wang N, Zhang W, Yang K, et al. Characterization of constitutive promoters for piggyBac transposon-mediated stable transgene expression in mesenchymal stem cells (MSCs). PLoS One. 2014;9:e94397. doi:10.1371/journal.pone.0094397.
15. Wang F, Okawa H, Kamano Y, Niibe K, Kayashima H, Osathanon T, et al. Controlled osteogenic differentiation of mouse mesenchymal stem cells by tetracycline-controlled transcriptional activation of amelogenin. PLoS One. 2015;10:e0145677. doi:10.1371/journal.pone.0145677.
16. Masui S, Shimosato D, Toyooka Y, Yagi R, Takahashi K, Niwa H. An efficient system to establish multiple embryonic stem cell lines carrying an inducible expression unit. Nucleic Acids Res. 2005;33:e43. doi:10.1093/nar/gni043.
17. Zhou BY, Ye Z, Chen G, Gao ZP, Zhang YA, Cheng L. Inducible and reversible transgene expression in human stem cells after efficient and stable gene transfer. Stem Cells. 2007;25:779–89. doi:10.1634/stemcells.2006-0128.

18. Qian K, Huang CT, Chen H, Blackbourn LW, Chen Y, Cao J, et al. A simple and efficient system for regulating gene expression in human pluripotent stem cells and derivatives. Stem Cells. 2014;32:1230–8. doi:10.1002/stem.1653.
19. Kay MA, Glorioso JC, Naldini L. Viral vectors for gene therapy: the art of turning infectious agents into vehicles of therapeutics. Nat Med. 2001;7:33–40. doi:10.1038/83324.
20. Boulaiz H, Marchal JA, Prados J, Melguizo C, Aranega A. Non-viral and viral vectors for gene therapy. Cell Mol Biol. 2005;51:3–22.
21. Evans CH, Robbins PD. Potential treatment of osteoarthritis by gene therapy. Rheum Dis Clin N Am. 1999;25:333–44. doi:10.1016/S0889-857x(05)70071-5.
22. Nixon AJ, Goodrich LR, Scimeca MS, Witte TH, Schnabel LV, Watts AE, et al. Gene therapy in musculoskeletal repair. Ann N Y Acad Sci. 2007;1117:310–27. doi:10.1196/annals.1402.065.
23. Lentz TB, Gray SJ, Samulski RJ. Viral vectors for gene delivery to the central nervous system. Neurobiol Dis. 2012;48:179–88. doi:10.1016/j.nbd.2011.09.014.
24. Bergelson JM, Cunningham JA, Droguett G, Kurt-Jones EA, Krithivas A, Hong JS, et al. Isolation of a common receptor for Coxsackie B viruses and adenoviruses 2 and 5. Science. 1997;275:1320–3. doi:10.1126/science.275.5304.1320.
25. Tashiro K, Kawabata K, Inamura M, Takayama K, Furukawa N, Sakurai F, et al. Adenovirus vector-mediated efficient transduction into human embryonic and induced pluripotent stem cells. Cell Reprogram. 2010;12:501–7. doi:10.1089/cell.2010.0023.
26. Tashiro K. Optimization of adenovirus vectors for transduction in embryonic stem cells and induced pluripotent stem cells. Yakugaku Zasshi. 2011;131:1333–8. doi:10.1248/yakushi.131.1333.
27. Brooks AR, Harkins RN, Wang P, Qian HS, Liu P, Rubanyi GM. Transcriptional silencing is associated with extensive methylation of the CMV promoter following adenoviral gene delivery to muscle. J Gene Med. 2004;6:395–404. doi:10.1002/jgm.516.
28. Meilinger D, Fellinger K, Bultmann S, Rothbauer U, Bonapace IM, Klinkert WE, et al. Np95 interacts with de novo DNA methyltransferases, Dnmt3a and Dnmt3b, and mediates epigenetic silencing of the viral CMV promoter in embryonic stem cells. EMBO Rep. 2009;10:1259–64. doi:10.1038/embor.2009.201.
29. Duan D, Sharma P, Yang J, Yue Y, Dudus L, Zhang Y, et al. Circular intermediates of recombinant adeno-associated virus have defined structural characteristics responsible for long-term episomal persistence in muscle tissue. J Virol. 1998;72:8568–77.
30. Schnepp BC, Jensen RL, Chen CL, Johnson PR, Clark KR. Characterization of adeno-associated virus genomes isolated from human tissues. J Virol. 2005;79:14793–803. doi:10.1128/JVI.79.23.14793-14803.2005.
31. McCarty DM, Young Jr SM, Samulski RJ. Integration of adeno-associated virus (AAV) and recombinant AAV vectors. Annu Rev Genet. 2004;38:819–45. doi:10.1146/annurev.genet.37.110801.143717.
32. Ratko TA, Cummings JP, Blebea J, Matuszewski KA. Clinical gene therapy for nonmalignant disease. Am J Med. 2003;115:560–9. doi:10.1016/S0002-9343(03)00447-9.
33. Russell DW, Hirata RK. Human gene targeting by viral vectors. Nat Genet. 1998;18:325–30. doi:10.1038/ng0498-325.
34. Mitsui K, Suzuki K, Aizawa E, Kawase E, Suemori H, Nakatsuji N, et al. Gene targeting in human pluripotent stem cells with adeno-associated virus vectors. Biochem Biophys Res Commun. 2009;388:711–7. doi:10.1016/j.bbrc.2009.08.075.
35. Khan IF, Hirata RK, Wang PR, Li Y, Kho J, Nelson A, et al. Engineering of human pluripotent stem cells by AAV-mediated gene targeting. Mol Ther. 2010;18:1192–9. doi:10.1038/mt.2010.55.
36. Asuri P, Bartel MA, Vazin T, Jang JH, Wong TB, Schaffer DV. Directed evolution of adeno-associated virus for enhanced gene delivery and gene targeting in human pluripotent stem cells. Mol Ther. 2012;20:329–38. doi:10.1038/mt.2011.255.
37. Damdindorj L, Karnan S, Ota A, Hossain E, Konishi Y, Hosokawa Y, et al. A comparative analysis of constitutive promoters located in adeno-associated viral vectors. PLoS One. 2014;9:e106472. doi:10.1371/journal.pone.0106472.

38. Cherry SR, Biniszkiewicz D, van Parijs L, Baltimore D, Jaenisch R. Retroviral expression in embryonic stem cells and hematopoietic stem cells. Mol Cell Biol. 2000;20:7419–26. doi:10.1128/MCB.20.20.7419-7426.2000.
39. Minoguchi S, Iba H. Instability of retroviral DNA methylation in embryonic stem cells. Stem Cells. 2008;26:1166–73. doi:10.1634/stemcells.2007-1106.
40. Wolf D, Goff SP. TRIM28 mediates primer binding site-targeted silencing of murine leukemia virus in embryonic cells. Cell. 2007;131:46–57. doi:10.1016/j.cell.2007.07.026.
41. Wolf D, Hug K, Goff SP. TRIM28 mediates primer binding site-targeted silencing of Lys1,2 tRNA-utilizing retroviruses in embryonic cells. Proc Natl Acad Sci U S A. 2008;105:12521–6. doi:10.1073/pnas.0805540105.
42. Grez M, Akgun E, Hilberg F, Ostertag W. Embryonic stem cell virus, a recombinant murine retrovirus with expression in embryonic stem cells. Proc Natl Acad Sci U S A. 1990;87:9202–6. doi:10.1073/pnas.87.23.9202.
43. Okita K, Nakagawa M, Hyenjong H, Ichisaka T, Yamanaka S. Generation of mouse induced pluripotent stem cells without viral vectors. Science. 2008;322:949–53. doi:10.1126/science.1164270.
44. Lewis PF, Emerman M. Passage through mitosis is required for oncoretroviruses but not for the human immunodeficiency virus. J Virol. 1994;68:510–6.
45. Lois C, Hong EJ, Pease S, Brown EJ, Baltimore D. Germline transmission and tissue-specific expression of transgenes delivered by lentiviral vectors. Science. 2002;295:868–72. doi:10.1126/science.1067081.
46. Hematti P, Hong BK, Ferguson C, Adler R, Hanawa H, Sellers S, et al. Distinct genomic integration of MLV and SIV vectors in primate hematopoietic stem and progenitor cells. PLoS Biol. 2004;2:e423. doi:10.1371/journal.pbio.0020423.
47. Modlich U, Navarro S, Zychlinski D, Maetzig T, Knoess S, Brugman MH, et al. Insertional transformation of hematopoietic cells by self-inactivating lentiviral and gammaretroviral vectors. Mol Ther. 2009;17:1919–28. doi:10.1038/mt.2009.179.
48. Apolonia L, Waddington SN, Fernandes C, Ward NJ, Bouma G, Blundell MP, et al. Stable gene transfer to muscle using non-integrating lentiviral vectors. Mol Ther. 2007;15:1947–54. doi:10.1038/sj.mt.6300281.
49. Philippe S, Sarkis C, Barkats M, Mammeri H, Ladroue C, Petit C, et al. Lentiviral vectors with a defective integrase allow efficient and sustained transgene expression in vitro and in vivo. Proc Natl Acad Sci U S A. 2006;103:17684–9. doi:10.1073/pnas.0606197103.
50. Sarkis C, Philippe S, Mallet J, Serguera C. Non-integrating lentiviral vectors. Curr Gene Ther. 2008;8:430–7. org/10.2174/156652308786848012.
51. Escors D, Breckpot K. Lentiviral vectors in gene therapy: their current status and future potential. Arch Immunol Ther Exp (Warsz). 2010;58:107–19. doi:10.1007/s00005-010-0063-4.
52. Pfeifer A, Ikawa M, Dayn Y, Verma IM. Transgenesis by lentiviral vectors: lack of gene silencing in mammalian embryonic stem cells and preimplantation embryos. Proc Natl Acad Sci U S A. 2002;99:2140–5. doi:10.1073/pnas.251682798.
53. Asano T, Hanazono Y, Ueda Y, Muramatsu S, Kume A, Suemori H, et al. Highly efficient gene transfer into primate embryonic stem cells with a simian lentivirus vector. Mol Ther. 2002;6:162–8. doi:10.1006/mthe.2002.0655.
54. Gropp M, Itsykson P, Singer O, Ben-Hur T, Reinhartz E, Galun E, et al. Stable genetic modification of human embryonic stem cells by lentiviral vectors. Mol Ther. 2003;7:281–7. doi:10.1016/S1525-0016(02)00047-3.
55. Kosaka Y, Kobayashi N, Fukazawa T, Totsugawa T, Maruyama M, Yong C, et al. Lentivirus-based gene delivery in mouse embryonic stem cells. Artif Organs. 2004;28:271–7. doi:10.1111/j.1525-1594.2004.47297.x.
56. Ma Y, Ramezani A, Lewis R, Hawley RG, Thomson JA. High-level sustained transgene expression in human embryonic stem cells using lentiviral vectors. Stem Cells. 2003;21:111–7. doi:10.1634/stemcells.21-1-111.
57. Tompers DM, Labosky PA. Electroporation of murine embryonic stem cells: a step-by-step guide. Stem Cells. 2004;22:243–9. doi:10.1634/stemcells.22-3-243.

58. Woodard LE, Wilson MH. piggyBac-ing models and new therapeutic strategies. Trends Biotechnol. 2015;33:525–33. doi:10.1016/j.tibtech.2015.06.009.
59. Cary LC, Goebel M, Corsaro BG, Wang HG, Rosen E, Fraser MJ. Transposon mutagenesis of baculoviruses: analysis of Trichoplusia ni transposon IFP2 insertions within the FP-locus of nuclear polyhedrosis viruses. Virology. 1989;172:156–69. doi:10.1016/0042-6822(89)90117-7.
60. Ding S, Wu X, Li G, Han M, Zhuang Y, Xu T. Efficient transposition of the piggyBac (PB) transposon in mammalian cells and mice. Cell. 2005;122:473–83. doi:10.1016/j.cell.2005.07.013.
61. Wu SC, Meir YJ, Coates CJ, Handler AM, Pelczar P, Moisyadi S, et al. piggyBac is a flexible and highly active transposon as compared to sleeping beauty, Tol2, and Mos1 in mammalian cells. Proc Natl Acad Sci U S A. 2006;103:15008–13. doi:10.1073/pnas.0606979103.
62. Wilson MH, Coates CJ, George Jr AL. PiggyBac transposon-mediated gene transfer in human cells. Mol Ther. 2007;15:139–45. doi:10.1038/sj.mt.6300028.
63. Fraser MJ, Cary L, Boonvisudhi K, Wang HG. Assay for movement of Lepidopteran transposon IFP2 in insect cells using a baculovirus genome as a target DNA. Virology. 1995;211:397–407. doi:10.1006/viro.1995.1422.
64. Mitra R, Fain-Thornton J, Craig NL. piggyBac can bypass DNA synthesis during cut and paste transposition. EMBO J. 2008;27:1097–109. doi:10.1038/emboj.2008.41.
65. Kaji K, Norrby K, Paca A, Mileikovsky M, Mohseni P, Woltjen K. Virus-free induction of pluripotency and subsequent excision of reprogramming factors. Nature. 2009;458:771–5. doi:10.1038/nature07864.
66. Woltjen K, Michael IP, Mohseni P, Desai R, Mileikovsky M, Hamalainen R, et al. piggyBac transposition reprograms fibroblasts to induced pluripotent stem cells. Nature. 2009;458:766–70. doi:10.1038/nature07863.
67. Yusa K, Rad R, Takeda J, Bradley A. Generation of transgene-free induced pluripotent mouse stem cells by the piggyBac transposon. Nat Methods. 2009;6:363–9. doi:10.1038/nmeth.1323.
68. Woltjen K, Kim SI, Nagy A. The piggyBac transposon as a platform technology for somatic cell reprogramming studies in mouse. Methods Mol Biol. 2016;1357:1–22. doi:10.1007/7651_2015_274.
69. Lacoste A, Berenshteyn F, Brivanlou AH. An efficient and reversible transposable system for gene delivery and lineage-specific differentiation in human embryonic stem cells. Cell Stem Cell. 2009;5:332–42. doi:10.1016/j.stem.2009.07.011.
70. Chen YT, Furushima K, Hou PS, Ku AT, Deng JM, Jang CW, et al. PiggyBac transposon-mediated, reversible gene transfer in human embryonic stem cells. Stem Cells Dev. 2010;19:763–71. doi:10.1089/scd.2009.0118.
71. Tanaka A, Woltjen K, Miyake K, Hotta A, Ikeya M, Yamamoto T, et al. Efficient and reproducible myogenic differentiation from human iPS cells: prospects for modeling Miyoshi Myopathy in vitro. PLoS One. 2013;8:e61540. doi:10.1371/journal.pone.0061540.
72. Kim SI, Oceguera-Yanez F, Sakurai C, Nakagawa M, Yamanaka S, Woltjen K. Inducible transgene expression in human iPS cells using versatile All-in-One piggyBac transposons. Methods Mol Biol. 2016;1357:111–31. doi:10.1007/7651_2015_251.
73. Shoji E, Woltjen K, Sakurai H. Directed myogenic differentiation of human induced pluripotent stem cells. Methods Mol Biol. 2016;1353:89–99. doi:10.1007/7651_2015_257.
74. Inada E, Saitoh I, Watanabe S, Aoki R, Miura H, Ohtsuka M, et al. PiggyBac transposon-mediated gene delivery efficiently generates stable transfectants derived from cultured primary human deciduous tooth dental pulp cells (HDDPCs) and HDDPC-derived iPS cells. Int J Oral Sci. 2015;7:144–54. doi:10.1038/ijos.2015.18.
75. Berens C, Hillen W. Gene regulation by tetracyclines. Genet Eng. 2004;26:255–77. doi:10.1007/978-0-306-48573-2_13.
76. Gossen M, Bujard H. Tight control of gene expression in mammalian cells by tetracycline-responsive promoters. Proc Natl Acad Sci U S A. 1992;89:5547–51.
77. Gossen M, Bujard H. Efficacy of tetracycline-controlled gene expression is influenced by cell type: commentary. Biotechniques. 1995;19:213–6.

78. Vieyra DS, Goodell MA. Pluripotentiality and conditional transgene regulation in human embryonic stem cells expressing insulated tetracycline-ON transactivator. Stem Cells. 2007;25:2559–66. doi:10.1634/stemcells.2007-0248.
79. Hockemeyer D, Soldner F, Cook EG, Gao Q, Mitalipova M, Jaenisch R. A drug-inducible system for direct reprogramming of human somatic cells to pluripotency. Cell Stem Cell. 2008;3:346–53. doi:10.1016/j.stem.2008.08.014.

Open Access   This chapter is distributed under the terms of the Creative Commons Attribution 4.0 International License (http://creativecommons.org/licenses/by/4.0/), which permits use, duplication, adaptation, distribution and reproduction in any medium or format, as long as you give appropriate credit to the original author(s) and the source, provide a link to the Creative Commons license and indicate if changes were made.

The images or other third party material in this chapter are included in the work's Creative Commons license, unless indicated otherwise in the credit line; if such material is not included in the work's Creative Commons license and the respective action is not permitted by statutory regulation, users will need to obtain permission from the license holder to duplicate, adapt or reproduce the material.

Chapter 12
Emerging Regenerative Approaches for Periodontal Regeneration: The Future Perspective of Cytokine Therapy and Stem Cell Therapy

Shinya Murakami

Abstract Cytokine therapy using basic fibroblast growth factor (FGF-2) has attracted attention as a next-generation periodontal tissue regenerative therapy. Clinical trial studies to date have shown that local application of 0.3 % FGF-2 induces statistically significant new alveolar bone formation. In vitro analyses using cultured periodontal ligament cells showed that FGF-2 maintains stem cells in an undifferentiated state and promotes the proliferation of these stem cells during the initial stages of wound healing. This increases the cell density of periodontal tissue stem cells at the site of healing, promotes angiogenesis, and produces specific extracellular matrix molecules, thereby preparing a local environment suitable for the regeneration of periodontal tissue. Additionally, based on an analysis in beagles, when adipose tissue-derived multi-lineage progenitor cells (ADMPCs) isolated from adipose tissue were transplanted together with fibrin gel to areas of periodontal tissue loss, significant regeneration of periodontal tissue was observed at the transplantation site. These results strongly suggest that adipose tissue, which is abundant in the human body and can be easily and safely collected, is a promising source of stem cells to promote the regeneration of periodontal tissue.

Keywords Periodontal regeneration • FGF-2 • ADSC

S. Murakami (✉)
Department of Periodontology, Osaka University Graduate School of Dentistry,
1-8 Yamadaoka, Suita, Osaka 565-0871, Japan
e-mail: ipshinya@dent.osaka-u.ac.jp

© The Author(s) 2017
K. Sasaki et al. (eds.), *Interface Oral Health Science 2016*,
DOI 10.1007/978-981-10-1560-1_12

12.1 Introduction

Periodontal disease occurs when a bacterial biofilm (dental plaque) adheres to the boundary between the teeth and gingiva, causing chronic inflammation and progressively destroying the periodontal tissue that supports the teeth. Therefore, periodontal treatment involves scaling and root planing, which mechanically removes the causative bacteria biofilm together with the necrotic cementum from the surface of the tooth root. Appropriate application of this therapy eliminates periodontal tissue inflammation and stops the process of destruction of the same tissue. However, removing the cause of the disease does not regenerate the lost periodontal tissue to its original state. Given the high prevalence of periodontal disease both in Japan and worldwide and the need to maintain or enhance "QOL supported by the teeth and the mouth" in middle-aged and elderly people, developing highly predictable periodontal tissue regenerative therapy that can be performed as a follow-on after treatment to remove the cause is urgently needed.

12.2 Periodontal Ligament as a Storage Site for Periodontal Tissue Stem Cells

Induction of tissue regeneration requires the presence of stem cells. Studies of regenerative medicine for periodontal tissue have shown that tissue stem cells that make the regeneration of periodontal tissue possible are present in the tissue surrounding the tooth root, known as the periodontal ligament, even in adults. The periodontal ligament is a ligament tissue located between the cementum and alveolar bone, with a thickness of 100–200 μm. Type I collagen fiber bundles form the main body of the periodontal ligament; the principal cellular components of this tissue include the periodontal ligament fibroblasts as well as characteristic cell groups such as osteoblasts and osteoclasts on the surface of alveolar bone and cementoblasts found on the cementum surface. Interestingly, the results of various molecular biological analyses of periodontal ligament tissue show that many of the cells in the periodontal ligament show consistently high expression of molecules such as RUNX-2 and alkaline phosphatase which are closely associated with cytodifferentiation into hard tissue-forming cells. These results suggest that the periodontal ligament is involved in remodeling hard tissues such as alveolar bone and cementum depending on environmental changes. Furthermore, undifferentiated mesenchymal stem cells, which can differentiate into a variety of cell types other than osteoblasts and cementoblasts, are present in the human periodontal ligament [1, 2].

12.3   Concept and Current Status of Periodontal Tissue Regenerative Therapy

Although mesenchymal stem cells that can differentiate into osteoblasts and cementoblasts are present in the adult periodontal ligament, periodontal tissue does not successfully regenerate when conventional periodontal treatment is performed to remove the cause of periodontal disease. Thus, in addition to conventional treatment, it is necessary to develop a method for inducing the periodontal tissue stem cells present in the periodontal ligament.

The specific biological processes required for the regeneration of periodontal tissue can be summarized into the following three processes: (1) selective and preferential induction of periodontal ligament-derived cells on the tooth root surface facing the areas of periodontal tissue loss, (2) proliferation and migration of undifferentiated mesenchymal stem cells (periodontal tissue stem cells) contained within the periodontal ligament-derived cells while retaining their differentiation potential and then achieving site-specific differentiation as hard tissue-forming cells (osteoblasts and cementoblasts) and periodontal ligament fibroblasts, and (3) collagen fiber bundles produced by periodontal ligament fibroblasts becoming embedded into bone tissue and cementum newly formed by osteoblasts and cementoblasts, resulting in regeneration of the fibrous connections between the teeth and alveolar bone (Fig. 12.1).

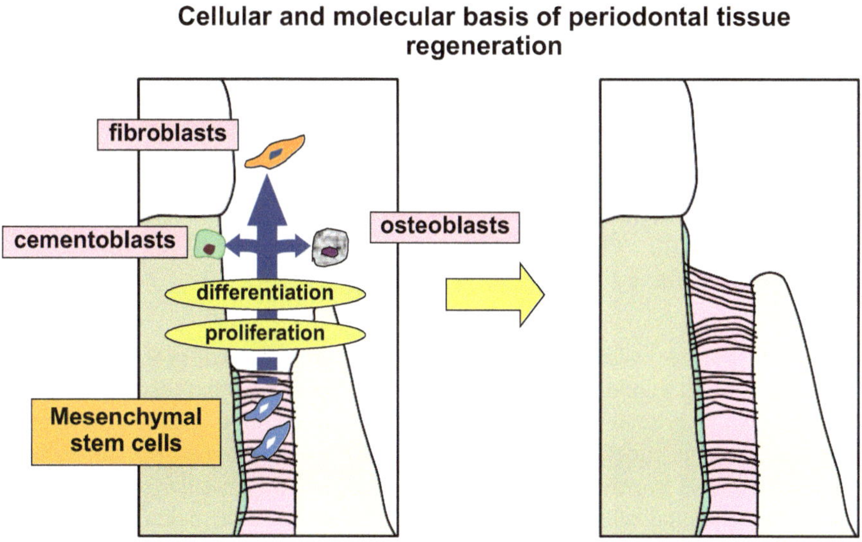

Cellular and molecular basis of periodontal tissue regeneration

Fig. 12.1 Cellular and molecular basis of periodontal tissue regeneration. Periodontal tissue stem cells, including undifferentiated mesenchymal stem cells, are present in the periodontal ligament, and regeneration of periodontal tissue can be achieved by activating these cells to proliferate, migrate, and site specifically differentiate at the site of periodontal tissue loss

Periodontal tissue regenerative therapies with clinical applications have been developed previously, including the "guided tissue regeneration method" and "enamel matrix protein method." These treatment methods have a long history of use in dental practice and have shown some degree of success. However, there are a number of issues remaining related to indicated conditions and predictability, and improvements to these methods are required. Therefore, the action mechanism should be determined and next-generation periodontal tissue regenerative therapy with stable predictability should be established.

12.4 Possibility of Cytokine Therapy

Human recombinant cytokines are currently used as therapeutic agents in a variety of diseases. In the field of periodontal tissue regenerative medicine, there have been many attempts to induce regeneration of periodontal tissue through the migration of periodontal tissue stem cells to the site of periodontal tissue loss and proliferation of these cells in the same area or differentiation to osteoblasts and cementoblasts, which are activated through local application of specific cytokines.

In the USA, a combination of platelet-derived growth factor (PDGF) and β-tricalcium phosphate (β-TCP) (β-TCP + 0.3 mg/mL PDGF) has been approved by the US Food and Drug Administration (FDA) to induce regeneration of periodontal tissue, which is marketed as GEM21S®. Cytokines are also applied in dentistry using a combination of bone morphogenic protein-2 (BMP-2) and bovine type I collagen, which has also been approved by the US FDA as a medical device used in alveolar ridge augmentation procedures and sinus elevation surgery. Thus, human recombinant cytokines are beginning to be applied in the field of dentistry.

12.5 Inducing Periodontal Tissue Regeneration with Basic Fibroblast Growth Factor (FGF-2)

Fibroblast growth factors (FGFs) are a group of proteins that promote fibroblast proliferation and were discovered in the brain and pituitary gland tissue. This family of proteins comprises FGF-1 to FGF-23. Among these, FGF-2 induces the proliferation of a wide variety of cells, including fibroblasts, vascular endothelial cells, neuroectodermal system cells, osteoblasts, chondrocytes, vascular smooth muscle cells, and epithelial cells. FGF-2 has attracted attention in the field of regenerative medicine because: (1) FGF-2 has potent angiogenesis-promoting action and (2) FGF-2 promotes the cellular proliferation of undifferentiated mesenchymal cells while retaining their pluripotency. An example of the clinical application of FGF-2 is as a therapeutic agent for intractable skin ulcers, such as decubital ulcers, and this therapy has been approved for manufacture in Japan.

In our laboratory, we verified the efficacy and safety and determined whether FGF-2 promotes the regeneration of periodontal tissue in animal studies [3, 4]. We experimentally prepared class II furcation involvement at the furcation sites of mandibular molars in beagles and nonhuman primate and filled the bone defect, on the experimental side, with 0.1–0.4 % FGF-2 with a gelatinous carrier and measured the histological morphology at 6 and 8 weeks after FGF-2 administration. The results showed that regeneration of periodontal tissue was associated with a statistically significant increase in new bone volume, new trabecular bone volume, and new cementum volume and was induced by local administration of FGF-2 (Fig. 12.2). At the same site, we observed the reappearance of Sharpey's fibers and rebuilding of fibrous attachments [3]. Extension of the peripheral nerves and Ruffini nerve endings was detected in the regenerated periodontal ligament. This strongly suggests that local administration of FGF-2 not only constructs periodontal tissue but also regenerates the function of sensory receptors in the tissue.

Furthermore, abnormal healing findings such as downgrowth of the gingival epithelium, ankylosis, and root resorption were not detected in cases on the side administered FGF-2.

We conducted an early phase II (PIIa) clinical trial (double-blind study with dose-response concurrent control, including placebo) with 13 participating facilities throughout Japan to investigate whether FGF-2 could induce the regeneration of periodontal tissue as well as the safety of FGF-2 (2002–2004). In this study, we investigated the efficacy and safety of FGF-2 as a drug to induce the regeneration of periodontal tissue using a placebo (3 % hydroxypropylcellulose, which was used as the carrier) and the investigational drug containing 0.03 %, 0.1 %, and 0.3 % FGF-2. The results revealed a statistically significant induction of new alveolar bone formation based on standardized x-ray images after local administration of 0.3 % FGF-2 to human 2- or 3-wall intrabony defects in the alveolar bone [5]. Next, we implemented a late phase II (PIIb) clinical trial (dose-response study) with 24 participating facilities throughout Japan to investigate the efficacy and safety of FGF-2 using placebo and the investigational drug containing 0.2 %, 0.3 %, and 0.4 % FGF-2 (2005–2007). The results showed that all investigational drugs containing FGF-2 induced statistically significant new alveolar bone formation, and 0.3 % FGF-2 was the clinically recommended dose [6] (Fig. 12.2a). Based on these results, we then implemented a phase III (PIII) clinical study (randomized double-blind parallel-group comparison study) with 23 participating facilities throughout Japan to investigate the efficacy and safety of FGF-2 using placebo and the investigational drug containing 0.3 % FGF-2 (2008–2010). The results confirmed that the investigational drug containing 0.3 % FGF-2 significantly increased alveolar bone ($p < 0.001$) (Fig. 12.2b) [7]. Additionally, no cases throughout the entire clinical study period were problematic in terms of safety.

We conducted detailed investigations into the effect of FGF-2 on cultured human periodontal ligament-derived cells to elucidate the mechanism of FGF-2-induced periodontal tissue regeneration; we predict the action mechanism to be as follows (Fig. 12.3). First, during the initial stages of wound healing, FGF-2 promotes the proliferation of stem cells while retaining their undifferentiated state and increases

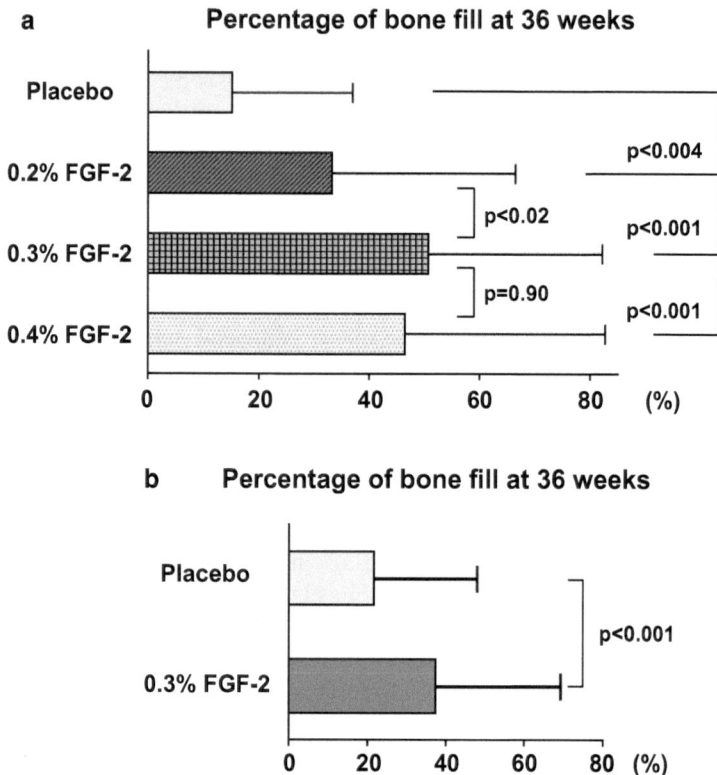

Fig. 12.2 Rate of increase in the height of alveolar bone 9 months after FGF-2 administration (clinical trial results on efficacy). (**a**) Late phase II clinical study: The subjects were divided into four groups (placebo group and 0.2%, 0.3%, and 0.4% FGF-2 treatment groups); each investigational drug was administered under double-blinded conditions. The figure shows the digitized results of a standardized dental x-ray taken 9 months after administration of the investigational drug and shows the percentage of the depth of the intrabony defect prior to treatment that had been filled with new bone (Adapted from literature [6]). (**b**) Phase III clinical study: The subjects were divided into two groups (placebo group and 0.3% FGF-2 treatment group); each investigational drug was administered under double-blinded conditions. The figure shows the digitized results of a standardized dental x-ray taken 9 months after administration of the investigational drug and shows the percentage of the depth of the intrabony defect compared to before treatment that had been filled with new bone (Adapted from literature [7])

the cell density of periodontal tissue stem cells at the wound site. Furthermore, FGF-2 promotes angiogenesis and the production of specific extracellular matrix molecules, creating a suitable local environment for the regeneration of periodontal tissue. After the effect of the locally administered FGF-2 is removed from the administration site through degradation or other actions, the periodontal tissue stem cells, which had increased in number, begin to differentiate into hard tissue-forming cells in the optimal environment created by FGF-2 administration. This promotes

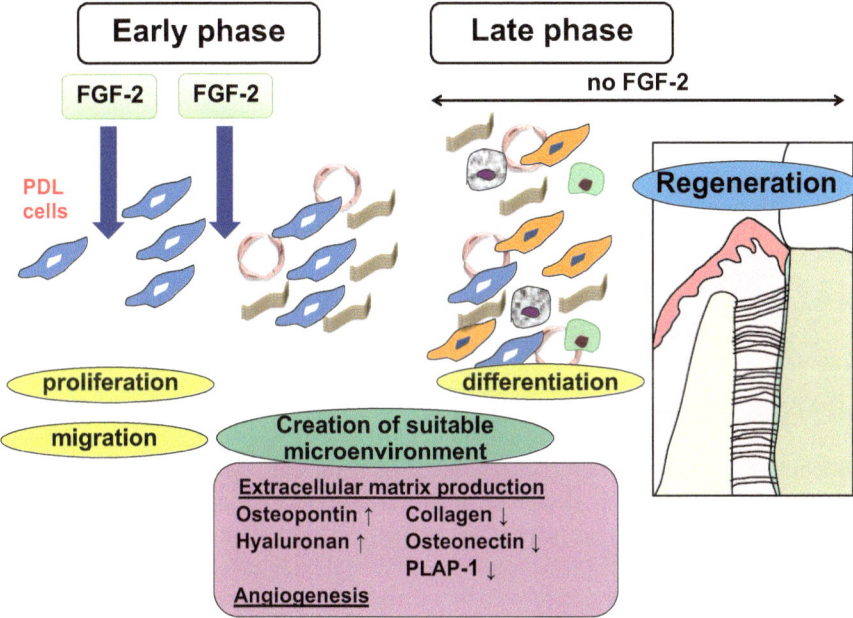

Fig. 12.3 Mechanism of induction of periodontal tissue regeneration with FGF-2. FGF-2 strongly promotes the migration and proliferation of immature human periodontal ligament (PDL) cells while retaining the cells' differentiation potential and increases the number of periodontal tissue stem cells at the site of periodontal tissue loss. Furthermore, FGF-2 promotes the production of angiogenesis and various extracellular matrices at the administration site, creating a suitable local environment for "periodontal tissue regeneration"

the regeneration of periodontal tissue, both in number and time period, including the formation of new alveolar bone and cementum [8].

12.6 Future Outlook of Cytokine Therapy Using FGF-2

The concept of "scaffold" has not been introduced for FGF-2 preparations, and clinical studies are currently underway. This is because the aim of both clinical studies was to first clarify the efficacy and safety of FGF-2 alone as a drug. However, in the next stage, we will examine tissue engineering-type methods for the FGF-2 carrier.

To date, in animal experiments using a severe periodontal tissue loss model (one wall intrabony defect model), we have found that following administration of 0.3% FGF-2 with β-TCP, there is no recession of the gingiva postoperatively and that significant periodontal tissue regeneration is induced at the same loss site [9]. If a new carrier for FGF-2 preparation is developed that can retain the space (space-making) at which the regeneration of periodontal tissue is expected, and with the appropriate formativeness and osteoconductive properties required by operators, the application of FGF-2 preparations is expected to be further broadened. Additionally, the results of animal experiments suggest that if the 0.3 % FGF-2 preparation is used concurrently during dental implant surgery, the healing time to achieve osseointegration will be shortened and robust integration will be achieved; further verification of these concepts is required in the future [10].

12.7 Possibility of Periodontal Tissue Regenerative Therapy with Stem Cell Transplantation

The periodontal tissue regenerative therapy described above promotes the regeneration of periodontal tissue by inducing the functions of stem cells present in all periodontal ligaments. However, the number of stem cells in periodontal ligament decreases with age [11]. Therefore, in elderly people and in cases of severe periodontal disease, we do not expect an adequate amount of regeneration simply by activating the stem cells present in the periodontal ligament. In these cases, it may be necessary to promote periodontal tissue regeneration by transplanting mesenchymal stem cells collected from other tissue in the same patient to the site of periodontal tissue loss.

The use of induced pluripotent stem cells is expected in the future, but it will still be some time before this technology can be clinically applied in the field of dentistry. However, it has been clarified that undifferentiated mesenchymal stem cells are present in a variety of tissues in our bodies, even after reaching adulthood. Studies are currently being conducted to regenerate periodontal tissue by transplanting stem cells collected from these tissues into the site of periodontal tissue loss together with suitable scaffolding materials.

Examples of bone marrow cells used to regenerate periodontal tissue are as follows. There is a report showing that mixing bone marrow cells collected from the ilium or other such sites with platelet-rich plasma and transplanting the mixture into the site of periodontal tissue loss is effective for inducing periodontal tissue regeneration [12], whereas another report found that after stimulating and proliferating collected bone marrow cells with FGF-2 in the laboratory, subsequent mixing the cells with collagen gel as scaffolding material promotes periodontal tissue regeneration [13]. In these cases, undifferentiated mesenchymal cells and osteoblast progenitor cells in the bone marrow were used as stem cell sources during periodontal tissue regeneration.

Another study investigated the promotion of periodontal tissue regeneration by collecting tissue slices from the maxilla periosteum, culturing the cells collected from the same tissue slice (periosteum-derived cells: expected to be cells with high osteogenic potential) in a sheet format, and transplanting the cultured cells together with platelet-rich plasma and hydroxyapatite into the site of periodontal tissue loss; this process stimulated new alveolar bone formation and promoted regeneration

Periodontal regeneration by transplantation of ADMPC

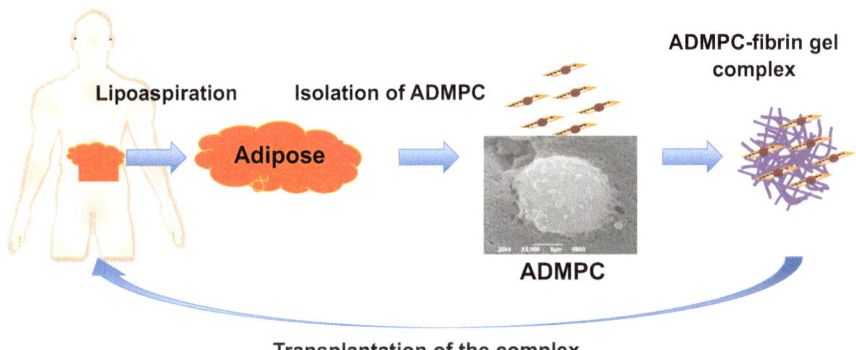

Fig. 12.4 Periodontal regeneration by transplantation of ADMPC. Approximately 30 mL of abdominal adipose tissue was suction sampled from the patient with periodontal disease, and ADMPCs were collected from the same tissue in the Cell Processing Center. These cells were then cultured until a sufficient volume of ADMPCs was acquired, which was then mixed with fibrin gel. Regeneration of periodontal tissue was induced by transplanting the same admixture into the site of the intrabony defect during periodontal surgery in the same patient

[14]. Furthermore, another study investigated culturing periodontal ligament cells in a sheet format, transplanting the cells onto the surface of a tooth root that had been exposed at the site of periodontal tissue loss, and transplanting β-TCP at the site of bone loss for total regeneration of periodontal tissue [15].

In our laboratory, we focused on adipose tissue, which we consider to be less of a burden on the patient during collection and a safer tissue. We investigated the use of undifferentiated mesenchymal stem cells present in another tissue as a source of stem cells (Fig. 12.4). We previously reported that adipose tissue-derived multi-lineage progenitor cells (ADMPCs) collected using our method differentiate into osteoblasts, myocardial cells, liver cells, and insulin-producing cells, as ADMPCs are pluripotent [16–19]. We also confirmed in in vitro that ADMPC isolated from human adipose tissue can differentiate into osteoblasts and periodontal ligament cells. Next, we investigated the effect of inducing periodontal tissue regeneration by ADMPC transplantation in a beagle periodontal disease model. We created artificial periodontal tissue loss (class II furcation involvement) at the furcation site of the fourth premolars in beagles and transplanted ADMPC isolated from adipose tissue collected from the abdominal area together with fibrin gel into the same site of the defect. The results confirmed significant regeneration of periodontal tissue at the transplantation site [20]. Interestingly, ADMPC had a remarkable trophic effect; insulin-like growth factor binding protein 6, which was secreted from ADMPC in large quantities, promoted the differentiation of periodontal tissue stem cells [21]. These results strongly suggest that adipose tissue, which is abundant in the body and can be easily and safely collected, can be used as a source of stem cells to promote the regeneration of periodontal tissue.

12.8 Conclusion

Periodontal tissue is a complex tissue comprised of alveolar bone, cementum, periodontal ligament, and gingiva, and the end target is to regenerate each type of tissue at will depending on the extent of periodontal tissue destruction. In the future, after ensuring the safety of this therapy, optimal scaffolding material can be customized for periodontal tissue regeneration to support stem cell migration and proliferation, as well as differentiation into osteoblasts and cementoblasts. The application of cytokine therapy using FGF-2 and periodontal tissue regenerative therapy through stem cell transplantation of ADMPC will not be restricted to the field of dentistry, but can also be expanded to other medical fields. Additionally, in the near future, we hope to establish patient-specific periodontal tissue regenerative therapy by combining optimal conditions with cytokines, stem cells, and scaffolding materials.

References

1. Seo BM, Miura M, Gronthos S, Bartold PM, Batouli S, Brahim J, et al. Investigation of multipotent postnatal stem cells from human periodontal ligament. Lancet. 2004;364:149–55.
2. Beertsen W, McCulloch CA, Sodek J. The periodontal ligament: a unique, multifunctional connective tissue. Periodontology. 1997;13:20–40.
3. Takayama S, Murakami S, Shimabukuro Y, Kitamura M, Okada H. Periodontal regeneration by FGF-2 (bFGF) in primate models. J Dent Res. 2001;81:2075–9.
4. Murakami S, Takayama S, Kitamura M, Shimabukuro Y, Yanagi K, Ikezawa K, et al. Recombinant human basic fibroblast growth factor (bFGF) stimulates periodontal regeneration in class II furcation defects created in beagle dogs. J Periodontal Res. 2003;38:97–103.
5. Kitamura M, Nakashima K, Kowashi Y, Fujii T, Shimauchi H, Sasano T, et al. Periodontal tissue regeneration using fibroblast growth factor-2 periodontal tissue regeneration using fibroblast growth factor-2: randomized controlled phase II clinical trial. PLoS One. 2008;3:e2611.
6. Kitamura M, Akamatsu M, Machigashira M, Hara Y, Sakagami R, Hirofuji T, et al. FGF-2 stimulates periodontal regeneration: results of a multi-center randomized clinical trial. J Dent Res. 2011;90:35–40.
7. Kitamura M, Akamatsu M, Kawanami M, Furuichi Y, Fujii T, Mori M, et al. Randomized placebo-controlled and controlled non-inferiority phase III trials comparing trafermin, a recombinant human fibroblast growth factor 2, and enamel matrix derivative in periodontal regeneration in intrabony defects. J Bone Miner Res. 2016;31:80.
8. Murakami S. Periodontal tissue regeneration by signalling molecule(s): what role does basic fibroblast growth factor (FGF-2) have in periodontal therapy? Periodontol 2000. 2011;56:188–208.
9. Anzai J, Kitamura M, Nozaki T, Nagayasu T, Terashima A, Asano T, et al. Effects of concomitant use of fibroblast growth factor (FGF)-2 with beta-tricalcium phosphate (β-TCP) on the beagle dog 1-wall periodontal defect model. Biochem Biophys Res Commun. 2010;403:345–50.

10. Nagayasu-Tanaka T, Nozaki T, Anzai J, Noriko S, Terashima A, Miki K, et al. FGF-2 promotes initial osseointegration and enhances stability of implants with low primary stability. Clin Oral Implants Res. 2016; in press.
11. Zheng W, Wang S, Ma D, Tang L, Duan Y, Jin Y. Loss of proliferation and differentiation capacity of aged human periodontal ligament stem cells and rejuvenation by exposure to the young extrinsic environment. Tissue Eng A. 2009;15:2363–71.
12. Yamada Y, Ueda M, Hibi H, Baba S. A novel approach to periodontal tissue regeneration with mesenchymal stem cells (MSCs) and platelet-rich plasma (PRP) using tissue engineering technology: a clinical case report. Int J Periodontics Restor Dent. 2006;26:363–9.
13. Kawaguchi H, Kurihara H. Clinical trial of periodontal tissue regeneration. Nippon Rinsho. 2008;66:948–54.
14. Okuda K, Yamamiya K, Kawase T, Mizuno H, Ueda M, Yoshie H. Treatment of human infrabony periodontal defects by grafting human cultured periosteum sheets combined with platelet-rich plasma and porous hydroxyapatite granules: case series. Int Acad Periodontol. 2009;11:206–13.
15. Iwata T, Yamato M, Tsuchioka H, Takagi R, Mukobata S, Washio K, et al. Periodontal regeneration with multi-layered periodontal ligament-derived cell sheets in a canine model. Biomaterials. 2009;30:2716–23.
16. Okura H, Komoda H, Fumimoto Y, Lee CM, Nishida T, Sawa Y, et al. Transdifferentiation of human adipose tissue-derived stromal cells into insulin-producing clusters. J Artif Organs. 2009;12:123–30.
17. Okura H, Matsuyama A, Lee CM, Saga A, Kakuta-Yamamoto A, Nagao A, et al. Cardiomyoblast-like cells differentiated from human adipose tissue-derived mesenchymal stem cells improve left ventricular dysfunction and survival in a rat myocardial infarction model. Tissue Eng C Methods. 2010;16:417–25.
18. Okura H, Komoda H, Saga A, Kakuta-Yamamoto A, Hamada Y, Fumimoto Y, et al. A properties of hepatocyte-like cell clusters from human adipose tissue-derived mesenchymal stem cells. Tissue Eng C Methods. 2010;16:761–70.
19. Komoda H, Okura H, Lee CM, Sougawa N, Iwayama T, Hashikawa T, et al. Reduction of N-glycolylneuraminic acid xenoantigen on human adipose tissue-derived stromal cells/mesenchymal stem cells leads to safer and more useful cell sources for various stem cell therapies. Tissue Eng A. 2010;16:1143–55.
20. Ozasa M, Sawada K, Iwayama T, Yamamoto S, Morimoto C, Okura H, et al. Periodontal tissue regeneration by transplantation of adipose tissue-derived multi-lineage progenitor cells. Inflamm Regen. 2014;34:109–16.
21. Sawada K, Takedachi M, Yamamoto S, Morimoto C, Ozasa M, Iwayama T, et al. Trophic factors from adipose tissue-derived multi-lineage progenitor cells promote cytodifferentiation of periodontal ligament cells. Biochem Biophys Res Commun. 2015;464:299–305.

Open Access This chapter is distributed under the terms of the Creative Commons Attribution 4.0 International License (http://creativecommons.org/licenses/by/4.0/), which permits use, duplication, adaptation, distribution and reproduction in any medium or format, as long as you give appropriate credit to the original author(s) and the source, provide a link to the Creative Commons license and indicate if changes were made.

The images or other third party material in this chapter are included in the work's Creative Commons license, unless indicated otherwise in the credit line; if such material is not included in the work's Creative Commons license and the respective action is not permitted by statutory regulation, users will need to obtain permission from the license holder to duplicate, adapt or reproduce the material.

Chapter 13
Molecular Mechanisms Regulating Tooth Number

Maiko Kawasaki, Katsushige Kawasaki, James Blackburn, and Atsushi Ohazama

Abstract Tooth number, shape, and position are consistent in mammals and are subject to strict genetic control. Multiple signaling pathways including Shh, Tgf, Bmp, Wnt, Fgf, Notch, and NF-kB are known to play critical roles in regulating tooth development. Recent studies show that these signaling pathways interact with each other through positive and negative feedback loops to regulate tooth number, shape, and spatial pattern. Teeth develop via a dynamic and complex reciprocal interaction between dental epithelium and cranial neural crest-derived mesenchyme. These interactions contain a series of inductive and permissive processes that lead to the determination, differentiation, and organization of odontogenic cells, which are controlled by these signaling pathways. It is believed that dozens of different molecules together form complex molecular networks that regulate tooth development. Studies of human congenital disease and transgenic mice suggest that disturbance of the molecular network results in abnormal tooth formation. Since molecular mechanisms involved in tooth development should be reproduced in tooth regeneration, knowledge of tooth development from both human and mouse studies is crucial for exploring tooth regenerative therapies. In this paper, we present an overview of the current literature covering the molecular mechanisms of tooth development, especially those regulating tooth number.

Keywords Tooth development • Tooth number • Missing teeth • Supernumerary teeth

M. Kawasaki • K. Kawasaki • A. Ohazama (✉)
Division of Oral Anatomy, Department of Oral Biological Science, Niigata University
Graduate School of Medical and Dental Sciences,
2-5274 Gakkocho-dori, Chuo-ku, Niigata 951-8514, Japan
e-mail: atsushiohazama@dent.niigata-u.ac.jp

J. Blackburn
Laboratory of Transcriptomic Research, The Kinghorn Cancer Centre, Garvan Institute of Medical Research, Sydney, Australia

UNSW Medicine, St. Vincent's Clinical School, Darlinghurst, Sydney, Australia

© The Author(s) 2017
K. Sasaki et al. (eds.), *Interface Oral Health Science 2016*,
DOI 10.1007/978-981-10-1560-1_13

13.1 Introduction

Tooth position, number, size, and shape are consistent in mammals and are determined genetically [6, 10, 13, 41, 80]. Teeth develop via sequential, reciprocal interactions between the oral epithelium and neural crest-derived mesenchyme. The first morphological sign of tooth development is an epithelial thickening (dental placode). The thickened tooth epithelium progressively takes the form of bud, cap, and bell configurations as differentiation and morphogenesis proceed. In addition to thickening of the epithelium, mesenchymal cells condense – a process that has been shown to be critical for organogenesis [44]. Subsequently, epithelial cells and mesenchymal cells (dental papilla) differentiate into enamel-producing ameloblasts and dentin-secreting odontoblasts, respectively. Since substantial research efforts over the last decade – using both human studies and transgenic mouse studies – have elucidated many aspects of the molecular mechanisms in tooth development, our understanding of the control of tooth shape diversity and location in the jaws has advanced considerably.

These research efforts have also elucidated that multiple signaling pathways including Shh, Tgf, Bmp, Wnt, Fgf, Notch, and NF-kB are known to play critical roles in regulating tooth development. The fine-tuning of these signaling pathways has been shown to be crucial in governing odontogenic precision. Recent studies show that crosstalk between these signaling pathways build complex molecular networks that regulate tooth development [6, 10, 13, 26, 28, 41, 80].

Regenerative medicine is one of the revolutionary future therapies in dentistry. Since regeneration of organs starts from initiation, knowledge of the molecular mechanisms involved at this stage is crucial for developing tooth regenerative therapies. Missing and extra teeth have been shown to be caused by disturbance of the developmental mechanisms during the initiation stage. Studying these numerical anomalies in both humans and mice therefore provides invaluable information to understanding the molecular mechanisms of tooth initiation and therefore tooth regeneration. In this paper, we present an overview covering the molecular mechanisms of tooth development, especially those determining tooth number.

13.2 Skin Appendage

The skin serves several functions, including thermoregulation, protection from the external environment, maintaining internal tissue fluid from evaporation, sensation, defense against infection, and supporting scaffold for hair. To fulfill these multiple functions, skin develops many structures as epidermal appendages such as nails, sweat glands, hair, sebaceous glands, tooth, and mammary glands. Although diverse in their structural phenotypes, these organs including teeth and hair share common morphological features in the early stages of development – i.e., epithelial components originate as thickenings that subsequently form buds around which the underlying mesenchymal cells condense. Interactions between the epithelial and mesenchymal tissues play central roles in regulating the morphogenesis of the skin appendages. When cultured alone, neither the epithelial nor mesenchymal components of these structures can differentiate into specific cells. It is also known that epithelial-mesenchymal interactions are sequential and reciprocal occurring in both directions between the epithelial and mesenchymal tissues.

Mice are the most highly studied mammals for investigating the mechanisms of tooth development. Initiation begins before the organ anlagen are morphologically visible. The first odontogenic signals derive from the tooth epithelium. Bmps-, Fgfs-, Wnts-, NF-kB-, and Shh-related genes are expressed in the presumptive tooth epithelium before the thickening process. Unlike humans, rodent incisors grow continuously throughout life and the stem cell niche is known to be located at the apical end of the incisor tooth. In the laboratory, hair can be initiated from murine dental stem cells, suggesting that tooth epithelium also retains the ability to form hair [84].

13.3 Missing Teeth

13.3.1 Missing Teeth in Humans

Congenital dental anomalies can occur either as non-syndromic familial cases or as part of a syndrome. Missing teeth are one of the most common human developmental anomalies. It has been shown that more than 20 % of humans lose at least one of the third molar teeth [36, 45, 53]. 1.6–9.6 % of the population suffers from hypodontia, with loss of one or two teeth (except the third molar). Oligodontia, with loss of more than six teeth (except the third molar), ranges from 0.0 % to 1.1 %, depending on the population studied. Loss of all teeth is referred to as anodontia. Missing teeth are more common in the permanent dentition than those in the primary dentition. However, it is believed that hypodontia in the primary dentition is correlated to hypodontia in the permanent dentition [36, 45]. Additionally, a higher prevalence ratio of hypodontia has been identified in women, with a 3:2 ratio [9].

13.3.1.1 Non-syndromic (Isolated) Familial Missing Teeth

Non-syndromic familial hypodontia has been reported to probably be inherited either as an autosomal dominant, autosomal recessive, or X-linked trait [1, 2, 15, 22, 65, 79].

It has been shown that mutation in *MSX1* is associated with hypodontia that predominantly affects second premolars and third molars which all have normal primary dentition [31, 45]. Mutation in *PAX9* results in hypodontia in the form of lost molars. Maxillary and/or mandibular second molars and central incisor are often affected in some individuals [37, 46]. The frequency of tooth loss is found to be

Table 13.1 Abnormal tooth numbers in human

Loss of tooth	
Syndrome	Mutation
Hypohidrotic/anhidrotic ectodermal dysplasia	*EDA, EDAR, and EDARADD*
Odonto-onycho-dermal dysplasia	*WNT10A*
Ectrodactyly ectodermal dysplasia cleft lip/palate syndrome	*P63*
Ellis-van Creveld syndrome	*EVC*
Incontinentia pigmenti	*IKKγ*
Van der Woude syndrome	*IRF6*
Rieger syndrome	*PITX2, FOXC1*
Oral facial digital syndrome type I	*OFD1*
Enamel renal gingival syndrome	*FAM20*
Extra tooth	
Cleidocranial dysplasia	*RUNX2*
SOX2 anophthalmia syndrome	*SOX2*
Gardner syndrome	*APC*
Opitz G/BBB syndrome	*MID1*
Tricho-rhino phalangic syndrome	*TRPS1*
Ehlers-Danlos syndrome type III	*Tenascin-XB, COL3A1*
Robinow syndrome	*ROR2*
Nance-Horan syndrome	*NHS*
Fabry syndrome	*a-galactosidase A*
Rothmund-Thomson syndrome	*RECQL4*
Hallermann-Streiff syndrome	*GJA1*

higher for second premolars and maxillary first premolars in association with *MSX1* mutation compared to *PAX9* mutation [53]. Each tooth type thus shows their different response to these gene mutations. Defects identified in *MSX1* and *PAX9* include gene deletion, as well as nonsense, frameshift, and missense mutation. In addition, it has been indicated that a nonsense mutation in *AXIN2* (a molecule essential for canonical Wnt signaling) leads to oligodontia [40]. Here, patients display an absence of at least eight permanent teeth, while primary dentition remains intact.

Mutation in *EDA* has been shown to be linked to non-syndromic hypodontia, although mutation in *EDA* can also lead to syndromic dental anomalies [83]. Mutation of *WNT10A* is found in patients with isolated hypodontia, even though mutation of the gene is also linked to odonto-onycho-dermal dysplasia [77]. Additionally, *GREM2* mutations have been found to exist within families with absent teeth [30].

13.3.1.2 Syndromic Missing Teeth

There are about eighty syndromes showing hypodontia (Table 13.1; [36]).

Ectodermal dysplasia consists of variable defects in the morphogenesis of ectodermal derivatives such as teeth, skin, hair, sweat glands, and nails, although more than 150 clinically distinct ectodermal dysplasias have been identified. Ectodermal dysplasia syndromes can be inherited in an autosomal dominant, autosomal recessive, or X-linked form. Disruption of the *EDA*, *EDAR*, and *EDARADD* loci has been causally identified for the onset of hypohidrotic/anhidrotic ectodermal dysplasia [56]. Odonto-onycho-dermal dysplasia is an autosomal recessive ectodermal syndrome, which is caused by mutation in *WNT10A*. It is characterized by hyperkeratosis, smooth tongue, dry hair, nail dysplasia, and hyperhidrosis of palms and soles. The dysplasia also presents with severe hypodontia [36]. Ectrodactyly ectodermal dysplasia cleft lip/palate syndrome is an autosomal dominant disorder characterized by ectrodactyly, ectodermal dysplasia, and cleft lip/palate. Patients with this syndrome also exhibit oligodontia/anodontia. Heterozygous mutation in *P63* has additionally been shown to be responsible for ectrodactyly ectodermal dysplasia cleft lip/palate syndrome [73]. Incontinentia pigmenti is an X-linked dominant disorder, which is characterized by abnormal skin pigmentation and hair loss. It is caused by mutation of *IKKγ* (*NEMO*) [56]. Incontinentia pigmenti is recognized as a form of ectodermal dysplasia by many societies, since ectodermal structures (hair, skin, nails, and teeth) as well as the eyes and nervous system are affected. Hypodontia is a common feature of the disorder [73].

Ellis-van Creveld syndrome is characterized by polydactyly, chondrodysplasia, nail dysplasia, and cardiac defects and is caused by mutation of *EVC*. Missing teeth are frequently observed in Ellis-van Creveld syndrome patients [5]. Van der Woude syndrome is characterized by cleft lip and/or cleft palate, paramedian lip pits and sinuses, and conical elevations of the lower lip. It is caused by mutation in the *IRF6* gene. Hypodontia is a common feature of the syndrome [55, 72]. Rieger syndrome is an autosomal dominant disorder, which consists of malformations in the anterior chamber of the eye and umbilical anomalies. Rieger syndrome patients also show hypodontia or anodontia. Mutations in *PITX2* and *FOXC1* have additionally been shown to be responsible for Rieger syndrome. Oral facial digital syndrome type I is an X-linked disorder which is caused by mutation of *OFD1*. The syndrome is characterized by malformation of the face, oral cavity, digits, central nervous system, and kidneys. Missing teeth are often observed in oral facial digital syndrome type I patients [23]. Enamel renal gingival syndrome is caused by the mutation of *FAM20*, and patients display impaired calcium metabolism and hypodontia [29].

It is generally accepted that hypodontia is often accompanied by clefts of the lip and palate. Hypodontia has been shown to be observed in 80 % of patients with clefts [36, 74]. The prevalence ratio of hypodontia is also correlated to the severity of the cleft [69].

13.3.1.3 Sporadic Missing Teeth

It is believed that sporadic hypodontia is caused by environmental factors such as trauma to the jaws, surgical procedures on the jaws, traumatic extraction of the primary teeth, chemotherapy, and radiation therapy [52, 71]. In common with environmental factors, mutation of *PAX9* has been shown to be associated with sporadic hypoplasia [62].

13.3.2 Missing Teeth in Mice

Mice are the most highly studied mammals for investigating the molecular mechanisms of tooth development, since the most common application of gene targeting is to produce knockout mice. Mice however only have incisor and molar teeth and as such their use is limited as a model for human tooth development.

The first odontogenic signal is derived from the epithelium. Prior to this, there is no prespecification of the cells into different populations within the mandibular arch. All ectomesenchyme cells in the arch are therefore equally responsive toward any signals, including those which instigate hair and limb development. After receiving appropriate signaling, the underlying mesenchyme becomes independent of the epithelial cues. From the bud stage, the direction of cellular communication is reversed and signals pass from the condensing mesenchyme to the epithelium.

Many genes have been shown to play a critical role in regulating tooth number (Table 13.2). In common with humans, mice with mutations in *Msx1* and *Pax9* mutation exhibit missing teeth, driven by an arrest of tooth development at the bud stage [64, 70]. Mutation of other molecules such as *Fgfr2*, *Lef1*, and *Runx2* also shows an arrest of tooth development at the bud stage [16, 18, 78]. Transition to cap stage from bud stage is thus a critical point in tooth development. In common with humans, *p63* and *Pitx2* mutant mice show missing teeth, but this is caused by an arrest of tooth development at stages preceding the bud stage [39, 42, 43]. Together, these data suggest that *p63* and *Pitx2* are essential for developmental signaling before the requirement of *Msx1*, *Pax9*, *Fgfr2*, *Lef1*, and *Runx2*. However *Msx1/2* mutants also show an arrest of tooth development at stages preceding the bud stage, suggesting that *Msx1* has a role in tooth initiation for which *Msx2* can compensate for gene loss in the mutant mouse [7].

It is believed that Shh, Tgf, Bmp, Wnt, and Fgf signaling are known to play critical roles in tooth initiation. *Gli2/3* (transcription factors mediated by Shh) mutants show an arrest of tooth development at stages prior to bud formation [24]. Mice with epithelial conditional deletion of *Bmpr1a* lead to an arrest of tooth development at the bud stage [4]. Additionally, *Fgf3/10* mutants show an arrest of tooth development preceding the bud stage, whereas mutation of *Fgfr2* results in an arrest of tooth development at the bud stage [16, 82]. Downregulation of Wnt signaling by overexpression of *Dkk1* (an antagonist of Wnt signaling) shows an arrest of tooth develop-

Table 13.2 Abnormal tooth numbers in mice

Loss of tooth		Extra tooth	
Tooth type	Mutation/Tg	Tooth type	Mutation/Tg
All tooth	*Msx1*	All tooth	*Apc*
	Pax9		*Sp6*
	Runx2		
	p63	Incisor	*Lrp4*
	Pitx2		*Wise*
	Msx1/Msx2		*K5-Ikkβ*
	Lef1		*Lhx6/Lhx7*
	Gli2/Gli3		
	K14-Dkk1	Diastema	*Ift88*
	Bmpr1a		*Lrp4*
	Fgf3/Fgf10		*Wise*
	Fgfr2		*Sprouty2*
	Dicer		*Sprouty4*
			R-Spondin2
Incisor	*Activinβa*		*Eda*
			K14-Eda
Molar	*Eda*		*K5-Edar*
	K14-Noggin		
	Bmp4	Molar	*Osr2*
	IkBα		
	Dlx1/Dlx2		

ment at stages prior to bud formation, whereas *Lef1* (a Wnt-related molecule) mutation results in an arrest of tooth development at the bud stage [3, 78]. In common with *Msx1* and *Msx2*, it is likely that there is redundancy between molecules within the same signaling pathway during tooth initiation events.

The term "epigenetics" refers to the covalent modification of DNA, protein, or RNA, resulting in changes to the function and/or regulation of these molecules without altering the genomic sequence. MicroRNAs (miRNAs) represent one of these epigenetic factors. These noncoding small single-stranded RNAs are 19–25 nucleotides in length and negatively regulate gene expression by binding target sequences in mRNA molecules. Absence of miRNAs in neural crest-derived cells, driven by deletion of *Dicer* (an essential molecule for microRNA processing), have been shown to lead to an absence of tooth development – although exact phenotypes vary between animals from lack of tooth development to almost normal tooth development [61]. This suggests that epigenetic factors also play a critical role in the regulation of tooth formation, but the exact mechanism is as yet unclear.

13.4 Supernumerary Teeth

Teeth are found in most vertebrates and have played a central role in their evolution. Change in tooth number is a significant evolutionary adaptation to accommodate novel feeding strategies. Reduction in tooth number is a well-known evolutionary trend of the dentition within eutherians. The total number of teeth per dentition has generally decreased, whereas tooth morphological complexity has increased during tooth evolution. Findings in tooth evolution are therefore a key feature to understanding the molecular mechanisms in regulating tooth number.

13.4.1 Supernumerary Teeth in Humans

The prevalence ratio of supernumerary teeth ranges from 0.2 % to 0.8 % in the deciduous dentition and from 0.3 % to 5.3 % in the permanent dentition [80]. The incidence of supernumerary teeth in men is higher than those in women (1.8:1–4.5:1, female/male) [53, 80].

13.4.1.1 Syndromic Extra Teeth in Humans

Multiple supernumerary teeth are usually a syndromic symptom, and the prevalence for non-syndromic multiple supernumerary teeth have been reported to be less than 1 % (Table 13.1; [68]).

Cleidocranial dysplasia (dysostosis) is characterized by general bone dysplasia, short stature, delayed closure of the cranial sutures, and hypoplastic or aplastic clavicles. Cleidocranial dysplasia is known to be associated with supernumerary dentition. Mutations in *RUNX2* are responsible for the dysplasia [36, 80]. *SOX2* anophthalmia syndrome is caused by mutations in the *SOX2* gene and is characterized by anophthalmia or microphthalmia, with various extraocular symptoms such as hypogonadotropic hypogonadism, brain anomaly, and esophageal abnormalities. Again, patients show supernumerary tooth formation [54]. Gardner syndrome is a rare autosomal dominant disorder which is caused by mutations in *APC*. The syndrome consists of gastrointestinal polyps, multiple osteomas, and skin and soft tissue tumors including a characteristic retinal lesion. Variant tooth anomalies including supernumerary teeth are observed in these patients. In addition, Opitz G/BBB syndrome, tricho-rhino phalangic syndrome, Ehlers-Danlos syndrome type III, Robinow syndrome, Nance-Horan syndrome, Fabry syndrome, Rothmund-Thomson syndrome, and Hallermann-Streiff syndrome are all known to be syndromes where supernumerary teeth occur [13, 73, 80].

As with hypodontia, syndromic extra tooth formation is often accompanied by cleft lip/palate. In this instance, splitting of the tooth germs due to cleft lip and/or palate is the cause of the formation of supernumerary teeth [38, 80].

13.4.1.2 Missing and Extra Teeth in Humans

The concomitant occurrence of hypodontia and supernumerary teeth is observed in Down syndrome, oral facial digital syndrome type I, Ellis-van Creveld syndrome, and Ehlers-Danlos syndrome [76]. These suggest that single gene mutation can lead to these opposite tooth phenotypes in same patients.

13.4.2 Supernumerary Teeth in Mice

It is widely accepted that modern eutherian mammals evolved from a common ancestor that had three incisors, one canine, four premolars, and three molars. Mice possess only one incisor and three molars in each jaw quadrant, separated by a toothless region called the "diastema." It is believed that mice lost the remaining teeth during evolution. Genetically modified mice therefore provide some limited information on the determinant of non-murine tooth types. However, it has been shown that mice have retained the genetic potential for the development of teeth lost during evolution, such as the premolars.

In wild-type mice, tooth germ-like structures are observed in the diastema at early stages of development, which disappear during later stages [63]. Both mutation and overexpression of *Eda* are known to lead to extra tooth formation in the diastema, suggesting that the precise signaling regulated by *Eda* is essential for controlling odontogenesis in the diastema [48, 63]. Extra tooth formation in the diastema is also observed in mice with mutation of *Gas1* (an inhibitor of Shh), *R-spondin2* (activators of Wnt signaling), *Ift88* (molecules present in the primary cilia where Shh signaling is activated), *Wise* (a secreted BMP antagonist and Wnt modulator), *Lrp4* (a negative Wnt co-receptor), *Sprouty2* (a negative feedback regulator of Fgf), and *Sprouty4* (a negative feedback regulator of Fgf) [32, 33, 35, 59, 60]. Evolutionary tooth loss in the diastema is likely to be associated with changes in these signaling pathways. These also suggest that odontogenic activity in the diastema is mainly controlled by inhibitors of signaling pathways.

Mammals have single-rowed dentitions, whereas many other vertebrates have dentition which consists of multiple rows. It is believed that mammals lost this additional dentition during evolution. *Osr2* is expressed in the molar tooth germ area with a lingual-to-buccal gradient and restricts expression of *Bmp4* – which is also mediated by *Msx1*. Mice with mutations in *Osr2* develop supernumerary teeth lingual to the endogenous molars due to expansion of *Bmp4* expression [85]. The interaction between *Osr2*, *Bmp4*, and *Msx1* thus plays a critical role in regulating molar tooth initiation in the buccolingual axis [41].

In common with the diastema and molar region, extra incisor tooth formation is also observed in mice with mutations in *Wise*, *Lrp4*, *Lhx6/7* and *Sprouty2/4*, and overexpressing *Ikkβ* [8, 11, 17, 32, 59]. Single extra incisor in a jaw quadrant is thus formed by changes in the Wnt, Bmp, Fgf, and NF-kB pathways.

Constitutive stabilization of β-catenin in the epithelium results in numerous supernumerary tooth formation in the mouse [25]. *Apc* is known to play a critical role in regulating the Wnt signaling pathway and conditional deletion of *Apc* in the epithelium also results in multiple supernumerary tooth formation [81]. Mice with overexpression of *Lef1* (a Wnt-related molecule) have been associated with ectopic tooth formation [86]. Numerous extra teeth are also observed in *Epiprophin* (*Sp6*)-deficient mice, which show upregulation of Wnt signaling [49]. Taken together, numerous extra tooth formation is thus directly related to over-activation of Wnt signaling. Wnt inhibitors, Dkks, are known to be expressed in wild-type developing tooth germs, suggesting that Wnt signaling activity is regulated by the balance between ligands and inhibitors [21].

13.5   Odontogenic Activity Between Tooth Germs

Each tooth type is known to show different developmental timing in humans. In mice, the second and third molars start to develop after the first molar reach the bell stage. It has been shown that the first molar tooth germs inhibit development of the second molar at an early stage [34]. The development of the second molar tooth initiates only when inhibitory factors from the first molar are sufficiently reduced. Odontogenic activity is thus regulated by the interaction between tooth germs. Bmp signaling is likely to be involved in this interaction. Indeed, mice with mesenchymal conditional deletion of *Bmp4* and mice overexpressing *Noggin* (Bmp antagonist) present with an absence of second and/or third molars [27, 67]. Mice with mutation of *Eda* and with reduction of NF-kB also exhibit loss of the second and/or third molars, suggesting that Eda-NF-kB cascade is also involved in the interaction between tooth germs [57, 66].

13.6   Tooth Initiation and Tooth Type

It has been shown that more than 20 % of humans lose at least one of the third molars [36, 45, 53]. Apart from the third molars, lower second premolars and/or upper lateral incisors are most commonly affected, followed by the upper second premolars [53]. Most of the common supernumerary teeth (46.9–92.8 % of supernumerary teeth) are observed between the upper central incisors – the so-called mesiodens [20, 68, 87]. Supernumerary teeth are also observed in the premolar region (10 % of the total supernumerary cases) and almost 75 % of those are in the mandible [80]. Supernumerary teeth are also found in the molar region as a distomolar (fourth molar). Furthermore, mice with targeted null mutations of both *Dlx1* and *Dlx2* homeobox genes have a tooth patterning phenotype where development of maxillary molar teeth is inhibited but development of all other teeth is normal. *ActivinβA* mutants show the opposite tooth phenotype, where maxillary molar teeth are present and other teeth are absent [19]. Odontogenic activities are thus differently regulated between regions of the maxillary and mandibular jaw.

It has been established that instructive signals are involved in the determination of tooth type. Homeobox genes have been found to regulate patterning in the development of many tissues including the maxillae and mandibles. In the developing jaws, several homeobox genes show highly restricted expression patterns in the ectomesenchyme along the proximodistal axis. For example, *Barx1* are expressed in mesenchymal cells of the presumptive molar region, whereas *Msx1* and *Msx2* are expressed in mesenchymal cells where the incisors develop. Spatially restricted gene expression is also observed in the epithelium. *Bmp4* is expressed in the presumptive incisor region, while *Fgf8* expression is restricted to the presumptive molar region. In the epithelium, these molecules have been shown to be responsible for regulating mesenchymal homeobox gene expression. Mis-expression of *Barx1* in the murine presumptive incisor mesenchyme results in a transformation of tooth shape, with molars developing from incisor primordia. Transposition of teeth (i.e., adjacent teeth switching positions) is an extremely rare dental anomalies in humans [12, 75]. However, examples of molar-like teeth in the maxillary central incisor region and premolar-like teeth in the maxillary lateral incisor region have previously been reported [28]. It is believed that the maxillary canine is most frequently involved in the transposition event [51]. Tooth identity is believed to be determined by complicated mechanisms, such as spatially restricted homeobox gene expression, and the gradient and overlap of signaling molecules [13, 14, 47]. Although the etiology of transposed teeth remains unclear, it is possible that changes in the balance of the determinant molecules results in the switching of tooth types.

13.7   The Midline and Tooth Development

Holoprosencephaly is a relatively common defect of the forebrain and midface in humans and is caused by impaired midline cleavage of the embryonic forebrain. It is believed that the holoprosencephaly spectrum is associated with the appearance of a solitary median maxillary central incisor. Single maxillary central incisors are also a feature of single median maxillary central incisor syndrome, which is caused by a failure in growth at the midline [10, 50]. Craniofacial development is thus linked to tooth development.

13.8   Conclusion

Tooth number, shape, and position are consistent in mammals and are subject to strict genetic control. Here, we highlight an overview covering the molecular mechanisms of tooth development, especially those regulating tooth number. Dozens of different molecules together form complex molecular networks creating positive and negative feedback loops, and a series of inductive and permissive processes that regulate tooth number. Multiple signaling pathways such as Shh, Tgf, Bmp, Wnt, Fgf, Notch, and NF-kB – and crosstalk between them – are known to play critical roles in regulating tooth development. Activity of signaling pathways is also regulated by the balance of ligands, activators, inhibitors, and receptors. Teeth develop via a dynamic and complex reciprocal interaction between dental epithelium and cranial neural crest-derived mesenchyme. It has been shown that transcription factors are involved in epithelial-mesenchymal interactions through the signaling loops between tissue layers by responding to inductive signals and regulating the expression of other signaling molecules. It has been shown that all these factors function as distinct role between tooth type, timing, location, and gender in mammals.

Studies of human congenital disease and transgenic mice suggest that disturbance of the molecular network results in abnormal tooth formation. Since molecular mechanisms involved in tooth development should be reproduced in tooth regeneration, knowledge of tooth development from both human and mouse studies provides crucial information for the advancement of tooth regenerative therapy. Among the molecular mechanisms involved in tooth development, those regulating tooth number are the most critical for tooth regeneration, as replacement dentition should start from tooth initiation. Rodent incisors grow continuously throughout life by utilizing a stem cell niche located at the apical end of the incisor tooth. As such, this structure is also able to provide crucial information pertinent to the study of tooth regeneration.

References

1. Ahmad W, Brancolini V, UL Faiyaz MF, Lam H, UL Haque S, Haider M, Maimon A, Aita VM, Owen J, Brown D, Zegarelli DJ, Ahmad M, Ott J, Christiano AM. A locus for autosomal recessive hypodontia with associated dental anomalies maps to chromosome 16q12.1. Am J Hum Genet. 1998;62:987–91.
2. Alvesalo L, Portin P. The inheritance pattern of missing, peg-shaped, and strongly mesio-distally reduced upper lateral incisors. Acta Odontol Scand. 1969;27:563–75.
3. Andl T, Reddy ST, Gaddapara T, Millar SE. WNT signals are required for the initiation of hair follicle development. Dev Cell. 2002;2:643–53.
4. Andl T, Ahn K, Kairo A, Chu EY, Wine-Lee L, Reddy ST, Croft NJ, Cebra-Thomas JA, Metzger D, Chambon P, Lyons KM, Mishina Y, Seykora JT, Crenshaw 3rd EB, Millar SE. Epithelial Bmpr1a regulates differentiation and proliferation in postnatal hair follicles and is essential for tooth development. Development. 2004;131:2257–68.
5. Baujat G, Le Merrer M. Ellis-van Creveld syndrome. Orphanet J Rare Dis. 2007;2:27.
6. Bei M. Molecular genetics of ameloblast cell lineage. J Exp Zool B Mol Dev Evol. 2009;312B(5):437–44.
7. Bei M, Maas R. FGFs and BMP4 induce both Msx1-independent and Msx1-dependent signaling pathways in early tooth development. Development. 1998;125:4325–33.
8. Blackburn J, Kawasaki K, Porntaveetus T, Kawasaki M, Otsuka-Tanaka Y, Miake Y, Ota MS, Watanabe M, Hishinuma M, Nomoto T, Oommen S, Ghafoor S, Harada F, Nozawa-Inoue K, Maeda T, Peterková R, Lesot H, Inoue J, Akiyama T, Schmidt-Ullrich R, Liu B, Hu Y, Page A, Ramírez A, Sharpe PT, Ohazama A*. Excess NF-kB induces ectopic odontogenesis in embryonic incisor epithelium. J Dent Res. 2015;94:121–8.

9. Brook AH. Dental anomalies of number, form and size: their prevalence in British schoolchildren. J Int Assoc Dent Child. 1974;5:37–53.
10. Catón J, Tucker AS. Current knowledge of tooth development: patterning and mineralization of the murine dentition. J Anat. 2009;214:502–15.
11. Charles C, Hovorakova M, Ahn Y, Lyons DB, Marangoni P, Churava S, Biehs B, Jheon A, Lesot H, Balooch G, Krumlauf R, Viriot L, Peterkova R, Klein OD. Regulation of tooth number by fine-tuning levels of receptor-tyrosine kinase signaling. Development. 2011;138:4063–73.
12. Chattopadhyay A, Srinivas K. Transposition of teeth and genetic etiology. Angle Orthod. 1996;66:147–52.
13. Cobourne MT, Sharpe PT. Making up the numbers: the molecular control of mammalian dental formula. Semin Cell Dev Biol. 2010;21:314–24.
14. Cobourne MT, Mitsiadis T. Neural crest cells and patterning of the mammalian dentition. J Exp Zool B Mol Dev Evol. 2006;306:251–60.
15. De Coster PJ, Marks LA, Martens LC, Huysseune A. Dental agenesis: genetic and clinical perspectives. J Oral Pathol Med. 2009;38:1–17.
16. De Moerlooze L, Spencer-Dene B, Revest JM, Hajihosseini M, Rosewell I, Dickson C. An important role for the IIIb isoform of fibroblast growth factor receptor 2 (FGFR2) in mesenchymal-epithelial signalling during mouse organogenesis. Development. 2000;127:483–92.
17. Denaxa M, Sharpe PT, Pachnis V. The LIM homeodomain transcription factors Lhx6 and Lhx7 are key regulators of mammalian dentition. Dev Biol. 2009;333:324–36.
18. D'Souza RN, Aberg T, Gaikwad J, Cavender A, Owen M, Karsenty G, Thesleff I. Cbfa1 is required for epithelial–mesenchymal interactions regulating tooth development in mice. Development. 1999;126:2911–20.
19. Ferguson CA, Tucker AS, Christensen L, Lau AL, Matzuk MM, Sharpe PT. Activin is an essential early mesenchymal signal in tooth development that is required for patterning of the murine dentition. Genes Dev. 1998;12:2636–49.
20. Fernández Montenegro P, Valmaseda Castellón E, Berini Aytés L, Gay Escoda C. Retrospective study of 145 supernumerary teeth. Med Oral Patol Oral Cir Bucal. 2006;11:E339–44.
21. Fjeld K, Kettunen P, Furmanek T, Kvinnsland IH, Luukko K. Dynamic expression of Wnt signaling-related Dickkopf1, -2, and -3 mRNAs in the developing mouse tooth. Dev Dyn. 2005;233(1):161–6.
22. Goldenberg M, Das P, Messersmith M, Stockton DW, Patel PI, D'Souza RN. Clinical, radiographic, and genetic evaluation of a novel form of autosomal-dominant oligodontia. J Dent Res. 2000;79:1469–75.
23. Gurrieri F, Franco B, Toriello H, Neri G. Oral-facial-digital syndromes: review and diagnostic guidelines. Am J Med Genet A. 2007;143A:3314–23.
24. Hardcastle Z, Mo R, Hui CC, Sharpe PT. The Shh signalling pathway in tooth development: defects in Gli2 and Gli3 mutants. Development. 1998;125:2803–11.
25. Järvinen E, Salazar-Ciudad I, Birchmeier W, Taketo MM, Jernvall J, Thesleff I. Continuous tooth generation in mouse is induced by activated epithelial Wnt/beta-catenin signaling. Proc Natl Acad Sci U S A. 2006;103:18627–32.
26. Jheon AH, Seidel K, Biehs B, Klein OD. From molecules to mastication: the development and evolution of teeth. Wiley Interdisc Rev Dev Biol. 2013;2:165–82.
27. Jia S, Zhou J, Gao Y, Baek JA, Martin JF, Lan Y, Jiang R. Roles of Bmp4 during tooth morphogenesis and sequential tooth formation. Development. 2013;140:423–32.
28. Kantaputra PN, Gorlin RJ. Double dens invaginatus of molarized maxillary central incisors, premolarization of maxillary lateral incisors, multituberculism of the mandibular incisors, canines and first premolar, and sensorineural hearing loss. Clin Dysmorphol. 1992;1:128–36.
29. Kantaputra PN, Bongkochwilawan C, Kaewgahya M, Ohazama A, Kayserili H, Erdem AP, Aktoren O, Guven Y. Enamel-renal-gingival syndrome, hypodontia, and a novel FAM20A mutation. Am J Med Genet A. 2014;164A:2124–8.

30. Kantaputra PN, Kaewgahya M, Hatsadaloi A, Vogel P, Kawasaki K, Ohazama A. Ketudat Cairns JR6. GREMLIN 2 mutations and dental anomalies. J Dent Res. 2015;94:1646–52.
31. Kapadia H, Mues G, D'Souza R. Genes affecting tooth morphogenesis. Orthod Craniofac Res. 2007;10:237–44.
32. Kassai Y, Munne P, Hotta Y, Penttila E, Kavanagh K, Ohbayashi N, Takada S, Thesleff I, Jernvall J, Itoh N. Regulation of mammalian tooth cusp patterning by ectodin. Science. 2005;309:2067–70.
33. Kawasaki M, Porntaveetus T, Kawasaki K, Oommen S, Otsuka-Tanaka Y, Hishinuma M, Nomoto T, Maeda T, Takubo K, Suda T, Sharpe PT, Ohazama A. R-spondins/Lgrs expression in tooth development. Dev Dyn. 2014;243:844–51.
34. Kavanagh KD, Evans AR, Jernvall J. Predicting evolutionary patterns of mammalian teeth from development. Nature. 2007;449:427–32.
35. Klein OD, Minowada G, Peterkova R, Kangas A, Yu BD, Lesot H, Peterka M, Jernvall J, Martin GR. Sprouty genes control diastema tooth development via bidirectional antagonism of epithelial-mesenchymal FGF signaling. Dev Cell. 2006;11:181–90.
36. Klein OD, Oberoi S, Huysseune A, Hovorakova M, Peterka M, Peterkova R. Developmental disorders of the dentition: an update. Am J Med Genet C Semin Med Genet. 2013;163C:318–32.
37. Kobielak A, Kobielak K, Wiśniewski AS, Mostowska A, Biedziak B, Trzeciak WH. The novel polymorphic variants within the paired box of the PAX9 gene are associated with selective tooth agenesis. Folia Histochem Cytobiol. 2001;39:111–2.
38. Kriangkrai R, Chareonvit S, Yahagi K, Fujiwara M, Eto K, Iseki S. Study of Pax6 mutant rat revealed the association between upper incisor formation and midface formation. Dev Dyn. 2006;235:2134–43.
39. Laurikkala J, Mikkola ML, James M, Tummers M, Mills AA, Thesleff I. p63 regulates multiple signalling pathways required for ectodermal organogenesis and differentiation. Development. 2006;133:1553–63.
40. Lammi L, Arte S, Somer M, Jarvinen H, Lahermo P, Thesleff I, Pirinen S, Nieminen P. Mutations in AXIN2 cause familial tooth agenesis and predispose to colorectal cancer. Am J Hum Genet. 2004;74:1043–50.
41. Lan Y, Jia S, Jiang R. Molecular patterning of the mammalian dentition. Semin Cell Dev Biol. 2014;25–26:61–70.
42. Lin CR, Kioussi C, O'Connell S, Briata P, Szeto D, Liu F, Izpisua-Belmonte JC, Rosenfeld MG. Pitx2 regulates lung asymmetry, cardiac positioning and pituitary and tooth morphogenesis. Nature. 1999;401:279–82.
43. Lu MF, Pressman C, Dyer R, Johnson RL, Martin JF. Function of Rieger syndrome gene in left-right asymmetry and craniofacial development. Nature. 1999;401:276–8.
44. Mammoto T, Mammoto A, Torisawa YS, Tat T, Gibbs A, Derda R, Mannix R, de Bruijn M, Yung CW, Huh D, Ingber DE. Mechanochemical control of mesenchymal condensation and embryonic tooth organ formation. Dev Cell. 2011;21:758–69.
45. Matalova E, Fleischmannova J, Sharpe PT, Tucker AS. Tooth agenesis: from molecular genetics to molecular dentistry. J Dent Res. 2008;87:617–23.
46. Mensah JK, Ogawa T, Kapadia H, Cavender AC, D'Souza RN. Functional analysis of a mutation in PAX9 associated with familial tooth agenesis in humans. J Biol Chem. 2004;279:5924–33.
47. Mitsiadis TA, Smith MM. How do genes make teeth to order through development? J Exp Zool B Mol Dev Evol. 2006;306:177–82.
48. Mustonen T, Pispa J, Mikkola ML, Pummila M, Kangas AT, Pakkasjärvi L, Jaatinen R, Thesleff I. Stimulation of ectodermal organ development by Ectodysplasin-A1. Dev Biol. 2003;259:123–36.
49. Nakamura T, de Vega S, Fukumoto S, Jimenez L, Unda F, Yamada Y. Transcription factor epiprofin is essential for tooth morphogenesis by regulating epithelial cell fate and tooth number. J Biol Chem. 2008;283:4825–33.

50. Nanni L, Ming JE, Du Y, Hall RK, Aldred M, Bankier A, Muenke M. SHH mutation is associated with solitary median maxillary central incisor: a study of 13 patients and review of the literature. Am J Med Genet. 2001;102:1–10.
51. Nambiar S, Mogra S, Shetty S. Transposition of teeth: a forensic perspective. J Forensic Dent Sci. 2014;6:151–3.
52. Näsman M, Forsberg CM, Dahllöf G. Long-term dental development in children after treatment for malignant disease. Eur J Orthod. 1997;19:151–9.
53. Nieminen P. Genetic basis of tooth agenesis. J Exp Zool B Mol Dev Evol. 2009;312B:320–42.
54. Numakura C, Kitanaka S, Kato M, Ishikawa S, Hamamoto Y, Katsushima Y, Kimura T, Hayasaka K. Supernumerary impacted teeth in a patient with SOX2 anophthalmia syndrome. Am J Med Genet A. 2010;152A:2355–9.
55. Oberoi S, Vargervik K. Hypoplasia and hypodontia in Van der Woude syndrome. Cleft Palate Craniofac J. 2005;42:459–66.
56. Ohazama A, Sharpe PT. TNF signalling in tooth development. Curr Opin Genet Dev. 2004;14:513–9.
57. Ohazama A, Hu Y, Schmidt-Ullrich R, Cao Y, Scheidereit C, Karin M, Sharpe. A dual role for Ikka in tooth development. Dev Cell. 2004;6:219–27.
58. Ohazama A, Sharpe PT. Development of epidermal appendages; teeth and hair. In: Epstein CJ, Erickson RP, Wynshaw-Boris A, editors. Inborn errors of development. The molecular basis of clinical disorders of morphogenesis. 2nd ed. Oxford: Oxford University Press; 2008. p. 245–62.
59. Ohazama A, Johnson EB, Ota MS, Choi HY, Porntaveetus T, Oommen S, Itoh N, Eto K, Gritli-Linde A, Herz J, Sharpe PT. Lrp4 modulates extracellular integration of cell signaling pathways in development. PLoS One. 2008;3, e4092.
60. Ohazama A, Haycraft CJ, Seppala M, Blackburn J, Ghafoor S, Cobourne M, Martinelli DC, Fan CM, Peterkova R, Lesot H, Yoder BK, Sharpe P. Primary cilia regulate Shh activity in the control of molar tooth number. Development. 2009;136:897–903.
61. Oommen S, Otsuka-Tanaka Y, Imam N, Kawasaki M, Kawasaki K, Jalani-Ghazani F, Anderegg A, Awatramani R, Hindges R, Sharpe PT, Ohazama A. Distinct roles of microRNAs in epithelium and mesenchyme during tooth development. Dev Dyn. 2012;241:1465–72.
62. Pawlowska E, Janik-Papis K, Poplawski T, Blasiak J, Szczepanska J. Mutations in the PAX9 gene in sporadic oligodontia. Orthod Craniofac Res. 2010;13:142–52.
63. Peterkova R, Lesot H, Peterka M. Phylogenetic memory of developing mammalian dentition. J Exp Zool B Mol Dev Evol. 2006;306:234–50.
64. Peters H, Neubuser A, Kratochwil K, Balling R. Pax9-deficient mice lack pharyngeal pouch derivatives and teeth and exhibit craniofacial and limb abnormalities. Genes Dev. 1998;12:2735–47.
65. Pirinen S, Kentala A, Nieminen P, Varilo T, Thesleff I, Arte S. Recessively inherited lower incisor hypodontia. J Med Genet. 2001;38:551–6.
66. Pispa J, Jung HS, Jernvall J, Kettunen P, Mustonen T, Tabata MJ, Kere J, Thesleff I. Cusp patterning defect in Tabby mouse teeth and its partial rescue by FGF. Dev Biol. 1999;216:521–34.
67. Plikus MV, Zeichner-David M, Mayer JA, Reyna J, Bringas P, Thewissen JG, Snead ML, Chai Y, Chuong CM. Morphoregulation of teeth: modulating the number, size, shape and differentiation by tuning Bmp activity. Evol Dev. 2005;7:440–57.
68. Rajab LD, Hamdan MA. Supernumerary teeth: review of the literature and a survey of 152 cases. Int J Paediatr Dent. 2002;12:244–54.
69. Ranta R. A review of tooth formation in children with cleft lip/palate. Am J Orthod Dentofacial Orthop. 1986;90:11–8.
70. Satokata I, Maas R. Msx1 deficient mice exhibit cleft palate and abnormalities of craniofacial and tooth development. Nat Genet. 1994;6:348–56.
71. Schalk-van der Weide Y, Steen WH, Bosman F. Taurodontism and length of teeth in patients with oligodontia. J Oral Rehabil. 1993;20:401–12.

72. Schneider EL. Lip pits and congenital absence of second premolars: varied expression of the Lip Pits syndrome. J Med Genet. 1973;10:346–9.
73. Schwabe GC, Opitz C, Tinschert S, Mundlos S, Sharpe PT. Molecular mechanisms of tooth development and malformations. Oral Biosci Med. 2004;1:77–91.
74. Shapira Y, Lubit E, Kuftinec MM. Congenitally missing second premolars in cleft lip and cleft palate children. Am J Orthod Dentofacial Orthop. 1999;115:396–400.
75. Shapira Y, Kuftinec MM. Tooth transpositions – a review of the literature and treatment considerations. Angle Orthod. 1989;59:271–6.
76. Van Buggenhout G, Bailleul-Forestier I. Mesiodens. Eur J Med Genet. 2008;51:178–81.
77. van den Boogaard MJ, Créton M, Bronkhorst Y, van der Hout A, Hennekam E, Lindhout D, Cune M, Ploos van Amstel HK. Mutations in WNT10A are present in more than half of isolated hypodontia cases. J Med Genet. 2012;49:327–31.
78. van Genderen C, Okamura RM, Farinas I, Quo RG, Parslow TG, Bruhn L, Grosschedl R. Development of several organs that require inductive epithelial–mesenchymal interactions is impaired in LEF-1-deficient mice. Genes Dev. 1994;8:2691–703.
79. Vastardis H, Karimbux N, Guthua SW, Seidman JG, Seidman CE. A human MSX1 homeodomain missense mutation causes selective tooth agenesis. Nat Genet. 1996;13:417–21.
80. Wang XP, Fan J. Molecular genetics of supernumerary tooth formation. Genesis. 2011;49:261–77.
81. Wang XP, O'Connell DJ, Lund JJ, Saadi I, Kuraguchi M, Turbe-Doan A, Cavallesco R, Kim H, Park PJ, Harada H, Kucherlapati R, Maas RL. Apc inhibition of Wnt signaling regulates supernumerary tooth formation during embryogenesis and throughout adulthood. Development. 2009;136:1939–49.
82. Wang XP, Suomalainen M, Felszeghy S, Zelarayan LC, Alonso MT, Plikus MV, Maas RL, Chuong CM, Schimmang T, Thesleff I. An integrated gene regulatory network controls stem cell proliferation in teeth. PLoS Biol. 2007;5:e159.
83. Yang Y, Luo L, Xu J, Zhu P, Xue W, Wang J, Li W, Wang M, Cheng K, Liu S, Tang Z, Ring BZ, Su L. Novel EDA p.Ile260Ser mutation linked to non-syndromic hypodontia. J Dent Res. 2013;92:500–6.
84. Yoshizaki K, Hu L, Nguyen T, Sakai K, He B, Fong C, Yamada Y, Bikle DD, Oda Y. Ablation of coactivator Med1 switches the cell fate of dental epithelia to that generating hair. PLoS One. 2014;9:e99991.
85. Zhang Z, Lan Y, Chai Y, Jiang R. Antagonistic actions of Msx1 and Osr2 pattern mammalian teeth into a single row. Science. 2009;323:1232–4.
86. Zhou P, Byrne C, Jacobs J, Fuchs E. Lymphoid enhancer factor 1 directs hair follicle patterning and epithelial cell fate. Genes Dev. 1995;9:700–13.
87. Zhu JF, Marcushamer M, King DL, Henry RJ. Supernumerary and congenitally absent teeth: a literature review. J Clin Pediatr Dent. 1996;20:87–95.

Open Access This chapter is distributed under the terms of the Creative Commons Attribution 4.0 International License (http://creativecommons.org/licenses/by/4.0/), which permits use, duplication, adaptation, distribution and reproduction in any medium or format, as long as you give appropriate credit to the original author(s) and the source, provide a link to the Creative Commons license and indicate if changes were made.

The images or other third party material in this chapter are included in the work's Creative Commons license, unless indicated otherwise in the credit line; if such material is not included in the work's Creative Commons license and the respective action is not permitted by statutory regulation, users will need to obtain permission from the license holder to duplicate, adapt or reproduce the material.

Part IV
Symposium IV: Medical Device Innovation for Diagnosis and Treatment of Biosis-Abiosis Interface

Chapter 14
Open-Source Technologies and Workflows in Digital Dentistry

Rong-Fu Kuo, Kwang-Ming Fang, and Fong-Chin Su

Abstract The Medical Device Innovation Center (MDIC) at the National Cheng Kung University has developed a complete and professional digital dentistry design workflow with cutting-edge equipment and software. MDIC certified with ISO13485 can provide total solutions in digital dentistry. An "Intelligent Manufacturing Systems Center" (IMSC) has been established using dental open technologies toward providing digital dentistry training and restoration design services. This digital dental laboratory is able to control the entire digital process from digital impressions to the CAD/CAM creation of the restoration and model milling. An intraoral scanner is used to make digital impressions for 3D geometric models from the chair side or from traditional impressions. In the design phase, three different commercial software packages are considered for the design portion. After importing the digital impression STL data file, one or more of these packages are used to design the restoration. The design is then sent to a five-axis milling machine for production. The CNC machines are chosen for machining or milling of the prosthetics from various materials including wax, PMMA, zirconia, chromium cobalt, resin nano-ceramics, glass ceramics, lithium disilicate, silicate ceramics, and titanium. As a cloud base solution, these design packages allow a connection to a remote manufacturing site. Our R&D team built a web-based cloud solution that can be deployed to each pertinent location. All data from the design package is stored on a private cloud which is then automatically synced to the remote public cloud. Work orders from various sources are then processed by any remote technician. Through the above settings, we may produce several common digital dentistry products including crowns and bridges, veneers, inlays and onlays, temporary crowns, and virtual diagnostic wax-ups. This digital dentistry laboratory is also equipped to handle advanced clinical cases such as implant planning, digital smile design analysis and customized surgical guides, custom abutments, implant bridges and bar designs, and orthodontics. Digital animation is applied for patient education and communication. For academics and training, a comprehensive digital technology training program has been developed to help dentists and dental technicians.

R.-F. Kuo • K.-M. Fang • F.-C. Su (✉)
Medical Device Innovation Center and Department of Biomedical Engineering, National Cheng Kung University, 1. Ta-Shueh Rd., Tainan 701, Taiwan
e-mail: fcsu@mail.ncku.edu.tw

© The Author(s) 2017
K. Sasaki et al. (eds.), *Interface Oral Health Science 2016*,
DOI 10.1007/978-981-10-1560-1_14

Keywords Intelligent manufacturing systems • Restoration • Digital dentistry •
Cloud base solution

14.1 Cloud Base Solution for Intelligent Dental Clinic and Manufacturing System

Industrial technologies are growing rapidly in recent years. Dental industry is toward a digital revolution as shown in Fig. 14.1 which causes traditional dental industry to face a serious impact and revolution [1, 2]. Dental clinics also transform gradually toward an intelligent environment illustrated in Fig. 14.2. In order to face this challenge, MDIC presents an innovative dental full-chain model focusing on digital design phase as a core concept plus the utilization of open-source technologies. This model based on an ecosystem platform provides a variety of services corresponding to the demand of dental clinics and dental labs. These needs may cover a variety of customizable dental products and services which are closely connected to dental design and manufacturing processes.

Integrating design and manufacturing for digital dentistry as a complete chain to provide custom service is the core in the model. In order to achieve high competence, MDIC established a mobile interface APP platform based on cloud computing and mobile ecommerce. Cloud computing provides a lot of advantages including high flexibility, low cost, mass storage, and parallel computing capacities to the

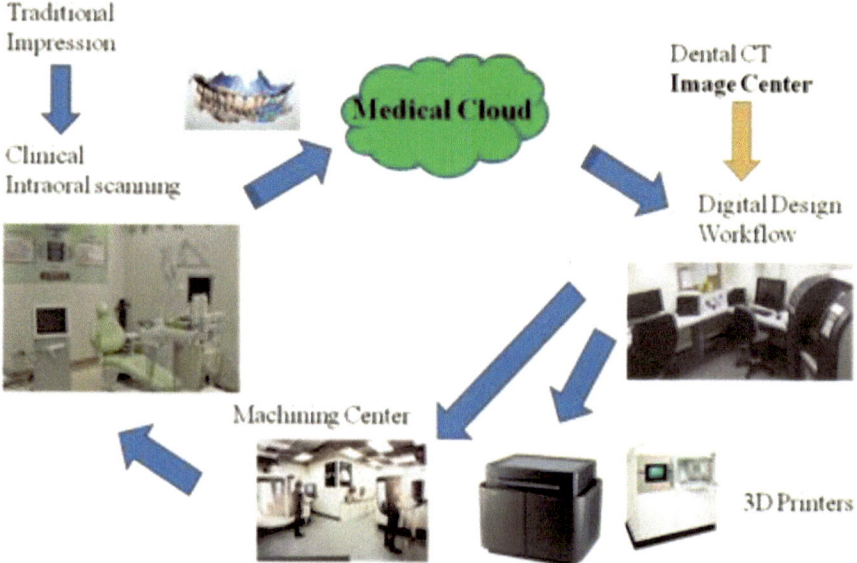

Fig. 14.1 Dental digital workflow

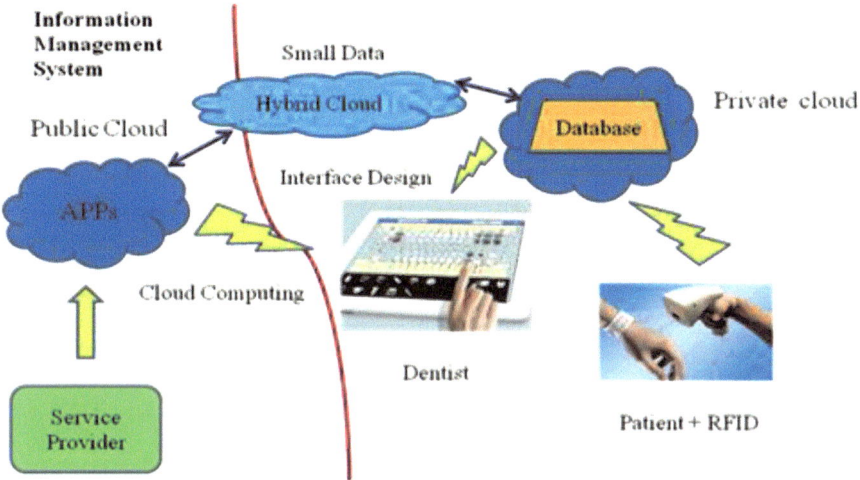

Fig. 14.2 Intelligent dental clinic

proposed architecture [3, 4]. Further, mobile services provide customers with ubiquitous services and shorter response time to quickly respond to customers' demand. Moreover, a lot of customer messages can be aware and collected by the sensors of mobile devices to continuously improve service quality [5].

The platform provides further features including flexible communications on geometry and color shading, the customizable dental order, and global data analysis. Moreover, the platform offers a variety of connecting services for various mobile devices such as smartphones and tablet PCs to connect to the system and communicate with global partners.

14.2   Integration of Digital Information

In the dental clinic, there are quite a lot of data produced on a regular base including information of extraoral impression, intraoral impression, CBCT dental medical image, occlusion and TMJ analysis, and tooth shade matching. The development of tooth shading device is discussed in the following.

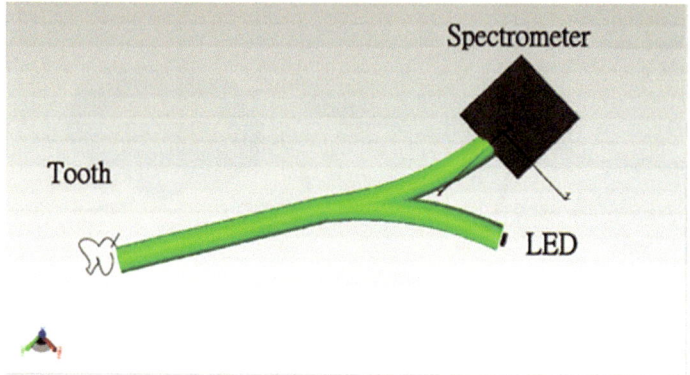

Fig. 14.3 Experiment setup using Y-shaped bifurcated fiber cables with a source leg to carry light from a light-emitting diode (*LED*) to a tooth surface and a spectrometer leg to carry light reflected by the tooth to a spectrometer

14.3 A Shade Guide Development Based on Open-Source Technology

Dental color matching is a challenging aspect of prosthetic dentistry. A dental shade guide is an essential tool used in dental labs and dental clinics to determine the color of natural teeth [6–13]. Color measurement apparatus consisted of a spectrometer and fiber bundles fixed on an optical table as shown in Fig. 14.3 with probe bundles with round bundle light source and single-fiber spectrometer legs with SMA connector. The bifurcated fiber cables with a source leg to carry light from a light-emitting diode (LED) to a tooth surface and a spectrometer leg to carry light reflected by the tooth to a spectrometer. The probe tip is in contact with the tooth surface and the height of the tip raised from tooth surface to 0.5 mm and 1 mm, respectively.

14.4 3D Printing Applications in Digital Dentistry

3D printing, also referred to as "additive manufacturing," has become one of the most well-known technologies since the twentieth century. Comparing to the conventional manufacturing, 3D printing has several remarkable advantages, such as overcoming the limitation of machining to make complex structures, high resolution, being simple for manipulation, and being easy for customization. Moreover, some 3D printers are small and economic enough for public to afford. Ideas and products can be made without sending into the workshop, which reduces the criteria of designing and fabrications at present; medical applications for 3D printings can be briefly catalogued into several uses, such as a means of manufacturing prototypes rapidly, or models for clinic, teaching, and surgical guides, or implants and prosthetics, or bioprinting [14, 15].

Fig. 14.4 RP model products from FDP and milling machine

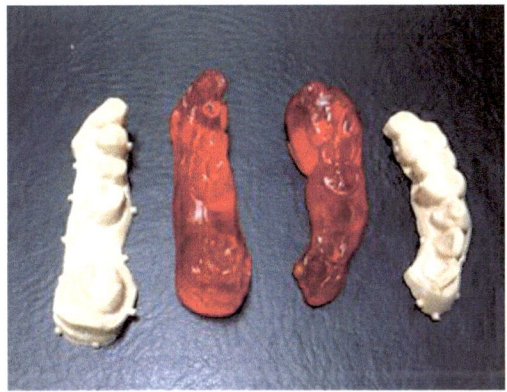

IMSC has established a 3D printing production line to assist industries and clinics in dental areas. Among different methods of 3D printing, "projection-based stereolithography apparatus" (PSLA) has additional advantages of economic and relatively smooth finished textures. In Fig. 14.4 which shows that two tooth models (colored in red) were manufactured for a dental technology company. In contrast to conventional manufacturing of making tooth models, the advantages of using PSLA manufacturing are higher resolution provided by digital files and less relaxation occurring in the PSLA material. The manufacturing procedure of PSLA requires .stl files to regenerate geometries, and these .stl files are often obtained by either intraoral or extraoral scanners [16]. Despite the improvement of resolution of scanners from macros to micros, one of the major reasons that this technique has not yet been largely used in practice is due to its loose accuracy and inconsistency, especially when it is applied to those assemblies requiring high precision. The accuracy of a finished component is influenced by many effects, from a preparation process [17], operating conditions [18, 19], a post-process [20], and other manufacturing processes with specific chemical reaction [21]. A standard operation procedure is defined to overcome these issues making loose accuracy and inconsistency to allow a printed model fitted with a manufactured zirconium denture, as shown in Fig. 14.5 below. It is believed that once accuracy and precision are repeatable, 3D printing cannot only be used in making models for teaching or guiding but also explores more applications, for instance, making orthodontic braces and quality control in the production line of tooth manufacturing.

14.5 Conclusions

Change toward digital era is an irreversible global trend. Due to business competition, most medical facilities are designed as a close system. With the gradual opening up of the technology and broad network connection, medical processes will become simplified and efficient. Thus, both dental lab and dental clinics work

Fig. 14.5 PMMA
restoration on FDP model

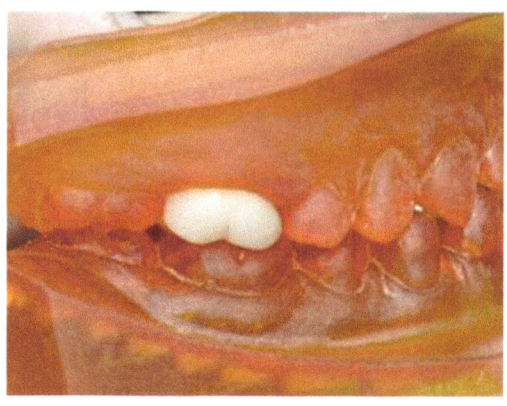

increasingly in a smart way. Based on the development and validation of new tech-nologies, MDIC continuously develops an international digital ecosystem platform that allows patients' clinical satisfaction even higher than before.

References

1. Miyazaki T, Hotta Y, Kunii J, Kuriyama S, Tamaki Y. A review of dental CAD/CAM: current status and future perspectives from 20 years of experience. Dent Mater J. 2009;28(1):44–56.
2. Van Noort R. The future of dental devices is digital. Dent Mater. 2012;28(1):3–12.
3. Hashem IAT, Yaqoob I, Anuar NB, Mokhtar S, Gani A, Khan SU. The rise of "big data" on cloud computing: review and open research issues. Inf Syst. 2015;47:98–115.
4. Qi H, Gani A. Research on mobile cloud computing: review, trend and perspectives. Paper presented at the Digital Information and Communication Technology and it's Applications (DICTAP). Second International Conference on. 2012.
5. Alqahtani AS, Goodwin R. E-commerce Smartphone application. IJACSA Int J Adv Comput Sci Appl. 2012;3(8):54–9.
6. Tung FF et al. The repeatability of an intraoral dental colorimeter. J Prosthet Dent. 2002;88(6):585–90.
7. Browning WD et al. Color differences: polymerized composite and corresponding Vitapan classical shade tab. J Dent. 2009;37:p. e34–9.
8. Vichi A et al. Color related to ceramic and zirconia restorations: a review. Dent Mater. 2011;27(1):p. 97–108.
9. Jarad FD, Russell MD, Moss BW. The use of digital imaging for colour matching and com-munication in restorative dentistry. Br Dent J. 2005;199(1):43–9.
10. Karaagaclioglu L, et al. In vivo and in vitro assessment of an intraoral dental colorimeter. J Prosthodont. 2010;19(4):279–85.
11. Pop-Ciutrila I-S, et al. Shade correspondence, color, and translucency differences between human dentine and a CAD/CAM hybrid ceramic system. J Esthet Restor Dent. 2016;28:S46–55.
12. Gómez-Polo C et al. Differences between the human eye and the spectrophotometer in the shade matching of tooth colour. J Dent. 2014;42(6):p. 742–5.
13. Schwabacher WB, Goodkind RJ. CIE (Commission Internationale del'Eclairage). Colorimetry – technical report. CIE Pub. No. 15, r.e.V.B.C.d.l.C. 2004.

14. Seol Y-J, et al. Bioprinting technology and its applications. Eur J Cardiothorac Surg. 2014;46(3):342–8.
15. Ventola CL. Medical applications for 3D printing: current and projected uses. Pharm Ther. 2014;39(10):704–11.
16. Gibson I, Rosen DW, Stucker B. Additive manufacturing technologies: rapid prototyping to direct digital manufacturing. New York: Springer US; 2009.
17. Canellidis V, et al. Pre-processing methodology for optimizing stereolithography apparatus build performance. Comput Ind. 2006;57(5):424–36.
18. Nee AYC, Fuh JYH, Miyazawa T. On the improvement of the stereolithography (SL) process. J Mater Process Technol. 2001;113(1–3):262–8.
19. Chartier T, et al. Stereolithography process: influence of the rheology of silica suspensions and of the medium on polymerization kinetics – cured depth and width. J Eur Ceram Soc. 2012;32(8):1625–34.
20. Kim H-C, Lee S-H. Reduction of post-processing for stereolithography systems by fabrication-direction optimization. Comput Aided Des. 2005;37(7):711–25.
21. Dufaud O, Le gall H, Corbel S. Application of stereolithography to chemical engineering: 'From Macro to Micro'. Chem Eng Res Des. 2005;83(2):133–8.

Open Access   This chapter is distributed under the terms of the Creative Commons Attribution 4.0 International License (http://creativecommons.org/licenses/by/4.0/), which permits use, duplication, adaptation, distribution and reproduction in any medium or format, as long as you give appropriate credit to the original author(s) and the source, provide a link to the Creative Commons license and indicate if changes were made.

The images or other third party material in this chapter are included in the work's Creative Commons license, unless indicated otherwise in the credit line; if such material is not included in the work's Creative Commons license and the respective action is not permitted by statutory regulation, users will need to obtain permission from the license holder to duplicate, adapt or reproduce the material.

Chapter 15
Detection of Early Caries by Laser-Induced Breakdown Spectroscopy

Yuji Matsuura

Abstract To improve sensitivity of dental caries detection by laser-induced breakdown spectroscopy (LIBS) analysis, it is proposed to utilize emission peaks in the ultraviolet. We newly focused on zinc whose emission peaks appear in ultraviolet because zinc exists at high concentration in the outer layer of enamel. It was shown that by using ratios between heights of an emission peak of Zn and that of Ca, the detection sensitivity and stability are largely improved. It is also shown that early caries are differentiated from healthy part by properly setting a threshold in the detected ratios. The proposed caries detection system can be applied to dental laser systems such as ones based on Er:YAG lasers. When ablating early caries part by laser light, the system notices the dentist that the ablation of caries part has been finished. We also show the intensity of emission peaks of zinc decreased with ablation with Er:YAG laser light.

Keywords Laser-induced breakdown spectroscopy • Dental caries detection • Ultraviolet spectroscopy • Hollow optical fiber

15.1 Introduction

Visual observation and contact methods using a dental probe have been applied for diagnosis of dental caries although they sometimes cause misdiagnosis and pain. To improve diagnosis accuracy without pain, many non-contact methods based on radiation of electromagnetic wave have been developed [1–5]. In contrast to these methods that are usually performed as comparison between decayed and healthy parts, caries detection methods based on laser-induced breakdown spectroscopy (LIBS) enable absolute and quantitative analysis of elements contained in teeth. It is a method for elemental analysis which is based on spectral analysis of plasma

Y. Matsuura (✉)
Graduate School of Biomedical Engineering, Tohoku University,
6-6-05 Aoba, Aramaki, Aoba-ku, Sendai, Miyagi 980-8579, Japan
e-mail: yuji@ecei.tohoku.ac.jp

© The Author(s) 2017   173
K. Sasaki et al. (eds.), *Interface Oral Health Science 2016*,
DOI 10.1007/978-981-10-1560-1_15

emission generated by irradiation of high-powered laser pulses [6]. LIBS is different from other element analysis methods such as inductively coupled, argon plasma-atomic emission spectroscopy (ICP-AES) [7], and LIBS needs no pretreatment of samples and, thus, it is capable of real-time analysis of very small amount. Recently, many groups applied LIBS methods to biomedical applications because they can be in vivo, less invasive diagnosis methods for a variety of soft and hard tissues [8–10]. For dental applications, many groups have proposed LIBS methods for caries detection by analyzing element contents in teeth [11, 12], and some groups have shown results of in vivo studies [13]. Enamels that are the outermost layer of the tooth are biochemical composite whose components are 96 % of inorganic materials including minerals, hydroxyapatites ($Ca_{10}(PO_4)_6(OH)_2$), and a small amount of metals, 3 % of water, and 1 % of organics such as protein and fat. In decayed parts of tooth, the amounts of the above contents vary in accordance with degree of caries progress, and therefore, by analyzing these elements, one can detect caries and execute accurate diagnosis of decaying stages. These methods usually detect relatively strong emission lines of elements such as Ca and P in hydroxyapatite and C, Mg, Cu, and Sr in visible wavelengths. However, the sensitivity and the accuracy were not sufficient for detection of early caries.

We have built an optical-fiber-based LIBS system for in vivo and real-time analysis of teeth enamels during laser dental treatment using a dental Er:YAG laser system. These two systems are combined by using a hollow optical fiber that transmits both of Q-switched Nd:YAG laser for LIBS and infrared Er:YAG laser for tooth ablation. In this paper, we expand the spectral region under analysis to ultraviolet light to improve the sensitivity of caries detection and show that, by analyzing emission peaks of zinc (Zn) in ultraviolet, early caries are detected in high accuracy.

15.2 Experimental Setup

Figure 15.1 shows the schematic of experimental setup. A Q-switched Nd:YAG laser with an operating wavelength of 1,064 nm, a pulse width of 7–8 ns, and a repetition rate of 10 pps was used as the light source for plasma generation. The laser light was coupled to a hollow optical fiber with an inner diameter of 700 μm by a convex lens with a focal length of 250 mm. By using a hollow optical fiber for delivery of laser light to the sample surface, a system is capable of both of diagnosis based on LIBS and caries removal because the hollow optical fibers deliver high-powered infrared laser light for tooth ablation as well [14, 15]. Plasma emission induced by laser radiation was detected by a step-index, pure-silica-glass optical fiber with a core diameter of 400 μm and numerical aperture of 0.22. Detected emission was delivered to a fiber-coupled spectrometer (Ocean Optics HR2000+, slit width 10 μm, 1,800 lines/mm) to measure the power spectra of emitted light from 200 to 340 nm wavelength with a resolution of 0.14 nm. Experiments were performed at atmospheric pressure, and argon gas was injected onto the sample via the

Fig. 15.1 Schematic of
experimental setup of
fiber-based LIBS system

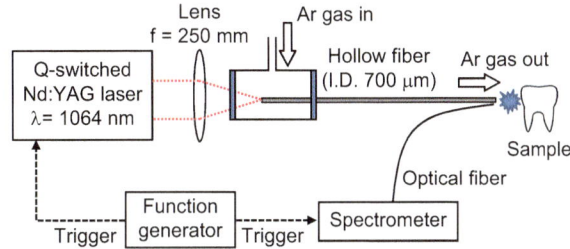

bore of hollow optical fiber to enhance emitted plasma intensities of low concentration elements [16].

As measurement samples, extracted human teeth in different levels of decay
were prepared. The samples were divided into three stages: early stages of decay
called "E1" and "E2" where the decays and cavities are stayed only in the tooth
enamel, advanced stages called "D1" and "D2" where the cavities reach into the
dentin, and healthy teeth without a decay or a cavity. More than ten samples in each
stages including both of incisors and molars were prepared, and the samples were
washed with brushes and pure water before being tested.

15.3   Results and Discussion

It is known that, usually, healthy teeth contain Ca and P that are main components
of hydroxyapatite in high concentration. As tooth decay progresses, other inorganic
elements such as Mg and Cu precipitate when crystals of hydroxyapatite are demineralized. Therefore, one can detect development of dental caries as increases of
densities of minor inorganic materials in LIBS spectra. However, in our preliminary
tests, we found that the differences in the intensities of these elements showing
strong emission peaks in visible and near infrared wavelengths are too small
between caries and healthy parts to detect early caries. Therefore, in this study, we
focus on ultraviolet region where characteristic emission peaks of various minor
components appear.

A measured LIBS spectrum of a healthy tooth in a wavelength region of 200–
340 nm is shown in Fig. 15.2. This is an averaged spectrum of emissions by 50 laser
shots, and the integration time of each emissions were 100 ms. Radiated pulse energy
was 21–22 mJ. In the LIBS spectrum in Fig. 15.2, we found that, in addition to components in hydroxyapatite, minor inorganic components such as Mg were detected.

Figure 15.3 shows LIBS spectra of a healthy tooth and a decayed tooth. In the
measurement of caries, radiated pulse energy was set to 15–16 mJ. We confirmed
from the figure that, for the caries tooth, the densities of C were higher and that of
Ca was lower than the healthy tooth. Based on this result, we firstly tried to set an
evaluation standard for diagnosis of caries progress by using the peak intensity of C

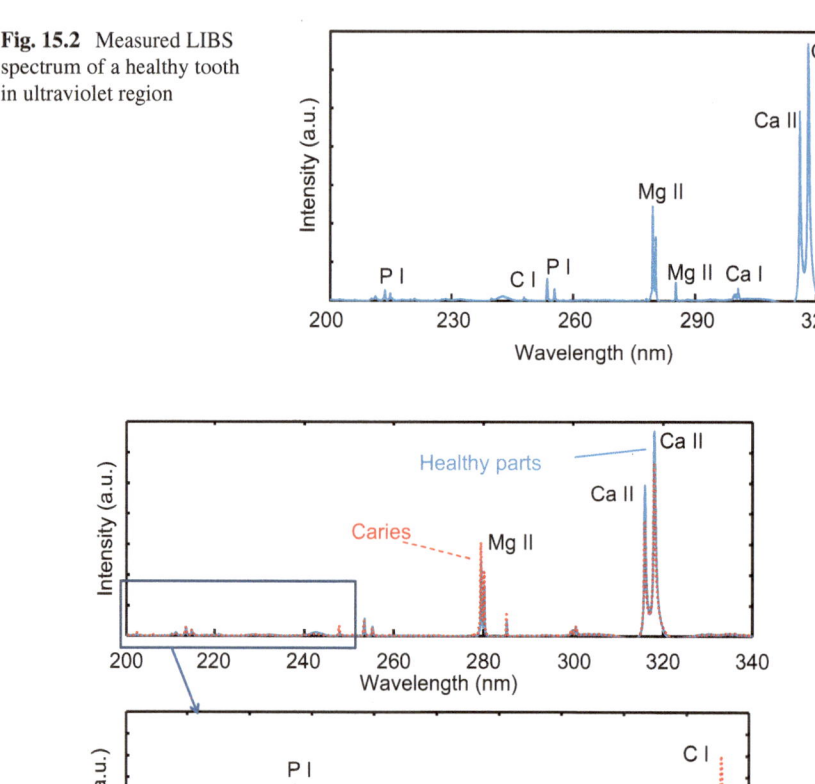

Fig. 15.2 Measured LIBS spectrum of a healthy tooth in ultraviolet region

Fig. 15.3 LIBS spectra of a healthy tooth and a decayed tooth

and Ca that show relatively high peak intensities and large differences between caries and healthy teeth. We prepared five samples from each of healthy teeth, early caries (E1 and E2), and dentin caries (D1 and D2). We performed 30 measurements for each samples and calculated intensity ratios of C at 247.7 nm and Ca at 317.9 nm for quantitative evaluation that was not affected by fluctuation of the emission intensities [17, 18]. Figure 15.4 shows a scatter diagram of the measured results. From this result, we found that dentin caries were differentiated from the other levels. However, averaged values of the healthy teeth and the early caries were 0.0050 and 0.0059, respectively, and the difference was too small to distinguish.

Based on the above result, we repeatedly performed similar experiments while changing the combination of focusing elements to choose the optimum target elements for detection of early caries. In the spectra shown in Fig. 15.5, we found that the

Fig. 15.4 Scattered diagram of calculated intensity ratios of C at 247.7 nm and Ca at 317.9 nm for three different stages of caries development

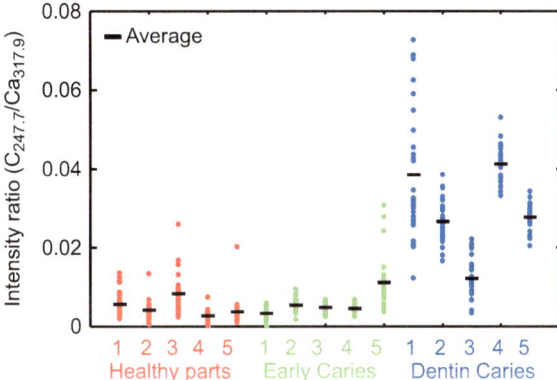

Fig. 15.5 LIBS spectra of healthy tooth and decayed tooth in short wavelength UV region

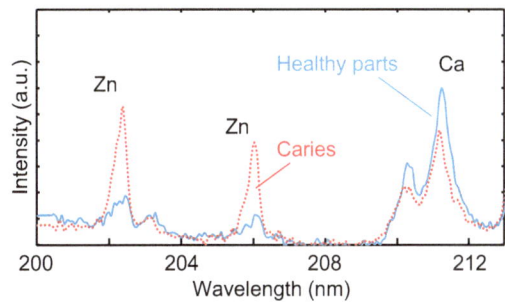

Fig. 15.6 Scatter diagram of measured intensity ratios between Zn at 202.5 nm and Ca at 317.9 nm

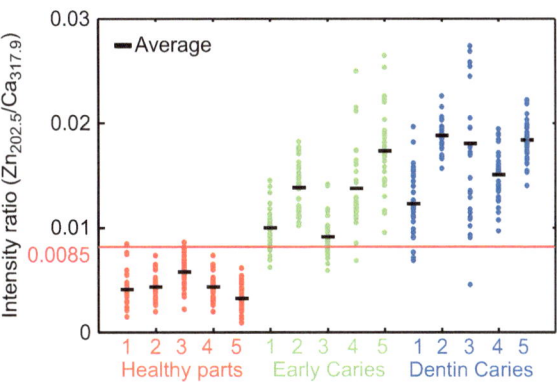

peak height of Zn increases in caries teeth as well. It is known that, compared with other inorganic materials, zinc exists at high concentration in the outer layer of enamel [19, 20]. Although various inorganic elements precipitate when hydroxyapatite is demineralized, change in the concentration due to tooth decay is seen more obviously in zinc because early caries stays only at the very surface of the enamel layer. Therefore, we assume that zinc was strongly detected in early stage of dental decay.

Figure 15.6 shows a scatter diagram of measured intensity ratios between Zn at 202.5 nm and Ca at 317.9 nm. In this figure, average values of intensity ratios are

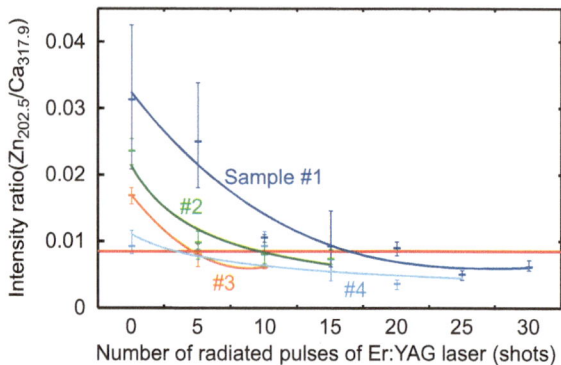

Fig. 15.7 Measured intensity ratio Zn/Ca as a function of number of radiated pulses of Er:YAG laser

0.0044 for the healthy teeth and 0.013 for the early caries. By setting a boundary at around 0.008, one can clearly distinguish between healthy teeth and early caries. Accuracies of diagnosis based on this result were as high as 98.2 % for the healthy teeth, 85.2 % for the early caries, and 96.6 % for the dentin caries. In additional experiments using more than five samples, the healthy parts and early caries are repeatedly tested. As a result, we confirmed that diagnosis accuracy for early caries was higher than 80 %.

Next, we evaluated feasibility of real-time analysis of the proposed method during laser treatment using an Er:YAG dental laser system. We utilized a dental laser system (J. MORITA, Erwin Adverl) and laser pulses with a wavelength of 2.94 μm and pulse energy of 100 mJ were radiated onto early caries by using a hollow optical fiber. Simultaneously, the LIBS analysis proposed above was performed. The results are shown in Fig. 15.7 as the measured intensity ratio Zn/Ca as a function of number of radiated pulses of the Er:YAG laser. For all of five samples that we tested, decreases in the intensity ratio were observed, and thus we confirmed that the proposed system is feasible for informing a practitioner finish of removal of caries parts.

15.4 Conclusion

We proposed a LIBS system for in vivo analysis of teeth enamels. The system utilizes a hollow optical fiber that transmits both of Q-switched Nd:YAG laser light for LIBS and infrared Er:YAG laser light for tooth ablation, and thus the system enables real-time analysis of teeth during laser dental treatment. By expanding the spectral region under analysis to ultraviolet light and focusing on emission peaks of Zn in the UV region, we largely improved the sensitivity of caries detection. We showed that, by using ratios of peak intensities of Zn and Ca, early caries were distinguished from healthy teeth with accuracies higher than 80 %. Then we applied this LIBS analysis to caries teeth while ablating the caries part with Er:YAG laser light and have shown that the intensity ratio Zn/Ca decreases with radiation of Er:YAG laser pulses. Therefore, the proposed system is feasible for informing a practitioner finish of removal of caries.

References

1. de Josselin de Jong E, ten Bosch JJ, Noordmans J. Optimized microcomputer-guided quantitative microradiography on dental mineralized tissue slices. Phys Med Biol. 1987;32:887–9.
2. Stookey GK, Gonzales-Cabezas C. Emerging methods of caries diagnosis. J Dent Educ. 2001;65:1001–6.
3. de Josselin de Jong E, Sundström F, Westerling H, Tranaeus S, ten Bosch JJ, Angmar-Månsson B. A new method for in vivo quantification of changes in initial enamel caries with laser fluorescence. Caries Res. 1995;29:2–7.
4. Takamori K, Hokari N, Okumura Y, Watanabe S. Detection of occlusal caries under sealants by use of a laser fluorescence system. J Clin Laser Med Surg. 2001;19:267271.
5. Subhash N, Thomas SS, Mallia RJ, Jose M. Tooth caries detection by curve fitting of laser-induced fluorescence emission: a comparative evaluation with reflectance spectroscopy. Laser Surg Med. 2005;37:320–8.
6. Cremers DA, Radziemski LJ. Handbook of laser-induced breakdown spectroscopy. Wes Sussex: Wiley; 2006.
7. Chew LT, Bradley DA, Mohd AY, Jamil MM. Zinc, lead and copper in human teeth measured by induced coupled argon plasma atomic emission spectroscopy (ICP-AES). Appl Radiat Isot. 2000;53:633–8.
8. Sun Q, Tran M, Smith BW, Winefordner JD. Zinc analysis in human skin by laser induced-breakdown spectroscopy. Talanta. 2000;52:293–300.
9. Hosseinimakarem Z, Tavassoli SH. Analysis of human nails by laser-induced breakdown spectroscopy. J Biomed Opt. 2011;16:057002.
10. Singh VK, Kumar V, Sharma J. Importance of laser-induced breakdown spectroscopy for hard tissues (bone, teeth) and other calcified tissue materials. Lasers Med Sci. 2014. doi:10.1007/s10103-014-1549-9.
11. Samek O, Beddows DCS, Telle HH, Morris GW, Liska M, Kaiser J. Quantitative analysis of trace metal accumulation in teeth using laser-induced breakdown spectroscopy. Appl Phys A. 1999;69:179–82.
12. Unnikrishnan VK, Choudhari KS, Kulkarni SD, Nayak R, Kartha VB, Santhosh C, Suri BM. Biomedical and environmental applications of laser-induced breakdown spectroscopy. Pramana J Phys. 2014;82:397–401.
13. Samek O, Telle HH, Beddows DCS. Laser-induced breakdown spectroscopy: a tool for real-time, in vitro and in vivo identification of carious teeth. BMC Oral Health. 2001. doi:10.1186/1472-6831-1-1.
14. Matsuura Y, Shi Y, Abe Y, Yaegashi M, Takada G, Mohri S, Miyagi M. Infrared-laser delivery system based on polymer-coated hollow fibers. Opt Laser Technol. 2001;33:279–83.
15. Matsuura Y, Hanamoto K, Sato S, Miyagi M. Hollow-fiber delivery of high-power pulsed Nd:YAG laser light. Opt Lett. 1998;23:1858–60.
16. Farid N, Bashir S, Mahmood K. Effect of ambient gas conditions on laser-induced copper plasma and surface morphology. Phys Scr. 2012;85:1–7.
17. Samek O, Beddows DCS, Telle HH, Kaiser J, Liska M, Caceres JO, Urena AG. Quantitative laser-induced breakdown spectroscopy analysis of calcified tissue samples. Appl Phys B. 2001;56:865–75.

18. Singh KV, Rai AK. Potential of laser-induced breakdown spectroscopy for the rapid identifica-
    tion of carious teeth. Lasers Med Sci. 2011;26:307–15.
19. Zipkin I. Biological mineralization. New York: Wiley; 1973.
20. Reitznerova E, Amarasiriwardena D, Kopcakova M, Barnes RM. Determination of some trace
    elements in human tooth enamel. Fresenius J Anal Chem. 2000;367:748–54.

Open Access This chapter is distributed under the terms of the Creative Commons Attribution 4.0
International License (http://creativecommons.org/licenses/by/4.0/), which permits use, duplication, adaptation, distribution and reproduction in any medium or format, as long as you give appropriate credit to the original author(s) and the source, provide a link to the Creative Commons
license and indicate if changes were made.

The images or other third party material in this chapter are included in the work's Creative
Commons license, unless indicated otherwise in the credit line; if such material is not included in
the work's Creative Commons license and the respective action is not permitted by statutory regulation, users will need to obtain permission from the license holder to duplicate, adapt or reproduce
the material.

Chapter 16
Acoustic Diagnosis Device for Dentistry

Kouki Hatori, Yoshifumi Saijo, Yoshihiro Hagiwara, Yukihiro Naganuma, Kazuko Igari, Masahiro Iikubo, Kazuto Kobayashi, and Keiichi Sasaki

Abstract There are a lot of diseases which show the abnormal elastic property. Although many medical doctors and dentists have noticed the change of tissue elasticity due to the disease, the diagnostic device to examine the tissue elastic property objectively has not well developed.

At Tohoku University, acoustic microscopy (AM) for medicine and biology has been developed and applied for more than 20 years. Application of AM has three major features and objectives. First, specific staining is not required for characterization or observation. Second, it provides the elastic property and information of

K. Hatori (✉)
Department of Prosthodontics, Matsumoto Dental University,
1780 Gobara Hirooka, Shiojiri, Nagano 399-0781, Japan

Division of Advanced Prosthetic Dentistry, Tohoku University Graduate School of Dentistry,
4-1 Seiryo-machi, Aoba-ku, Sendai, Miyagi, Japan
e-mail: khat810@yahoo.co.jp

Y. Saijo
Biomedical Imaging Laboratory, Graduate Schools of Biomedical Engineering and Medical
Sciences, Tohoku University, Sendai, Miyagi, Japan

Y. Hagiwara
Department of Orthopaedic Surgery, Tohoku University, Tohoku University School of
Medicine, Sendai, Miyagi, Japan

Y. Naganuma
Division of Advanced Prosthetic Dentistry, Tohoku University Graduate School of Dentistry,
4-1 Seiryo-machi, Aoba-ku, Sendai, Miyagi, Japan

Dentistry for the Disabled, Tohoku University Hospital, Sendai, Miyagi, Japan

K. Igari
Dentistry for the Disabled, Tohoku University Hospital, Sendai, Miyagi, Japan

M. Iikubo
Department of Oral Diagnosis, Tohoku University Graduate School of Dentistry,
Sendai, Miyagi, Japan

K. Kobayashi
Honda Electronics Co. Ltd., Toyohashi, Aichi, Japan

K. Sasaki
Division of Advanced Prosthetic Dentistry, Tohoku University Graduate School of Dentistry,
4-1 Seiryo-machi, Aoba-ku, Sendai, Japan

© The Author(s) 2017
K. Sasaki et al. (eds.), *Interface Oral Health Science 2016*,
DOI 10.1007/978-981-10-1560-1_16

the subject, because acoustic properties have close correlation with the mechanical property of the subject. Third, it makes the observation easy and rapid. The purpose of this study is to observe carious dentin and periodontal ligament (PDL) using AM. Moreover, I would like to propose the prototype of the portable caries detector.

AM clearly visualizes the color distribution of the acoustic properties of the subject.

The acoustic examination device may be a powerful apparatus not only to visualize the morphological appearance but also to diagnose the disease.

Keywords Acoustic imaging • Human teeth • Dentin caries • Periodontal ligament • Diagnosis

16.1   Introduction

We have been developing a scanning acoustic microscope (SAM) system at Tohoku University since 1985. SAM has been applied to observe the acoustic properties of various cells, organs, and disease state: vein endothelial cells, gastric cancer, normal kidney and renal cancer, infarcted myocardium, atherosclerosis of aorta and carotid arterial plaques, etc. In biomedical study, the application of SAM is useful for the intraoperative pathological examination, the observation of low-frequency ultrasonic images, and the assessment of biomechanical properties at a microscopic level [1–11]. The application of SAM in biomedical study has two major features and objectives. First, the specific staining is unnecessary for observation. Second, SAM provides the acoustic properties and information such as the acoustic intensity, the sound speed propagating in the sample, and the acoustic attenuation. Therefore, SAM enables to acquire the elastic property of the subject and the quantitative values of the acoustic properties, because the acoustic properties have close correlation with the mechanical property of the subject. After 2004, the third advantage to use SAM was added, which is that SAM makes the observation easy and rapid. One scanning takes 2 min approximately using new concept SAM, because new concept SAM system was developed using a single-pulsed wave instead of continuous waves used in the conventional SAM system in 2004 [12, 13]. Sano et al. reported that the decalcification during tissue preparation did not alter the sound speed propagating in fibrocartilage or bone [14]. Hagiwara et al. reported that the sound speed propagating in the posterior synovial membrane after knee joint contracture increased significantly using decalcified specimen [13]. These reports indicate that the decalcified specimen is also appropriate to observe the acoustic properties using SAM.

Periodontal ligament (PDL) is the dense fibrous connective tissue which binds the root cementum to the alveolar bony socket. This tissue is especially cellular and vascular tissue and plays pivotal role to develop and maintain periodontal membrane. PDL supports and dissipates the occlusal force and regulates the tooth eruption. Thus, PDL performs these important oral functions [15, 16]. Since many studies of the mechanical properties of PDL have been reported using animal model

[17], it was pointed out that mechanical properties of PDL showed the values (0.9–1.6 N/mm²) in the extrusion testing in rat mandibular first molar [18, 19]. These reports were the dynamic tests for the measurement of the mechanical properties of PDL, but little is known about the static test for the measurement of the mechanical properties of PDL in vivo. In this paper, the first purpose is to observe the rat PDL using SAM.

Acoustic impedance, which is given as a product of sound speed and density, has a close correlation with the elastic modulus, when the value of density of the embedding material was constant. Based on the above background, HONDA ELECTRONICS Co., Ltd. (Toyohashi, Japan) and Tohoku University Graduate School of Biomedical Engineering have developed the acoustic impedance microscope (AIM) which enables to image the two-dimensional distribution of the acoustic intensity and impedance of the sample [20]. AIM has been applied to observe cerebellum tissue [21] and cultured glial and glioma cells [22]; these studies reported that AIM also provided the acoustic intensity and impedance image at a microscopic level.

Dental caries is a common and chronic disease in oral cavity and results from the production of acid by caries-inducing bacterial fermentation on the tooth surface. The acid by caries-inducing bacterial metabolism causes the decalcification of the hard tissues of the teeth such as enamel and dentin [23]. Finally the decalcified lesion becomes soft and is recognized as dental caries. Dental caries cause the change of hardness of the teeth [24, 25]. The change of acoustic properties of dentin caries has not been observed using AIM yet. In this paper, the second purpose is to observe the human dentin caries using AIM and to observe the microstructure of the human dentin caries using the scanning electron microscope (SEM).

Based on these the technology of producing such acoustic imaging devices, HONDA ELECTRONICS Co., Ltd. and Tohoku University Graduate School of Biomedical Engineering fabricated a prototype of portable acoustic stiffness checker. In this paper, the last purpose is to introduce about this acoustic stiffness checker.

16.2 Materials and Methods

16.2.1 Observation of Rat Periodontal Ligament Using Scanning Acoustic Microscopy

16.2.1.1 Scanning Acoustic Microscope (SAM)

The sound speed propagating in a solid substance is defined as:

$$C = \sqrt{\frac{E(1-\delta)}{\rho(1-2\delta)(1+\delta)}} \tag{16.1}$$

where C is the sound speed propagating in the object, E is the Young's modulus of the object, ρ is the Poisson's ratio, and δ is the density of the object. Thus, it is implied that the change of the sound speed propagating in the tissue is the change of the tissue elastic property [26]. The Young's modulus (E) is in proportion to the square of the sound speed; in the other words, the change of the sound speed provides with the information of the change of material hardness. The observation of the sound speed propagating in the thin sliced sample enables us to understand the proper hardness of the thin sliced sample [14].

Our SAM is mainly composed of five units: (1) ultrasonic transducer with acoustic lens, (2) pulse generator, (3) digital oscilloscope with system control PC (Windows), (4) stage control microcomputer, and (5) display monitor (Figs. 16.1 and 16.2). The emission of the electric pulse was within 400 ps from excitation, and the pulse voltage was 40 V. A single-pulsed ultrasonic impulse with a pulse width of 5 ns was generated and received by the same ultrasonic transducer above the section. The aperture diameter of the ultrasonic transducer was 1.2 mm, and the focal length was 1.5 mm (Fig. 16.3). The central frequency was 80 MHz, the nominal bandwidth was 50–105 MHz (−6 dB), and the pulse repetition rate was 10 kHz. Taking the focal distance and the sectional area of the transducer into consideration, the diameter of the focal spot was estimated as 20 μm at 80 MHz. Physiologic saline was used as the coupling medium between the transducer and the section. An ultrasonic wave with a wide frequency component was generated by applying the voltage pulse of 40 V and was irradiated to the section on the glass slide. The reflections from the section surface and those from interface between the section and the glass slide were detected by the transducer and were introduced into the digital oscilloscope (TDS 5052, Tektronix Inc., Portland, USA). The band limit was 300 MHz, and the sampling rate was 2.5 GS/s. In order to reduce random noise, four times of pulse responses at the same point were averaged in the digital oscilloscope prior to being introduced into the computer.

The transducer was mounted on an X-Y stage driven by the stage control microcomputer which was operated by the system control PC via an RS232C. The X- and Y-scan were operated by a linear servo motors by the incoming signal from stage control circuit.

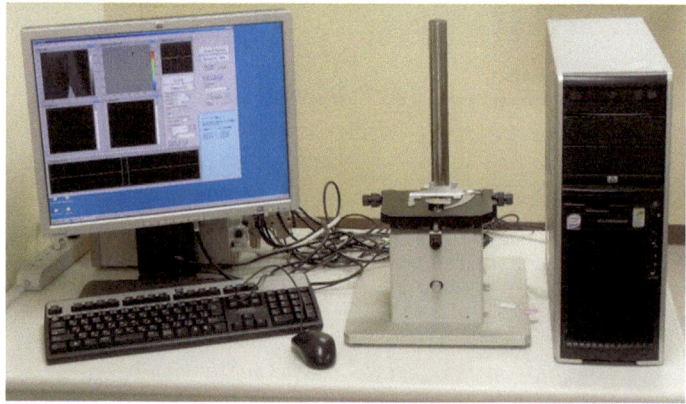

Fig. 16.1 System of scanning acoustic microscopy. This system can display a two-dimensional distribution of sound speed and acoustic impedance

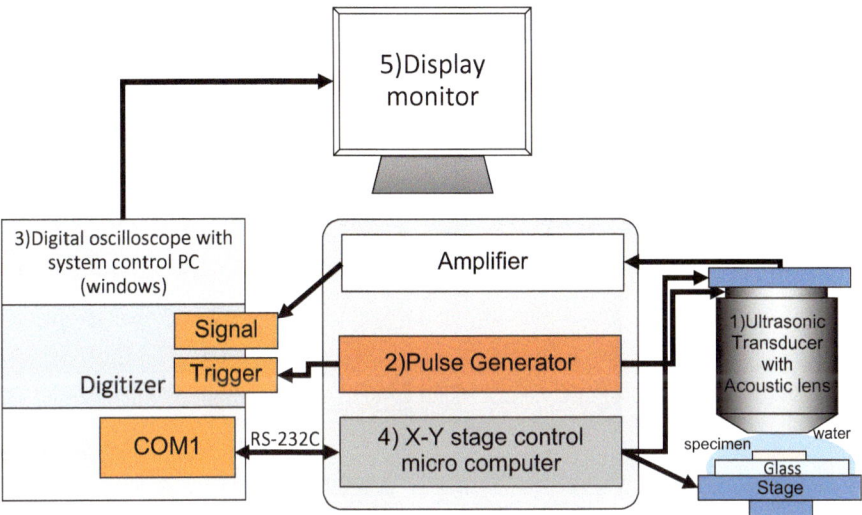

Fig. 16.2 Block diagram of scanning acoustic microscopy

Fig. 16.3 Schematic illustration of SAM transducer. The aperture diameter of the ultrasonic transducer was 1.2 mm, and the focal length was 1.5 mm. The central frequency was 80 MHz

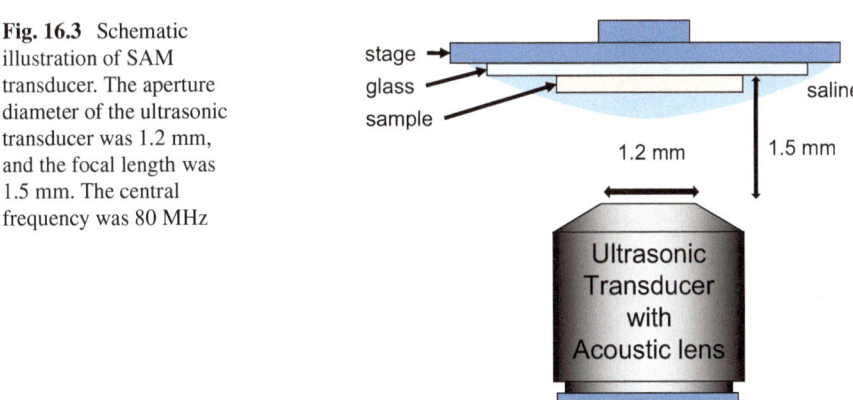

The distance from the transducer to the glass surface was 1.5 mm. Reflected ultrasonic wave included three types of waveforms, which were from the surface of the section (*Sfront*), from the interface between the section and the glass (*Srear*), and the surface from the glass (*Sref*) (Fig. 16.4). The ultrasonic phases of these waveforms were standardized and recognized as a reflection wave from the glass in system control PC. Finally, two-dimensional distributions of the acoustic intensity, the sound speed, the acoustic attenuation, and the thickness of the sample of the 2.4×2.4 mm sample area were visualized with 300×300 pixels and calculated by Fourier transforming the waveform [12–14].

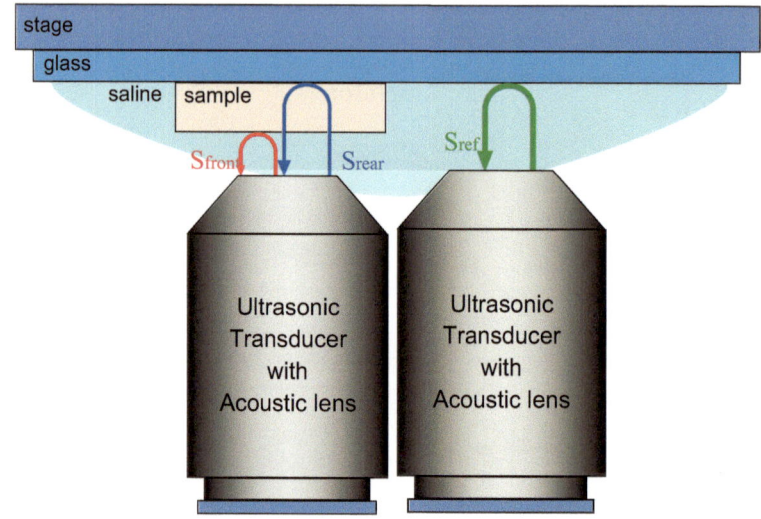

Fig. 16.4 Schematic illustration of scanning image of sound speed microscope. A single-pulsed ultrasonic impulse was generated and received by the same transducer. *Sfront* is the reflection from the surface of the section, *Srear* is the reflection from the interface between the section and the glass, and *Sref* is the reflection the surface from the glass

16.2.1.2 Tissue Preparation

The protocol for this experiment was approved by the Animal Research Committee of Tohoku University (approval number: 22–24).

A total of 28 male Wistar rats at 3-, 5-, 7- and 10-week-old postnatally were used in this study. Twenty-eight rats were prepared for SAM analysis (3-, 5-, 7- and 10-week-old; $n = 7$/each age).

The rats were generally anesthetized with intraperitoneal injection of pentobarbital sodium (30 mg/kg body weight) (Somnopentyl®; Kyoritsu Seiyaku, Tokyo, Japan) and fixed with 4 % paraformaldehyde (PFA) in 0.1 M phosphate buffered saline (PBS) (pH7.4) by perfusion through the aorta. The mandibles were resected and immersed in the same fixative overnight at 4 °C. The fixed specimens were decalcified in 10 % ethylenediaminetetraacetic acid (EDTA) in 0.01 M PBS (pH7.4) for 6 weeks at 4 °C. After dehydration through a graded series of ethanol solution, the specimens were embedded in paraffin and finally cut into 5-μm sagittal (mesiodistal) sections from the buccal to the lingual side of the mandible. Standardized serial sections were made in the medial periodontal space region of the lower first molar.

16.2.1.3 Image Analysis

Normal light microscope (DM3000, LEICA, Wetzlar, Germany) was used to take the microscopic images corresponding to the stored SAM images. A region of observation by SAM was set in the interradicular region (IRR) and the mesial radicular region (MRR) of PDL in each section (Fig. 16.5). SAM images with a gradation color scale were acquired to visualize the clear distribution of the sound speed. In order to calculate the sound speed propagating in PDL, the color SAM images were processed for the gray scale images by the same SAM system. Finally, in the observed region, the sound speed propagating in PDL, excluding cementum, dentin, and alveolar bone, was calculated using commercially available image analysis software (PhotoShop 7.0, Adobe Systems Inc., San Jose, CA, USA).

16.2.1.4   Statistics

Data in each PDL region were analyzed using one-way factorial analysis of variance (ANOVA) with Tukey's post hoc multiple comparisons. Differences between IRR and MRR at each rat age were analyzed using paired t-test. All data were expressed as the mean ± SD. A value of $p < 0.05$ was accepted as statistically significant.

16.2.2   Observation of Human Carious Dentin Using Acoustic Impedance Microscopy

16.2.2.1   Acoustic Impedance Microscope (AIM)

Acoustic impedance mode was applied for visualization and observation of sample surface. The acoustic impedance of a material is defined as a product of the density and the sound speed. The acoustic impedance of physiological saline is used to verify the calibration curve in the following equation in its simplest form [22]:

$$Z = \rho c \qquad (16.2)$$

Fig. 16.5 Histological appearance of PDL of the first molar of rat mandible. Interradicular region (*IRR*) area is encircled by the *black solid rectangle*. Mesial radicular region (*MRR*) is encircled by the *black dotted rectangle*. 5× original magnification. Bar = 500 μm

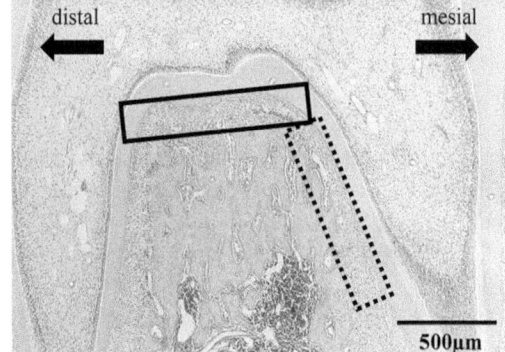

where Z is acoustic impedance, ρ is density, and c is sound speed.

The acoustic approach should provide the information of the elastic bulk modulus because the relationship among the sound speed, the density, and the elastic bulk modulus of the material is described in the following equation in its simplest form:

$$C = \sqrt{\frac{K}{\rho}} \qquad (16.3)$$

where C is sound speed, K is elastic bulk modulus, and ρ is density [27].

Taken together these equations, the elastic bulk modulus is described in the following equation:

$$K = Z^2 / \rho \qquad (16.4)$$

where K is elastic bulk modulus, Z is acoustic impedance, and ρ is density. Therefore, if the density of embedding material is constant, the change of acoustic impedance will indicate the change of tissue elasticity. This equation indicates that the elastic bulk modulus is in proportion to the square of the acoustic impedance. In other words, the change of the acoustic impedance indicates the change of the elastic property.

The acoustic impedance microscope (AIM) (AMS-50SI, HONDA ELECTRONICS CO., LTD, Toyohashi, Japan) in the present study was shown in Figs. 16.1 and 16.6.

An electric impulse was generated by a high-speed switching semiconductor. The start of the pulse was within 400 ps, the pulse width was 2 ns, and the pulse voltage was 40 V. The frequency of the impulse covered up to 380 MHz. The electric pulse was used to excite a transducer that had a central frequency of 300 MHz and a sapphire rod as an acoustic lens. The ultrasound spectrum of the reflected ultrasound was broad enough to cover 220–380 MHz (−6 dB). The reflections from the sample were received by the transducer and were introduced into a Windows-based PC (Pentium D, 3.0 GHz, 2 GB RAM, 250 GB HDD) via digital oscilloscope (Tektronix TDS7154B, Beaverton, USA). The frequency range was 1 GHz, and the sampling rate was 20 GS/S. Four pulse echo sequences were averaged for each scan point in order to increase the signal-to-noise ratio.

The transducer was mounted on an X-Y stage with a microcomputer board that was driven by the PC through RS232C. The both X-scan and Y-scan were driven by linear servo motors. Finally, two-dimensional distribution of the acoustic impedance was visualized and the image area was 2.4×2.4 mm. The total scanning takes 121 s [20].

16.2.2.2 Tissue Preparation

The protocols for the experiments were approved by the Ethical Committee of Tohoku University (approval number: 26–28).

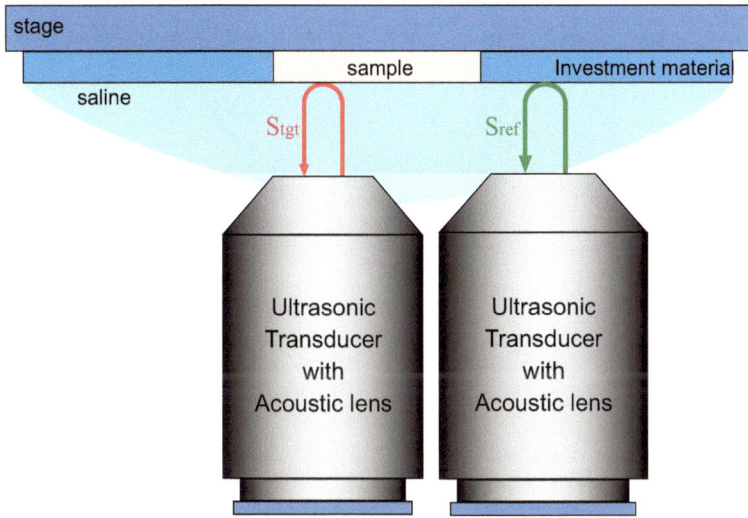

Fig. 16.6 Schematic illustration of scanning image of the acoustic impedance microscope. *Stgt* is the reflection from the sample; *Sref* is the reflection from the investment material. In this case, substrate is water

Human molar teeth with dental caries ($n = 10$) were used in this study. The teeth were extracted under the diagnosis of severe periodontitis, and the extracted teeth were immediately fixed with 0.1 % glutaraldehyde and kept into the same fixative for 24 h. The teeth samples were embedded in the self-curing resin (UNIFAST III CLEAR, GC, Tokyo, Japan). The embedded samples were cut sagittally by saw microtome (SP1600, LEICA, Wetzlar, Germany) perpendicular to the caries surface, and the sliced sections (sagittal sections) were polished and smoothened with waterproof abrasive papers (#1000 and #2000) (Fig. 16.7). The thickness of polished section was 400–600 μm. The prepared sections were assessed by AIM and the scanning electron microscope (SEM), respectively.

16.2.2.3 Scanning Electron Microscopy (SEM)

After the acoustic microscopic observation, the scanning electron microscopic investigation for the carious dentin lesion and the sound dentin area was carried out in order to observe the microstructural surface of both the carious lesion and the sound dentin area. The polished samples for SEM observation were initially air-dried and mounted on aluminum stubs. The sample was sputtered with a 20-μm-thickness of carbon film layer (JFC-1100, JEOL Ltd., Tokyo, Japan) and then analyzed with SEM (JSM-6060, JEOL Ltd., Tokyo, Japan). The accelerating voltage was 15 kV and the specimen irradiation current was 1.0 nA at SEM examination.

Fig. 16.7 Sliced human molar teeth with dental caries. Dentin caries is encircled by the *red dotted rectangle*, and sound dentin is encircled by the *black dotted rectangle*

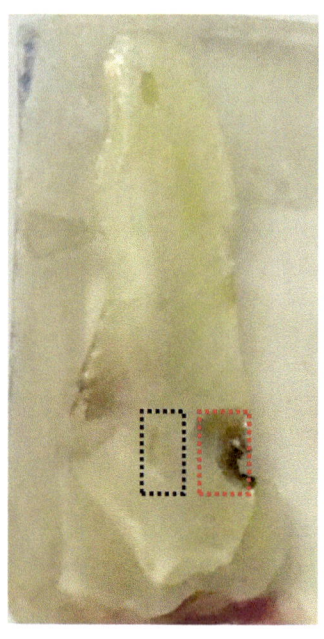

16.2.2.4 Statistics

Differences between the dentin caries and the sound dentin were analyzed using paired *t*-test. All data were expressed as the mean ± SD. A value of $p < 0.05$ was accepted as statistically significant.

16.2.3 Portable Acoustic Stiffness Checker (PASC)

A prototype of portable acoustic stiffness checker was shown in Fig. 16.8. Our PACS is mainly composed of three units: (1) acoustic probe, (2) pulsar receiver, and (3) tablet PC (Windows) (Fig. 16.9).

The target signal is compared with the reference signal and interpreted into acoustic impedance as:

$$Z(target) = \frac{1 - \dfrac{S(\text{target})}{S0}}{1 + \dfrac{S(\text{target})}{S0}} Z(sub) = \frac{1 - \dfrac{S(\text{target})}{S(\text{ref})} \cdot \dfrac{Z(\text{sub}) - Z(\text{ref})}{Z(\text{sub}) + Z(\text{ref})}}{1 + \dfrac{S(\text{target})}{S(\text{ref})} \cdot \dfrac{Z(\text{sub}) - Z(\text{ref})}{Z(\text{sub}) + Z(\text{ref})}} Z(sub) \quad (16.5)$$

where $S0$ is the transmitted signal, S_{target} and S_{ref} are reflections from the target and reference, respectively, and Z_{target}, Z_{ref}, and Z_{sub} are the acoustic impedances of the target, reference, and substrate, respectively [20].

Fig. 16.8 A prototype of ultrasound stiffness checker. Figure is the whole system of the ultrasound stiffness checker

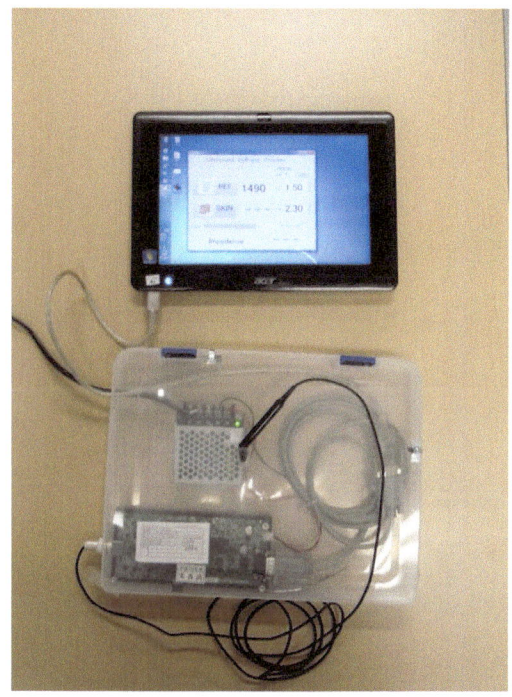

Finally, we can acquire the acoustic impedance of the targeted object on the tablet PC, based on the acoustic impedance of the reference.

16.3   Results

16.3.1   SAM Observation of PDL

Figure 16.10 is the result of the PC screen of SAM showing rat PDL. This result provided four visual information such as the sound speed intensity, the sound speed, the attenuation, and the sample thickness (Fig. 16.10a–d). SAM clearly visualized two-dimensional color distribution of the sound speed of both IRR and MRR (Fig. 16.11a, b). SAM also clearly visualized two-dimensional color distribution of the dentin and the alveolar bone. The gradation color table indicated the sound speed propagating in the sample. In the gradation color table, the lowest sound speed was 1,500 m/s, and the highest sound speed was 1,800 m/s (Fig. 16.11c).

IRR and MRR were composed of very low sound speed areas (blue) in 3-week-old postnatal group (Fig. 16.12a, e). In 5-week-old postnatal group, IRR and MRR were composed of low to middle sound speed areas (blue to green) (Fig. 16.12b, f).

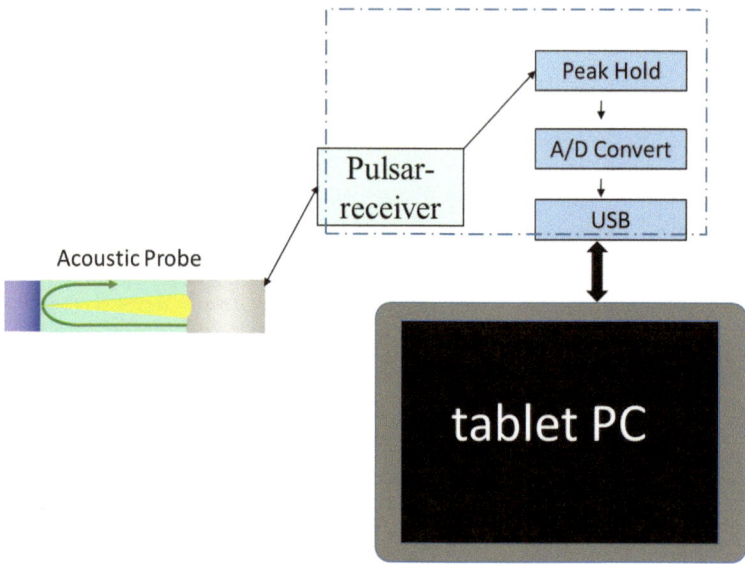

Fig. 16.9 Block diagram of ultrasound stiffness checker

In a 7-week-old postnatal group, IRR and MRR were composed of middle sound speed areas (yellow to green) (Fig. 16.12c, g). In 10 weeks postnatal group, IRR and MRR were composed of middle to high sound speed areas (green) (Fig. 16.12d, h). The low sound speed area decreased, and middle sound speed areas gradually increased in both IRR and MRR depending on the rat growth.

16.3.2   Sound Speed Analysis

The mean value of sound speed propagating in each region was calculated. The sound speed of IRR and MRR is shown in Table 16.1. In IRR, there was no statistical difference between 3-week group and the 5-week group (3w 1,529 m/s $\pm$ 15.2 m/s vs. 5w 1,541 m/s $\pm$ 14.1 m/s; $p=0.605$), between 5-week group and 7-week group (5w 1,541 m/s $\pm$ 14.1 m/s vs. 7w 1,563 m/s $\pm$ 18.2 m/s; $p=0.169$), between 5-week group and 10-week group (5w 1,541 m/s $\pm$ 14.1 m/s vs. 10w 1,564 m/s $\pm$ 21.4 m/s; $p=0.054$), and between 7-week group and 10-week group (7w 1,563 m/s $\pm$ 18.2 m/s vs. 10w 1,564 m/s $\pm$ 21.4 m/s; $p=0.999$). However, the sound speed of IRR was significantly increased depending on the development (3w 1,529 m/s $\pm$ 15.2 m/s vs. 7w 1,563 m/s $\pm$ 18.2 m/s; $p=0.012$ and 3w 1,529 m/s $\pm$ 15.2 m/s vs. 10w 1,564 m/s $\pm$ 21.4 m/s; $p=0.002$). In MRR, there was no statistical difference between 3-week group and 5-week group (3w 1,536 m/s $\pm$ 16.1 m/s vs. 5w 1,556 m/s $\pm$ 18.4 m/s; $p=0.125$), between 3-week group and 7-week group (3w 1,536 m/s $\pm$ 16.1 m/s vs. 7w 1,560 m/s $\pm$ 15.7 m/s; $p=0.063$), between 5-week group and 7-week group

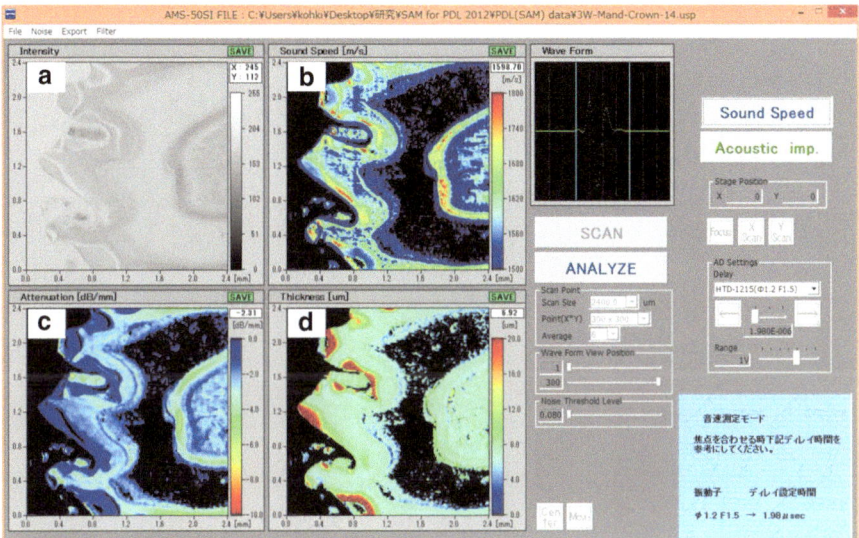

Fig. 16.10 PC screen of SAM image of PDL. Two-dimensional color distribution with the resolution of the 2.4×2.4 mm sample area is visualized with 300×300 pixels. (**a**) sound speed intensity image, (**b**) sound speed image, (**c**) attenuation image, and (**d**) sample thickness image

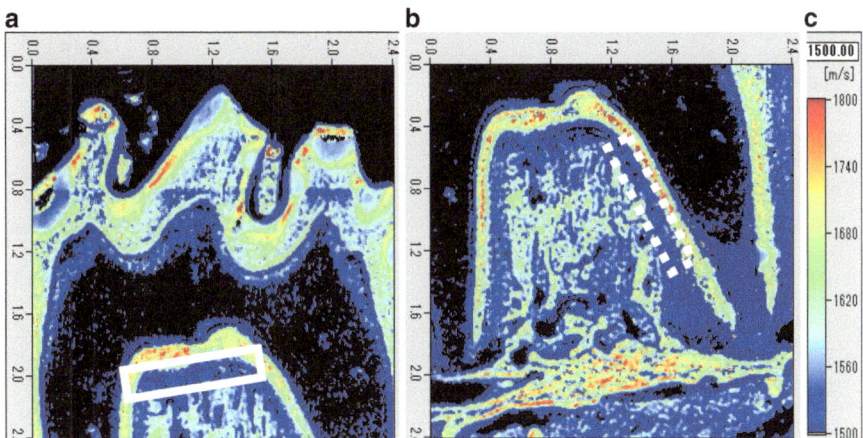

Fig. 16.11 SAM image of IRR. IRR is encircled by the *solid rectangle* (**a**). SAM image of MRR. MRR is encircled by the *dotted rectangle* (**b**). The color graduation table of the sound speed (**c**)

(5w 1,556 m/s ± 18.4 m/s vs. 7w 1,560 m/s ± 15.7 m/s; $p = 0.974$), between 5-week group and 10-week group (5w 1,556 m/s ± 18.4 m/s vs. 10w 1,566 m/s ± 16.9 m/s; $p = 0.583$), and between 7-week group and 10-week group (7w 1,560 m/s ± 15.7 m/s vs. 10w 1,566 m/s ± 16.9 m/s; $p = 0.881$). However, the sound speed of MRR was

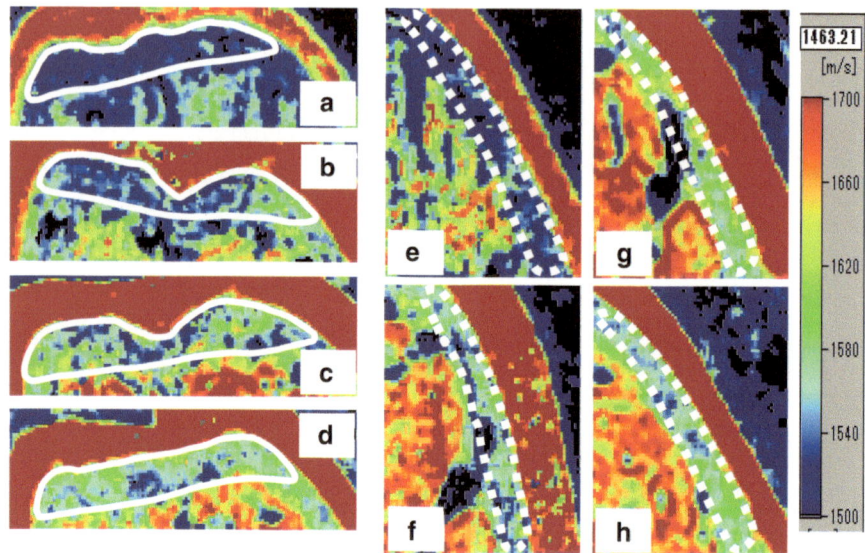

Fig. 16.12 SAM images of IRR (**a–d**) and MRR (**e–h**). 3-week-old (**a, e**). 5-week-old (**b, f**). 7-week-old (**c, g**). 10-week-old (**d, h**)

Table 16.1 Mean sound speed propagating in IRR and MRR, respectively, and standard deviations between IRR and MRR. Statistic evaluation was performed by one-way ANOVA and Tukey's method, or by paired t-test

	IRR(m/s)	sMRR(m/s)
3w	$1,529 \pm 15.2^{a,b}$	$1,536 \pm 16.1^{c}$
5w	$1,541 \pm 14.1^{d}$	$1,556 \pm 18.4^{d}$
7w	$1,563 \pm 18.2^{a}$	$1,560 \pm 15.7$
10w	$1,564 \pm 21.4^{b}$	$1,566 \pm 16.9^{c}$

[a,b]$P < 0.01$ by one-way ANOVA and Tukey's method
[c]$P < 0.05$ by one-way ANOVA and Tukey's method
[d]$P < 0.05$ by paired t-test

significantly increased depending on the development (3w 1,536 m/s ± 16.1 m/s vs. 10w 1,566 m/s ± 16.9 m/s; $p = 0.003$).

There was significant difference between IRR and MRR in 5-week group (IRR 1,541 m/s ± 14.1 m/s vs. MRR 1,556 m/s ± 18.4 m/s; $p = 0.003$).

16.3.2.1 AIM Observation of Dentin Caries

Figure 16.13 is the result of the PC screen of AIM showing dentin caries. This result provided two visual information, the acoustic intensity (Fig. 16.13a) and the acoustic impedance (Fig. 16.13b). In the acoustic impedance image, AIM clearly visualized two-dimensional color distribution of the acoustic impedance of the sound

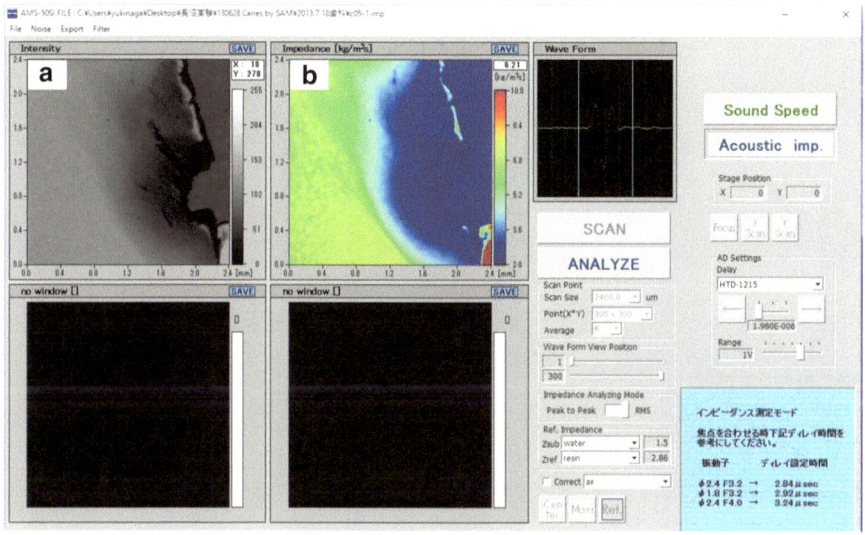

Fig. 16.13 PC screen of AIM image of dentin caries. Two-dimensional color distribution with the resolution of the 2.4×2.4 mm sample area is visualized with 300×300 pixels. Acoustic intensity image (**a**), acoustic impedance (**b**) image

dentin as well as the dentin caries. The gradation color of the dentin caries (Fig.16.14a) differed from that of the sound dentin (Fig. 16.14b). The dentin caries was composed of low acoustic impedance areas (blue to light blue). However, the sound dentin was composed of higher acoustic impedance areas (green to yellow) than the carious dentin lesion. The comparison of the acoustic impedance of both sound dentin and dentin caries is shown in Fig. 16.15. The acoustic impedance of the dentin caries was significantly lower than that of the sound dentin (sound dentin vs. dentin caries; 6.75 ± 0.14 kg/m^{2}s vs. 2.47 ± 0.41 kg/m^{2}s; $p = 2.29 \times 10^{-12}$).

16.3.2.2 SEM Observation of Dentin Caries

SEM images are shown in Fig. 16.16. The sound dentin areas showed smooth surface, and the transverse sections of dentinal tubules were aligned regularly (Fig. 16.16a). However, surface erosion and destruction were detected on the dentin caries. The dentinal tubules were aligned irregularly, and the cross sections of dentinal tubules showed wider and narrower on the dentin caries than those on the sound dentin (Fig. 16.16b).

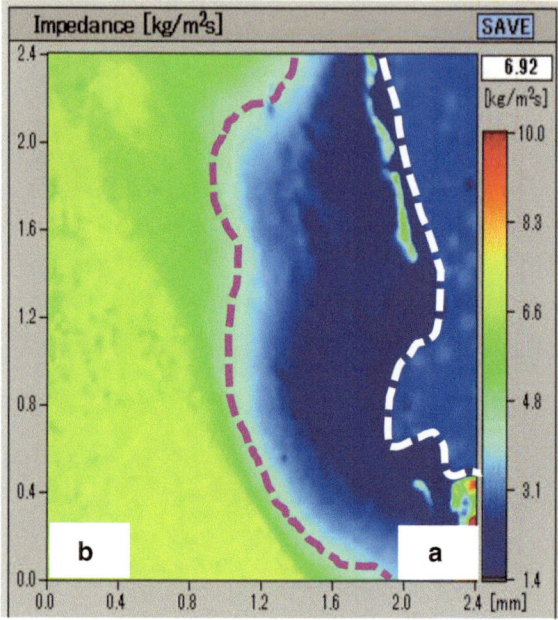

Fig. 16.14 Acoustic impedance image of the carious dentin lesion. Sound dentin area consists of higher acoustic impedance area (*green* to *yellow*) (**a**). Dentin caries consists of low acoustic impedance area (*light blue* to *blue*) (**b**)

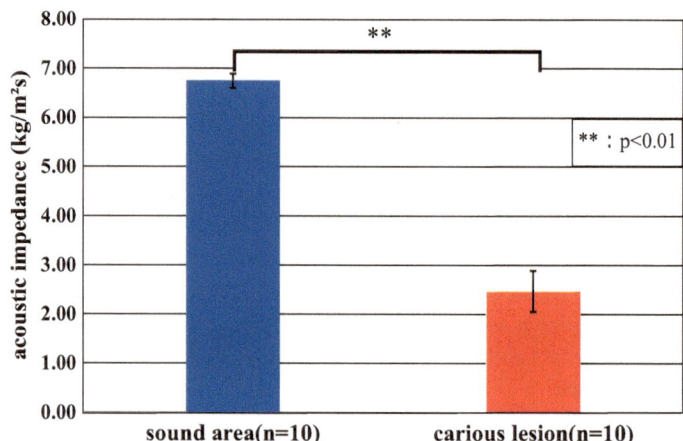

Fig. 16.15 Bar graph of the acoustic impedance between sound dentin area and carious dentin lesion. *Blue bar* is the mean acoustic impedance of sound dentin. *Red bar* is the mean acoustic impedance of dentin caries

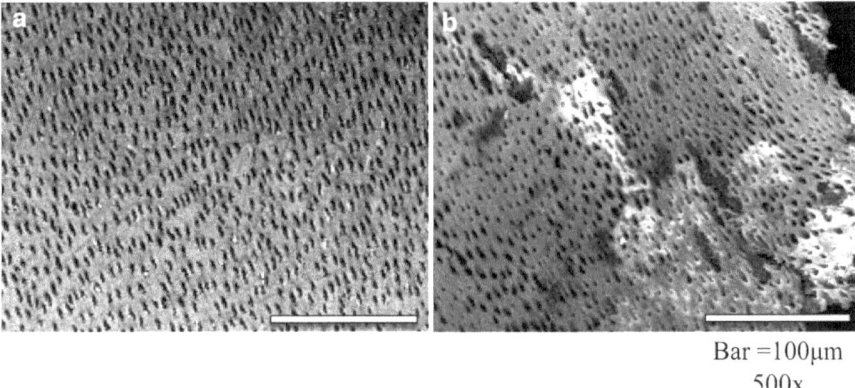

Bar =100μm
500x

Fig. 16.16 SEM images of sound dentin (**a**) and dentin caries (**b**). 500×original magnification. Bar=100 μm

16.3.3 Ultrasound Stiffness Checker

Figure 16.17 shows the actual measurement of human sound tooth. Since the probe surface is flat, it is difficult to measure the dental carious cavity. Many improvements are necessary to apply the diagnosis for dental caries.

16.4 Discussion

16.4.1 SAM Observation of PDL

In the present study, our SAM clearly visualized the gradation color images and distribution of rat PDL. SAM images were equal to histological appearances obtained by light microscopy. These results are the first time that SAM is applied to investigate the elastic property of PDL.

Previous studies reported that the elastic property of PDL was decreased along with development [28–31]. In *in vivo* studies, the elastic property of PDL were investigated by the displacement of the rat molar under the mechanical loads [28, 31] and the measurement of the load by pushing the rat incisor out of alveolar socket [29, 30]. However, these methods measured not only the elastic property of PDL but also the elastic property of alveolar bone. Meanwhile, in *in vitro* studies, numerical analyses using finite element model (FEM) study did not precisely reflect the elastic property of PDL [32, 33], because the occlusal conditions are not fully considered.

In the present study, the sound speed propagating in PDL was significantly increased in a course of development. These results indicate that the elasticity of PDL is decreased along with rat development; in other words PDL becomes harder during development. Previous studies reported that collagens such as type I, III, and

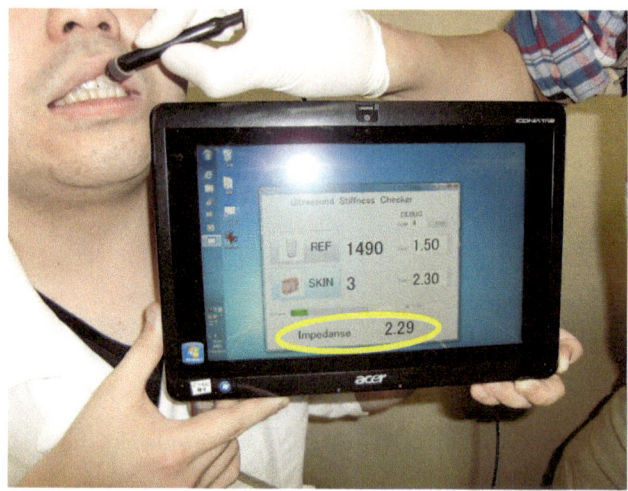

Fig. 16.17 Actual examination of human teeth. Ultrasound stiffness checker shows the ultrasound impedance of the tooth in the *yellow circle*

XII collagen were implicated in the regulation of the development and/or maturation of PDL [34–36]. And other previous studies reported that non-collagens such as fibroblast growth factor (FGF), alkaline phosphatase, and connective tissue growth factor (CCN/CTGF) were implicated in the regulation of the development and/or maturation of PDL [36–38]. Therefore, the present study indicates that the extracellular matrix molecules may play a pivotal role to regulate the elasticity of PDL.

Previous studies reported that occlusal force influence to mature PDL [35, 39]. In a study, the occlusal forces induced to promote the collagens synthesis [35]. In another study, the occlusal forces induced to express insulin-like growth factor-1. These collagens and growth factor had influence on the development and/or maturation of PDL [39]. In the present study, MRR elasticity was lower than IRR elasticity in 5-week-old. We speculate that in order to resist the lateral loading of occlusal force MRR became harder to bind firmly between the tooth root and the alveolar bony socket.

The present study indicates that the decreased elasticity of PDL is induced by the morphological development such as the production of collagen(s) and/or proteoglycan(s) and the functional development such as the occlusal construction and/or the development of the masticatory muscles.

SAM is a powerful tool for evaluating the elasticity and its distribution in the targeted tissues *in situ*. This study suggests that the increased elasticity of PDL is one of the developmental processes of the periodontal tissues.

16.4.2 AIM Observation of Dentin Caries

To date, AIM has been mainly applied for the observation of soft tissues such as synovial membrane [13] and coronary artery [20]. In the present study, AIM was able to clearly visualize the color distribution of the acoustic impedance of both the sound and carious dentin. Moreover, AIM enabled to evaluate the elastic property calculating the acoustic impedance of the sample. Taken together these results, we propose that AIM may be useful to diagnose the dental caries visually and numerically.

Previous study reported that the Vickers hardness of the carious dentin lesion was significantly lower than that of the sound dentin areas [40]. In the present study, the acoustic impedance of the carious dentin lesion was significant lower than in the sound dentin area. The Vickers hardness test is a destructive examination; therefore, there are limitations to apply the measurement of the Vickers hardness to diagnose human caries. The ultrasonic testing is a nondestructive examination, so the calculation of the acoustic impedance of the dentin caries may be a noninvasive and safe method to diagnose human caries.

SEM observation revealed that caries induced microstructural deterioration of dentin. This result suggested that acid produced by oral bacterial metabolism decalcified the Ca element included in dentin.

To understand the elastic property due to the pathological change is important information to diagnose the disease in medicine and dentistry. It has been well known that caries make dentin soft. But, unfortunately, it has not established to evaluate precisely the elasticity of the carious dentin. In the present study, AIM enabled to visualize the dentin caries clearly and evaluate it numerically. Based on the principle of AIM, we are going to complete a novel portable acoustic caries detector.

Acknowledgments We wish to thank Mr. Katsuyoshi Shoji, Department of Orthopaedic Surgery, Tohoku University Graduate School of Medicine, for his excellent assistance. This work was supported by MEXT KAKENHI Grant Number JP25462982.

References

1. Saijo Y, Tanaka M, Okawai H, Dunn F. The ultrasonic properties of gastric cancer tissues obtained with a scanning acoustic microscope system. Ultrasound Med Biol. 1991;17:709–14.
2. Sasaki H, Tanaka M, Saijo Y, et al. Ultrasonic tissue characterization of renal cell carcinoma tissue. Nephron. 1996;74:125–30.
3. Saijo Y, Tanaka M, Okawai H, Sasaki H, Nitta SI, Dunn F. Ultrasonic tissue characterization of infarcted myocardium by scanning acoustic microscopy. Ultrasound Med Biol. 1997;23:77–85.
4. Sasaki H, Saijo Y, Tanaka M, et al. Acoustic properties of dialysed kidney by scanning acoustic microscopy. Nephrol Dial Transplant Off Publ Eur Dial Transplant Assoc – Eur Ren Assoc. 1997;12:2151–4.

5. Saijo Y, Sasaki H, Okawai H, Nitta S, Tanaka M. Acoustic properties of atherosclerosis of human aorta obtained with high-frequency ultrasound. Ultrasound Med Biol. 1998;24:1061–4.
6. Saijo Y, Sasaki H, Sato M, Nitta S, Tanaka M. Visualization of human umbilical vein endothelial cells by acoustic microscopy. Ultrasonics. 2000;38:396–9.
7. Saijo Y, Ohashi T, Sasaki H, Sato M, Jorgensen CS, Nitta S. Application of scanning acoustic microscopy for assessing stress distribution in atherosclerotic plaque. Ann Biomed Eng. 2001;29:1048–53.
8. Sasaki H, Saijo Y, Tanaka M, Nitta S. Influence of tissue preparation on the acoustic properties of tissue sections at high frequencies. Ultrasound Med Biol. 2003;29:1367–72.
9. Saijo Y, Miyakawa T, Sasaki H, Tanaka M, Nitta S. Acoustic properties of aortic aneurysm obtained with scanning acoustic microscopy. Ultrasonics. 2004;42:695–8.
10. Sano H, Saijo Y, Kokubun S. Material properties of the supraspinatus tendon at its insertion- a measurement with the scanning acoustic microscopy. J Musculoskelet Res. 2004;8:29–34.
11. Saijo Y, Sasaki H, Hozumi N, Kobayashi K, Tanaka M, Yambe T. Sound speed scanning acoustic microscopy for biomedical applications. Technol Health Care: Off J Eur Soc Eng Med. 2005;13:261–7.
12. Hozumi N, Yamashita R, Lee CK, et al. Time-frequency analysis for pulse driven ultrasonic microscopy for biological tissue characterization. Ultrasonics. 2004;42:717–22.
13. Hagiwara Y, Saijo Y, Chimoto E, et al. Increased elasticity of capsule after immobilization in a rat knee experimental model assessed by scanning acoustic microscopy. Ups J Med Sci. 2006;111(3):303–13.
14. Sano H, Hattori K, Saijo Y, Kokubun S. Does decalcification alter the tissue sound speed of rabbit supraspinatus tendon insertion? In vitro measurement using scanning acoustic microscopy. Ultrasonics. 2006;44:297–301.
15. Berkowitz BKB, Holland GR, Moxham BJ. Periodontal ligament. In: Berkowitz BKB, Holland GR, Moxham BJ, editors. Oral anatomy, histology and embryology. 3rd ed. London: Mosby; 2007. p. 180–204.
16. Kaku M, Yamauchi M. Mechano-regulation of collagen biosynthesis in periodontal ligament. J Prosthodont Res. 2014;58:193–207.
17. Komatsu K. Mechanical strength and viscoelastic response of the periodontal ligament in relation to structure. J Dent Biomech. 2010;2010:502318.
18. Komatsu K, Chiba M. The effect of velocity of loading on the biomechanical responses of the periodontal ligament in transverse sections of the rat molar in vitro. Arch Oral Biol. 1993;38(5):369–75.
19. Komatsu K, Kanazashi M, Shimada A, Shibata T, Viidik A, Chiba M. Effects of age on the stress-strain and stress-relaxation properties of the rat molar periodontal ligament. Arch Oral Biol. 2004;49(10):817–24.
20. Saijo Y, Hozumi N, Kobayashi K, et al. Ultrasound speed and impedance microscopy for in vivo imaging. Conference proceedings: annual international conference of the IEEE engineering in medicine and biology society IEEE engineering in medicine and biology society conference 2007. 2007:1350–3.
21. Kobayashi K, Yoshida S, Saijo Y, Hozumi N. Acoustic impedance microscopy for biological tissue characterization. Ultrasonics. 2014;54:1922–8.
22. Gunawan AI, Hozumi N, Yoshida S, Saijo Y, Kobayashi K, Yamamoto S. Numerical analysis of ultrasound propagation and reflection intensity for biological acoustic impedance microscope. Ultrasonics. 2015;61:79–87.
23. Usha CSR. Dental caries – a complete changeover (part I). J Conserv Dent. 2009;12(2):46–54.
24. Shimizu A, Nakashima S, Nikaido T, Sugawara T, Yamamoto T, Momoi Y. Newly developed hardness testing system, "Cariotester": measurement principles and development of a program for measuring knoop hardness of carious dentin. Dent Mater J. 2013;32(4):643–7.
25. Kodaka T, Debari K, Yamada M. Correlation between microhardness and mineral content in sound human dentin. J Showa Univ Dent Soc. 1998;18(2):199–201.

26. O'Brien Jr WD, Olerud J, Shung KK, Reid JM. Quantitative acoustical assessment of wound maturation with acoustic microscopy. J Acoust Soc Am. 1981;69(2):575–9.
27. Saijo Y, Okawai H, Sasaki H, et al. Evaluation of the inner-surface morphology of an artificial heart by acoustic microscopy. Artif Organs. 2000;24(1):64–9.
28. Picton DC, Wills DJ. Viscoelastic properties of the periodontal ligament and mucous membrane. J Prosthet Dent. 1978;40(3):263–72.
29. Yamane A. The effect of age on the mechanical properties of the periodontal ligament in the incisor teeth of growing young rats. Gerodontology. 1990;9(1):9–16.
30. Pini M, Wiscott HW, Scherrer SS, Botsis J, Belser UC. Mechanical characterization of bovine periodontal ligament. J Periodontal Res. 2002;37(4):237–44.
31. Kawarizadeh A, Bourauel C, Jager A. Experimental and numerical determination of initial tooth mobility and material properties of the periodontal ligament in rat molar specimens. Eur J Orthod. 2003;25(6):569–78.
32. Komatsu K. In vitro mechanics of the periodontal ligament in impeded and unimpeded rat mandibular incisors. Arch Oral Biol. 1988;32(4):249–55.
33. Chiba M, Yamane A, Oshima S, Komatsu K. In vitro measurement of regional differences in the mechanical properties of the periodontal ligament in the rat incisor. Arch Oral Biol. 1990;35(2):153–61.
34. Huang YH, Ohsaki Y, Kurisu K. Distribution of type I and III collagen in the developing periodontal ligament of mice. Matrix. 1991;11(1):25–35.
35. Karimbux NY, Rosenblum ND, Nishimura I. Site-specific expression of collagen I and XII mRNA in the rat periodontal ligament at two developmental stages. J Dent Res. 1992;71(7):1355–62.
36. Ozaki S, Kaneko S, Podyma-Inoue KA, Yanagishita M, Soma K. Modulation of extracellular matrix synthesis and alkaline phosphatase activity of periodontal ligament cells by mechanical stress. J Periodontal Res. 2005;40(2):110–7.
37. McCulloch CA, Bordin S. Role of fibroblast subpopulations in periodontal physiology and pathology. J Periodontal Res. 1991;26(3):144–54.
38. Asano M, Kubota S, Nakanishi T, Nishida T, Yamaai T, Yosimichi G, Ohyama K, Sugimoto T, Murayama Y, Takigawa M. Effect of connective tissue growth factor (CCN/CTGF) on proliferation and differentiation of mouse periodontal ligament-derived cells. Cell Commun Signal. 2005;5(3):11.
39. Termsuknirandorn S, Hosomichi J, Soma K. Occlusal stimuli influence on the expression of IGF-1 and the IGF-1 receptor in the rat periodontal ligament. Angle Orthod. 2008;78(4):610–6.
40. Oikawa M, Itoh K, Kusunoki M, Kitanaka N, Hasegawa T. Comparison of rotary and ultrasonic caries removal determined by two fluorescent caries detection devices. Jpn J Conserv Dent. 2013;56(1):78–84.

Open Access This chapter is distributed under the terms of the Creative Commons Attribution 4.0 International License (http://creativecommons.org/licenses/by/4.0/), which permits use, duplication, adaptation, distribution and reproduction in any medium or format, as long as you give appropriate credit to the original author(s) and the source, provide a link to the Creative Commons license and indicate if changes were made.

The images or other third party material in this chapter are included in the work's Creative Commons license, unless indicated otherwise in the credit line; if such material is not included in the work's Creative Commons license and the respective action is not permitted by statutory regulation, users will need to obtain permission from the license holder to duplicate, adapt or reproduce the material.

Part V
Poster Presentation Award Winners

Chapter 17
Activation of TLR3 Enhance Stemness and Immunomodulatory Properties of Periodontal Ligament Stem Cells (PDLSCs)

Nuttha Klincumhom, Daneeya Chaikeawkaew, Supanniga Adulheem, and Prasit Pavasant

Abstract Periodontal ligament stem cells (PDLSCs) have been served as a cell reservoir for tissue regeneration during adulthood. For clinical applications, the challenging steps are to maintain the stem cell properties and to improve the regeneration capacity of PDLSCs during culture. Toll-like receptor 3 (TLR3) signaling has been shown to enhance therapeutic potential in several cell types including mesenchymal stem cells (MSCs) by inducing the secretion of multifunctional trophic factors. However, the role of TLR3 in PDLSCs is still unknown. The aim of this study was to investigate the responses of PDLSCs after TLR3 engagement using TLR3 agonist, poly(I:C). The result indicated that stimulation of TLR3 significantly enhanced pluripotent stem cell gene expression (e.g., REX-1 and SOX2) as well as immunomodulatory molecules (e.g., IFNγ and IDO). Interestingly, inhibition of NF-kB signaling decreased the TLR3-activated IFNγ but increased the TLR3-activated IDO expression, suggesting the multiple pathways in the inductive mechanism. Our finding supports the concept that activated TLR3 could encourage the stem cell and immunosuppressive properties of PDLSCs. Since immunosuppressive properties of stem cells could support tissue healing and regeneration, activation of TLR3 in PDL cells may trigger the effective PDL tissue regeneration.

Keywords TLR3 • Immunosuppressive property • PDLSCs • Stem cell markers

N. Klincumhom • P. Pavasant (✉)
Mineralized Tissue Research Unit, Faculty of Dentistry, Chulalongkorn University, 34 Henri-Dunant road, Pathumwan, Bangkok 10330, Thailand

Department of Anatomy, Faculty of Dentistry, Chulalongkorn University, 34 Henri-Dunant road, Pathumwan, Bangkok 10330, Thailand
e-mail: prasit215@gmail.com

D. Chaikeawkaew • S. Adulheem
Department of Anatomy, Faculty of Dentistry, Chulalongkorn University, 34 Henri-Dunant road, Pathumwan, Bangkok 10330, Thailand

© The Author(s) 2017
K. Sasaki et al. (eds.), *Interface Oral Health Science 2016*,
DOI 10.1007/978-981-10-1560-1_17

17.1 Introduction

Toll-like receptor 3 (TLR3) is a member of TLR family, the group of pattern recognition receptors (PRRs) that recognizes double-stranded RNA (dsRNA) produced by positive-strand RNA viruses and DNA viruses, resulting in pathogen clearance and recruitment of adaptive immune response [1]. TLR3-dsRNA interaction generally induces the expression and secretion of IFNγ and other proinflammatory cytokines through the Toll-interleukin-1 receptor domain-containing adaptor molecule-1 (TICAM-1, also known as TRIF) [2]. Recent study revealed that stimulation of TLR3 signaling by poly(I:C), the specific TLR3 agonist, enhanced immunosuppressive ability of human umbilical cord-derived mesenchymal stem cells (UC-MSCs) through microRNA-143 inhibition [3]. Nonetheless, the therapeutic advantages of TLR3 activation are not only limited to its immunoregulatory response but also to its regenerative capacity by secretion of multifunctional trophic factors [4]. In bone marrow-derived MSCs (BM-MSCs), treatment with poly(I:C) potentially augmented the therapeutic capability for hamster heart failure through induction of relevant trophic factors such as interleukin-6 (IL-6), STO-1, hepatic growth factor (HGF), and vascular endothelial growth factor (VEGF) [5].

There is evidence suggesting that TLR3 may function as sensor to perceive the signal from damaged tissues upon injury. Literatures indicate that dsRNA released from damaged and necrotic cells can activate TLR3, resulting in an acceleration of the reepithelialization for wound closure. Moreover, dsRNA-induced TLR3 activation could promote the release of several cytokines including IL-6, IL-11, LIF, IL-10, SDF-1, VEGF, and HGF [6, 7]. However, the release of these cytokines is dependent on the type of cells and the concentration of ligand.

Periodontal ligament stem cells (PDLSCs) are the MSC-like cell population resided in periodontal ligament tissue, the connective tissue connected tooth root to the surrounding alveolar bone. Thus, PDLSCs have the potential to regenerate multiple cells comprising periodontal structures such as cementum, bone, and ligament and may pave the way for periodontal regenerative treatment via cell-based transplantation [8–10]. Apart from regenerative properties, the potential of PDLSCs for therapeutic application regarding to their immune privilege and immunosuppressive capability has been reported recently [11]. The immunomodulatory effect of DAMP-activated TLR3 has been indicated in periodontopathic gingival cells by stimulating TLR2-mediated inflammatory response to infected pathogen [12].

It has been proposed that for tissue regeneration, three interrelated events must be occurred. First, our body must sense the loss of tissue integrity, followed by the signals to induce the migration of precursor cells to reconstitute missing structures. Finally, these migrating cells must be signaled for an appropriate differentiation or functions [13]. However, upon infection or injury of PDLSCs, the signal that triggers PDLSC function for tissue regeneration has not been dignified yet. In this study, to imitate TLR3-dsRNA recognition, poly(I:C) was used for stimulating TLR3 in PDLSCs. The responses of PDLSCs after poly(I:C) treatment were investigated regarding to their stemness and immunosuppressive properties.

17.2 Materials and Methods

17.2.1 PDLSC Culture

Human PDLSCs were collected from healthy third molars as previously described [14]. The protocol was approved by the Ethical Committee, Faculty of Dentistry, Chulalongkorn University. PDLSCs were expanded in growth medium containing Dulbecco's Modified Eagle's Medium supplemented with 10% fetal bovine serum, 2 mM L-Glutamine, 100 U/ml penicillin, 100 µg/ml streptomycin, and 5 µg/ml amphotericin B. All cell culture reagents were purchased from Gibco BRL (Carls-bad, CA). Human PDLSCs were characterized for MSC properties as our previous report [15].

For osteogenic differentiation, PDLSCs were cultured in osteogenic medium which was normal growth medium supplemented with 50 µg/ml ascorbic acid, 100 nM dexamethasone, and 10 mM β-glycerophosphate.

17.2.2 Poly(I:C) Treatment

For TLR3 activation, PDLSCs were seeded at a density of 1.2×10^5 cells/well in 12-well plate in growth medium for 24 h and then treated with poly(I:C) at concentration of 5, 10 and 25 µg/ml for 24 h (InvivoGen, San Diego, CA) in fresh growth medium. PDLSCs cultured in medium without poly(I:C) were used as control.

17.2.3 Mineralization Assay

Detection of calcium deposition was evaluated on day 14 after cultured in osteogenic medium using the Alizarin red S staining assay. Briefly, cells were washed with phosphate-buffered saline (PBS) prior to being fixed with cold methanol for 10 min and stained with 1% Alizarin red S staining solution (Sigma-Aldrich, St. Louis, MO).

17.2.4 Gene Expression Analysis

Total amount of RNA was extracted using Trizol reagent and 1 µg of RNA was converted into complementary DNA (cDNA) using a reverse transcriptase enzyme kit (Promega, Madison, WI). Conventional PCR was performed using Taq polymerase (Biotechrabbit Gmbh, Hennigsdorf, Germany) in a DNA thermal cycler (Biometra GmH, Göttingen, Germany). The products were electrophoresed on a 2% agarose gel and visualized using ethidium bromide (EtBr; Bio-Rad, Hercules, CA).

Table 17.1 Primer sequences for mRNA expression analysis (conventional PCR, qPCR)

Gene		Sequence 5–3′	Product size (bp)	Sequence ID
GAPDH	F	5′-TGAAGGTCGGAGTCAACGGAT-3′	396	NM_002046.5
	R	5′-TCACACCCATGACGAACATGG-3′		
OC	F	5′-CTTTGTGTCCAAGCAGGAGG-3′	165	NM_199173.4
	R	5′-CTGAAAGCCGATGTGGTCAG-3′		
NANOG	F	5′- CAGCCCCGATTCTTCCACCAGTC-3′	389	NM_024865.3
	R	5′- CGGAAGATTCCCAGTCGGGTTCA-3′		
OCT4	F	5′-GCAACCTGAGAATTTGTTCCT-3′	173	NM_002701.5
	R	5′-AGACCCAGCAGCCTCAAAATC-3′		
REX1	F	5′-CAGATCCTAAACAGCTCGCAGAA-3′	305	NM_174900.4
	R	5′-GCGTACGCAAATTAAAGTCCAGA-3′		
SOX2	F	5′-ACCAGCTCGCAGACCTACAT-3′	319	NM_003106.3
	R	5′-ATGTGTGAGAGGGGCAGTGT-3′		
TLR3	F	5′-AAATTGGGCAAGAACTCACAGG-3′	319	NM_003265.2
	R	5′-GTGTTTCCAGAGCCGTGCTAA-3′		
qOCT4	F	5′-TCGAGAACCGAGTGAGAGG-3′	124	NM_002701.5
	R	5′-GAACCACACTCGGACCACA-3′		
qNANOG	F	5′-ATGCCTCACACGGAGACTGT-3′	102	NM_024865.3
	R	5′-AAGTGGGTTGTTTGCCTTTG-3′		
qREX1	F	5′-TGGGAAAGCGTTCGTTGAGA-3′	89	NM_174900.4
	R	5′-CACCCTTCAAAAGTGCACCG-3′		
CD73	F	5′-ACACTTGGCCAGTAAAATAGGG-3′	122	NM_002526.3
	R	5′-ATTGCAAAGTGGTTCAAAGTCA-3′		
CD90	F	5′-GAAGACCCCAGTCCAGATCCA-3′	172	NM_006288.3
	R	5′-TGCTGGTATTCTCATGGCGG-3′		
CD105	F	5′-CATCACCTTTGGTGCCTTCC-3′	195	NM_001114753.2
	R	5′-CTATGCCATGCTGCTGGTGGA-3′		
IFNγ	F	5′-CTAGGCAGCCAACCTAAGCA-3′	179	NM_000619.2
	R	5′-CAGGGTCACCTGACACATTC-3′		
IDO	F	5′-CATCTGCAAATCGTGACTAAG-3′	187	NM_002164.5
	R	5′-GTTGGGTTACATTAACCTTCCTT-3′		

qPCR was performed using SYBR GREEN FastStart Essential DNA Green Master which was detected by the LightCycler 96 (Roche Diagnostic GmbH, Mannheim, Germany). GAPDH was used as a reference gene for all experimental quantification. The primer sequences for qPCR can be reviewed in Table 17.1.

17.2.5 Flow Cytometry

PDLSCs (2×10^6 cells/sample) were dissociated and resuspended in 200 μl of FAC buffer and stained with MSC surface marker comprised of FITC-conjugated anti-human CD90 (Abcam, Cambridge, MA), PE-conjugated anti-human CD105 (BD Biosciences

Pharmingen, San Diego, CA), and purified anti-human CD73 (Abcam). For CD73 detection, after incubating in primary antibody, cells were stained with biotinylated anti-mouse IgG2A followed by APC streptavidin. Stained cells will be analyzed on a FACSCalibur™ using the CellQuest™ software (BD Bioscience, San Jose, CA).

17.2.6 *Statistical Analyses*

Data were reported as Mean ± S.D. Statistical analyses of significance were evaluated using a two-tailed Student's *t*-test or analysis of variance where values of $P < 0.05$ were considered significant. The analysis was performed by the statistical software (SPSS, Chicago, IL). A minimum of three replicates were analyzed for each experiment.

17.3 Results and Discussion

In periodontitis, the unhealthy cells of sulcular epithelium as well as gingival tissue were damaged either from the effect of proinflammatory cytokines or from mechanical stimulation such as chewing or brushing [16]. As in skin wound healing that was reported previously, the damaged or dying cells could release dsRNA, resulting in the activation of TLR3 and subsequently initiating tissue regeneration [17]. In this case, damaged cells in periodontitis may activate TLR3 in PDLSCs leading to the healing and regeneration of periodontal tissue.

17.4 TLR3 Activation Promote Stemness of PDLSCs

Human PDLSCs isolated from PDL tissue expressed the MSC markers including CD73, CD90 and CD105, as well as pluripotent markers such as REX-1, OCT4, NANOG, and SOX2. They also exhibited osteogenic characteristic as presented in upregulation of osteocalcin (OC) mRNA level and calcium deposition after osteogenic induction (Fig. 17.1). These results support that our isolated PDL cells possess stem cell characteristics, referred as PDLSCs.

Next, we examined the expressions of TLR2-4 in PDL tissue and PDLSCs. PDL tissue and dental pulp and gingival tissues from healthy patients were obtained with informed consent and subjected to RT-PCR analysis for the detection of TLR2-4 expressions. The results indicated that the expression levels of TLR2-4 were detected differently depended on cell types. Our study found that PDL tissue expressed the highest level of TLR3 compared to cells from dental pulp and gingiva. Moreover, gingival tissue showed the highest level of TLR2. No difference in expression of TLR4 was observed among tissue types (Fig. 17.2a).

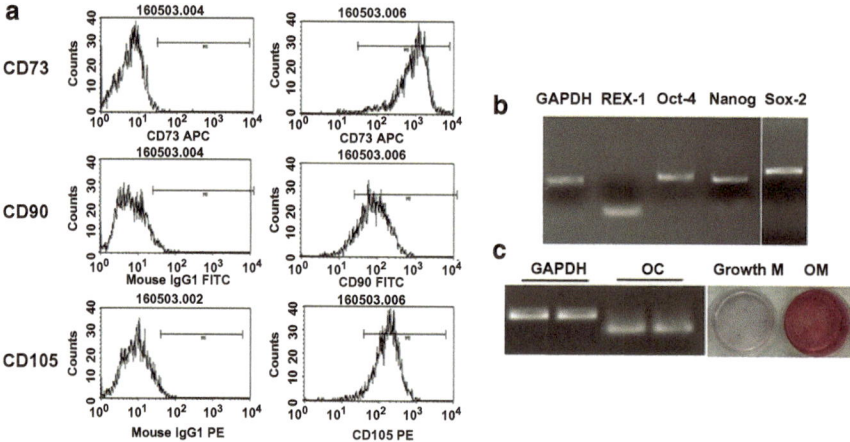

Fig. 17.1 Human PDL cells possessed the MSC characteristics. Human PDL cells were established from periodontal tissue obtained from extracted third molars. Cells from passage 3 were analyzed for the expression of mesenchymal stem cell markers. The results from FACS analysis showed that more than 90 % of cells expressed CD73, CD90, and CD105 (**a**). HPDL cells also expressed pluripotent markers such as Rex-1, Oct4, nanog, and Sox2 as judged by RT-PCR (**b**). For osteogenic differentiation ability, cells were cultured in osteogenic medium for 14 days. The increased expression of osteocalcin (*OC*) was detected. Moreover, in vitro calcification was observed as judged by alizarin red S staining (**c**). The figure was the representative results from three established PDL cell lines

As shown in Fig. 17.2, PDLSC itself promptly expressed TLR3. Poly(I:C) was used as specific TLR3 ligand to stimulate TLR3 function in PDLSCs. Cells were treated with poly(I:C) at concentration of 5, 10, and 25 µg/ml for 24 h. Our result showed the significant increased in mRNA levels of MSC-specific surface markers, CD73 and CD105, and the self-renewal markers REX-1 and SOX2 as compared to untreated control. No significant difference among poly(I:C)-treated groups was presented (Fig. 17.3a, b). Our result which corresponds with recent study indicated that activation of TLR3 by poly(I:C) promoted the cancer stem cell phenotype by upregulating the pluripotent genes including SOX2 in breast cancer [18]. Moreover, the increased mRNA level of hair progenitor-specific marker has been observed upon TLR3 activation in keratinocyte which contributed to hair follicle regeneration [17]. Attractively, TLR3 activation was required for induction of pluripotency in fibroblast by modification of epigenetic mechanism [19].

We imply that TLR3 may serve as a trigger receptor for damaged/infected response by enhancing the stemness of PDLSCs. Further determining of the signaling that involved in TLR3-acivated stem cell properties in PDLSCs would be needed to provide the better understanding of underlying mechanism of regeneration upon infection or injury. Besides, there are studies showing that the proliferation and differentiation capacities of PDLSCs had been weakened during in vitro culture expansion or in aged donors-derived, resulting in loss of proliferative and regenerative capacity [20, 21]. Here, we suggest that enhancement of PDLSC stemness by poly(I:C) treatment could concrete the way for PDLSC application in regenerative medicine.

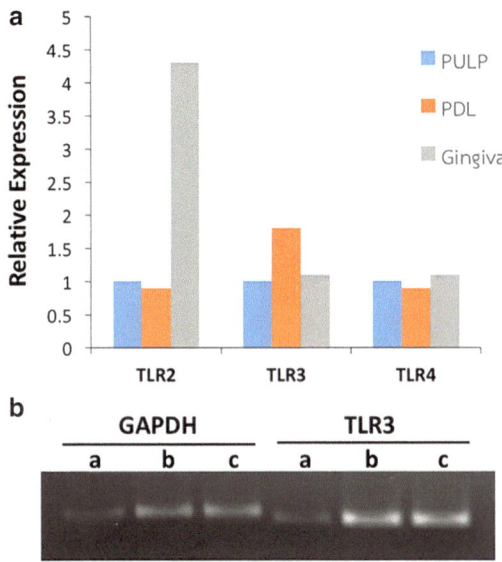

Fig. 17.2 Expression of TLR3 in periodontal tissue and PDL cells. Periodontal (*PDL*), dental pulp (pulp), and gingiva tissues were obtained from healthy patient and subject for RT-PCR analysis. The results (**a**) showed that all three tissues expressed TLR2, TLR3, and TLR4 with different levels. The highest expression of TLR3 was found in PDL tissue while gingiva expressed the highest level of TLR2. (**b**) showed the expression of TLR3 from three PDL cell lines (**a–c**). GAPDH was used as the internal control

17.5   TLR3 Activation Enhance Immunosuppressive Properties of PDLSCs

The MSC-like phenotypes of PDLSCs are not only the expression of stem cell markers, but also the immunosuppressive activity that can suppress immune responses after allogeneic transplantation [11]. The immune response interference of PDLSCs is generally processed through a cell-cell contact and secretion of immunomodulatory factors resulting in inhibition of T and B cell proliferation. Additionally, recent study indicated the immunosuppressive activity of PDLSCs is resisted even after osteogenic differentiation [22].

The role of TLR3 regarding the immunomodulation properties has been examined extensively in MSCs derived from several origins such as BM-MSCs [5], AD-MSCs [23], and UC-MSCs [3]; however, no comprehensive study in dental stem cells has been studied. In present study, after exposing to 25 µg/ml of poly(I:C) for 24 h, the upregulation of immunomodulatory mRNA levels, IFNγ, and IDO in PDLSCs was detected by qPCR (Fig. 17.4a). IFNγ is known as an upstream regulator of immunomodulatory molecules in MSCs. These cytokines promote an IDO production, resulting in restoring immunosuppressive ability of dental pulp stem cells (DPSCs) derived from irreversible pulpitis [24]. IDO is a catalytic enzyme of tryptophan, a crucial

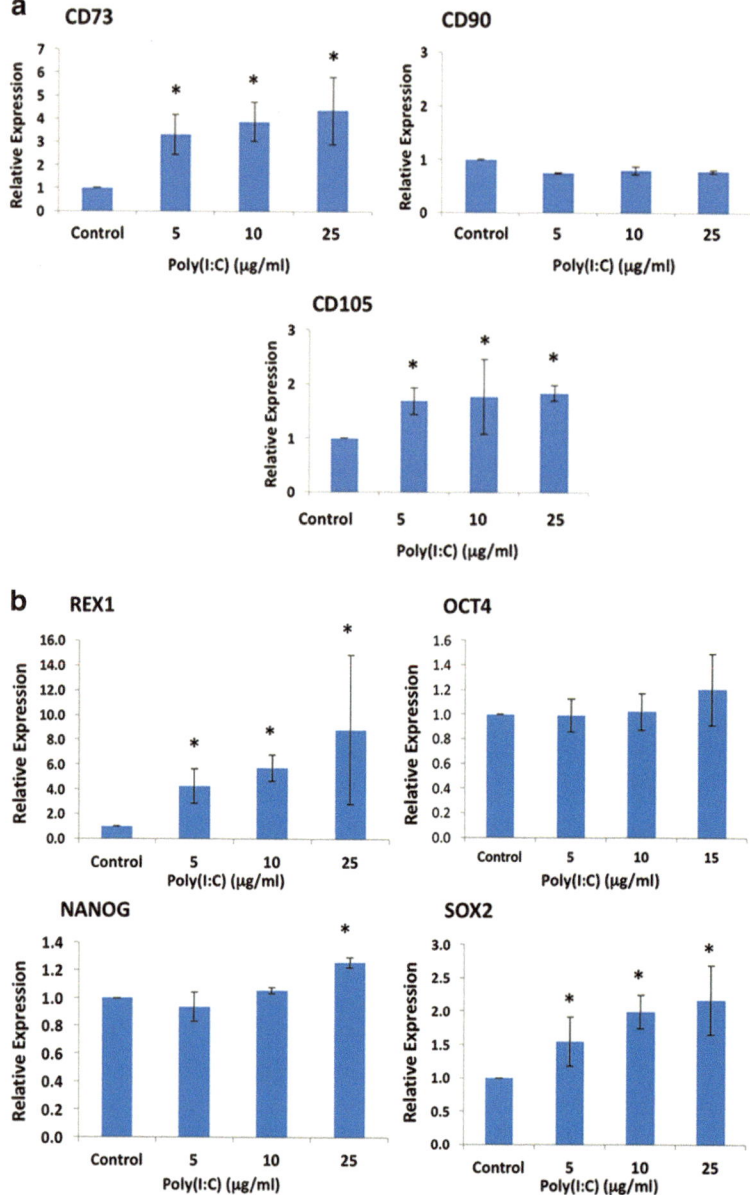

Fig. 17.3 Poly(I:C) enhanced the expression of stem cell markers in human PDL cells. Human PDL cells were treated with poly(I:C) at concentration of 5, 10 and 25 µg/ml for 24 h. Real-time RT-PCR analysis showed that poly(I:C) enhanced the expression of MSC markers; CD73, CD105 (**a**), and pluripotent markers; Rex-1, Sox2 (**b**). * indicated the significant difference ($p < 0.05$). The results were shown as mean ± S.D. from three experiments

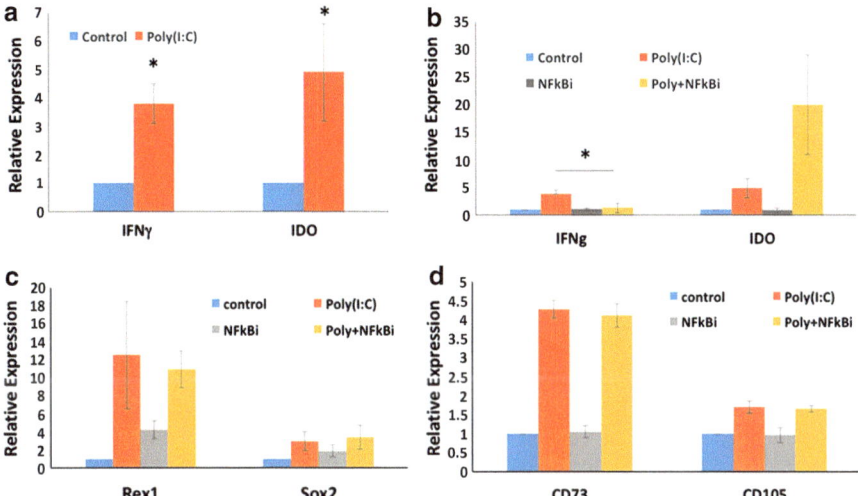

Fig. 17.4 TLR3 activation induced immunosuppressive property of human PDL cells. Human PDL cells were treated with 25 µg/ml poly(I:C) for 24 h. RT-PCR analysis revealed that poly(I:C) significantly induced the expression of IFNγ and IDO (**a**). Addition of NF-kB inhibitor inhibited the poly(I:C)-induced IFNγ but could not inhibit the induction of IDO (**b**). Moreover, poly(I:C)-induced stem cell markers were also NF-kB independent (**c, d**). * indicated the significant difference ($p < 0.05$). The results were shown as mean ± S.D. from three experiments

amino acid for T cell growth. Upregulation of IDO thus leads to the inhibition of T cell proliferation [25, 26]. In AD-MSCs, IDO is secreted in conditioned supernatant by IFNγ and potentially inhibited PBMC proliferation. Previous study indicated immunosuppressive mechanism of TLR3 in UC-MSCs through inhibiting microRNA-143 that results in IDO production [23]. We therefore denote that activation of TLR3 promotes immunosuppressive properties of PDLSCs via stimulating IFNγ and IDO. It has been shown that upon injury or inflammation of tissue, the immunosuppressive cytokines released from MSCs indicated to play a role in tissue homeostasis by inhibiting inflammatory response as well as stimulating the proliferation and differentiation of stem/progenitor cells [27]. By this cell protective mechanism, we prospect that activation of PDLSCs to secrete these immunosuppressive cytokines could augment the therapeutic potential of MSCs for tissue regeneration.

17.6   TLR3 Activation Induces IFNγ Production in PDLSCs Via NF-kB Pathway

Normally, released dsRNA after viral infection is recognized by TLR3, and then activate NF-kB signaling, through TIR domain-containing adaptor TRIF [28, 29]. To examine whether activation of TLR3 in PDLSCs regulates production of IFNγ and its downstream IDO via NF-kB signaling, the NF-kB inhibitor was added into normal medium with and without poly(I:C) treatment. The mRNA expression of immunomodulatory molecules IFNγ and IDO was determined using qPCR. The result showed that, without poly(I:C) treatment, NF-kB inhibitor did not affect the endogenous IFNγ and IDO in PDLSCs. Interestingly, pretreatment of NF-kB inhibitor could attenuate the induction of IFNγ while further increasing IDO expression when compared to those in poly(I:C) treatment alone (Fig. 17.4b). Typically, IDO is induced by IFNγ in dendritic cells and macrophages. However, the regulation of IDO expression via IFNγ-independent pathways has been reported previously in astrocyte [30]. The result revealed that TLR3 activates IFNγ production in PDLSCs via NF-kB signaling, while TLR3-activated IDO expression may be possessed by other NF-kB-independent pathways.

Additionally, TLR3-activated NF-kB nuclear translocation has been reported to play an important role in epigenetic mechanisms during reprogramming of fibroblast into induced pluripotent stem cells (iPSCs) via retrovirus encoding reprogramming factors [19]. Therefore, it is possible that TLR3-induced stemness of PDLSCs was regulated via NF-kB signaling. To test this idea, the mRNA expression of pluripotent stem cell markers (e.g., REX-1 and SOX2) and MSC markers (e.g., CD73 and CD105) of TLR3-activated PDLSCs was determined after pretreatment with NF-kB inhibitor using qPCR. We found that addition of NF-kB inhibitor had no effect on the induction of those stem cell markers in either with poly(I:C) activation or NF-kB inhibitor alone in PDLSCs (Fig. 17.4c, d), though more studies are required for better understanding on how TLR3 signaling regulates stemness in PDLSCs.

17.7 Conclusion

In summary, our study discovered that TLR3 engagement by its specific ligand poly(I:C) could stimulate the stem cell properties of PDLSCs including self-renewal and immunomodulation. In periodontitis, it is possible that PDLSCs might generate new cells and prevent the immune cell-initiated inflammation for repairing the damaged structure through these mechanisms. Hence, TLR3 agonist may be a novel approach to enhance stem cell properties of PDLSCs for tissue regeneration, especially allogeneic cell-based therapy.

Acknowledgments This work was supported by Research Chair Grant 2012, the National Science and Technology Development Agency (NSTDA), Thailand. N.K. was supported by Grants for Development of New Faculty Staff, Ratchadaphiseksomphot Endowment Fund, Chulalongkorn University and the ASAHI Glass Foundation.

References

1. Alexopoulou L, Holt AC, Medzhitov R, Flavell RA. Recognition of double-stranded RNA and activation of NF-kappaB by toll-like receptor 3. Nature. 2001;413(6857):732–8.
2. O'Neill LA, Golenbock D, Bowie AG. The history of toll-like receptors – redefining innate immunity. Nat Rev Immunol. 2013;13(6):453–60.
3. Zhao X, Liu D, Gong W, Zhao G, Liu L, Yang L, Hou Y. The toll-like receptor 3 ligand, poly(I:C), improves immunosuppressive function and therapeutic effect of mesenchymal stem cells on sepsis via inhibiting MiR-143. Stem Cells. 2014;32(2):521–33.
4. Delarosa O, Dalemans W, Lombardo E. Toll-like receptors as modulators of mesenchymal stem cells. Front Immunol. 2012;3:182.
5. Mastri M, Shah Z, McLaughlin T, Greene CJ, Baum L, Suzuki G, Lee T. Activation of toll-like receptor 3 amplifies mesenchymal stem cell trophic factors and enhances therapeutic potency. Am J Physiol Cell Physiol. 2012;303(10):C1021–33.
6. Kariko K, Bhuyan P, Capodici J, Weissman D. Small interfering RNAs mediate sequence-independent gene suppression and induce immune activation by signaling through toll-like receptor 3. J Immunol. 2004;172(11):6545–9.
7. Lai Y, Di Nardo A, Nakatsuji T, Leichtle A, Yang Y, Cogen AL, Wu ZR, Hooper LV, Schmidt RR, von Aulock S, et al. Commensal bacteria regulate toll-like receptor 3-dependent inflammation after skin injury. Nat Med. 2009;15(12):1377–82.
8. Seo BM, Miura M, Gronthos S, Bartold PM, Batouli S, Brahim J, Young M, Robey PG, Wang CY, Shi S. Investigation of multipotent postnatal stem cells from human periodontal ligament. Lancet. 2004;364(9429):149–55.
9. Gronthos S, Mrozik K, Shi S, Bartold PM. Ovine periodontal ligament stem cells: isolation, characterization, and differentiation potential. Calcif Tissue Int. 2006;79(5):310–7.
10. Shi S, Bartold PM, Miura M, Seo BM, Robey PG, Gronthos S. The efficacy of mesenchymal stem cells to regenerate and repair dental structures. Orthod Craniofac Res. 2005;8(3):191–9.
11. Wada N, Menicanin D, Shi S, Bartold PM, Gronthos S. Immunomodulatory properties of human periodontal ligament stem cells. J Cell Physiol. 2009;219(3):667–76.
12. Mori K, Yanagita M, Hasegawa S, Kubota M, Yamashita M, Yamada S, Kitamura M, Murakami S. Necrosis-induced TLR3 activation promotes TLR2 expression in gingival cells. J Dent Res. 2015;94(8):1149–57.
13. Brockes JP, Kumar A, Velloso CP. Regeneration as an evolutionary variable. J Anat. 2001;199(Pt 1–2):3–11.
14. Pattamapun K, Tiranathanagul S, Yongchaitrakul T, Kuwatanasuchat J, Pavasant P. Activation of MMP-2 by Porphyromonas gingivalis in human periodontal ligament cells. J Periodontal Res. 2003;38(2):115–21.
15. Sawangmake C, Nowwarote N, Pavasant P, Chansiripornchai P, Osathanon T. A feasibility study of an in vitro differentiation potential toward insulin-producing cells by dental tissue-derived mesenchymal stem cells. Biochem Biophys Res Commun. 2014;452(3):581–7.
16. Cekici A, Kantarci A, Hasturk H, Van Dyke TE. Inflammatory and immune pathways in the pathogenesis of periodontal disease. Periodontol 2000. 2014;64(1):57–80.
17. Nelson AM, Reddy SK, Ratliff TS, Hossain MZ, Katseff AS, Zhu AS, Chang E, Resnik SR, Page C, Kim D, et al. dsRNA released by tissue damage activates TLR3 to drive skin regeneration. Cell Stem Cell. 2015;17(2):139–51.
18. Jia D, Yang W, Li L, Liu H, Tan Y, Ooi S, Chi L, Filion LG, Figeys D, Wang L. beta-Catenin and NF-kappaB co-activation triggered by TLR3 stimulation facilitates stem cell-like phenotypes in breast cancer. Cell Death Differ. 2015;22(2):298–310.
19. Lee J, Sayed N, Hunter A, Au KF, Wong WH, Mocarski ES, Pera RR, Yakubov E, Cooke JP. Activation of innate immunity is required for efficient nuclear reprogramming. Cell. 2012;151(3):547–58.

20. Baxter MA, Wynn RF, Jowitt SN, Wraith JE, Fairbairn LJ, Bellantuono I. Study of telomere length reveals rapid aging of human marrow stromal cells following in vitro expansion. Stem Cells. 2004;22(5):675–82.
21. Sawa Y, Phillips A, Hollard J, Yoshida S, Braithwaite MW. The in vitro life-span of human periodontal ligament fibroblasts. Tissue Cell. 2000;32(2):163–70.
22. Tang R, Wei F, Wei L, Wang S, Ding G. Osteogenic differentiated periodontal ligament stem cells maintain their immunomodulatory capacity. J Tissue Eng Regen Med. 2014;8(3):226–32.
23. Lombardo E, DelaRosa O, Mancheno-Corvo P, Menta R, Ramirez C, Buscher D. Toll-like receptor-mediated signaling in human adipose-derived stem cells: implications for immunogenicity and immunosuppressive potential. Tissue Eng Part A. 2009;15(7):1579–89.
24. Sonoda S, Yamaza H, Ma L, Tanaka Y, Tomoda E, Aijima R, Nonaka K, Kukita T, Shi S, Nishimura F, et al. Interferon-gamma improves impaired dentinogenic and immunosuppressive functions of irreversible pulpitis-derived human dental pulp stem cells. Sci Rep. 2016;6:19286.
25. Meisel R, Zibert A, Laryea M, Gobel U, Daubener W, Dilloo D. Human bone marrow stromal cells inhibit allogeneic T-cell responses by indoleamine 2,3-dioxygenase-mediated tryptophan degradation. Blood. 2004;103(12):4619–21.
26. Haddad R, Saldanha-Araujo F. Mechanisms of T-cell immunosuppression by mesenchymal stromal cells: what do we know so far? Biomed Res Int. 2014;2014:216806.
27. Wang Y, Chen X, Cao W, Shi Y. Plasticity of mesenchymal stem cells in immunomodulation: pathological and therapeutic implications. Nat Immunol. 2014;15(11):1009–16.
28. Kawai T, Akira S. Signaling to NF-kappaB by toll-like receptors. Trends Mol Med. 2007;13(11):460–9.
29. Sen GC, Sarkar SN. Transcriptional signaling by double-stranded RNA: role of TLR3. Cytokine Growth Factor Rev. 2005;16(1):1–14.
30. Suh HS, Zhao ML, Rivieccio M, Choi S, Connolly E, Zhao Y, Takikawa O, Brosnan CF, Lee SC. Astrocyte indoleamine 2,3-dioxygenase is induced by the TLR3 ligand poly(I:C): mechanism of induction and role in antiviral response. J Virol. 2007;81(18):9838–50.

Open Access This chapter is distributed under the terms of the Creative Commons Attribution 4.0 International License (http://creativecommons.org/licenses/by/4.0/), which permits use, duplication, adaptation, distribution and reproduction in any medium or format, as long as you give appropriate credit to the original author(s) and the source, provide a link to the Creative Commons license and indicate if changes were made.

The images or other third party material in this chapter are included in the work's Creative Commons license, unless indicated otherwise in the credit line; if such material is not included in the work's Creative Commons license and the respective action is not permitted by statutory regulation, users will need to obtain permission from the license holder to duplicate, adapt or reproduce the material.

Chapter 18
Influence of Exogenous IL-12 on Human Periodontal Ligament Cells

Benjar Issaranggun Na Ayuthaya and Prasit Pavasant

Abstract Periodontal disease is the most prevalent oral disease. The pathogenesis of this disease is mostly due to the robust host immune response, leading to the destruction of tooth-supporting tissue. Several cytokines have been shown to play roles in pathogenesis of periodontal disease; however, inflammatory cytokines can also activate the immunomodulatory properties of mesenchymal stem cells (MSCs), the mechanism to protect and maintain cell survival under inflammatory environment. Therefore, inflammation can exert both negative and positive effects for regulating tissue homeostasis. Interleukin 12 (IL-12) is one of the potent destructive stimulators in pathogenesis of many inflammatory diseases. In periodontitis, the increased level of IL-12 in serum and gingival crevicular fluid was found associated with the severity of the periodontal disease. However, the exact role of IL-12 in periodontitis is still unclear. The aim of this study was to investigate the responses of human periodontal ligament (PDL) cells to exogenous IL-12, especially on the immunomodulatory effects of IL-12. The results demonstrated the presence of IL-12 and IL-12 receptor (IL-12R) in periodontal tissues, and the expression was enhanced in tissues from periodontitis patients. Exogenous IL-12 stimulated the expression of some inflammatory cytokines as well as the immunomodulatory molecules, such as interferon gamma (IFNγ), human leukocyte antigen (HLA), and indoleamine-pyrrole-2,3-dioxygenase (IDO) enzyme. In conclusion, the data suggested the influence of increased IL-12, during periodontal inflammation, on controlling tissue's homeostasis by upregulating the inflammatory cytokines and modulating the function of immune cells through the expression of immunosuppressive molecules.

Keywords Interleukin 12 • Immunomodulation • Periodontal ligament cell • Interferon gamma (IFNγ) • Indoleamine-pyrrole-2,3-dioxygenase (IDO) • Human leukocyte antigen (HLA)

B. Issaranggun Na Ayuthaya • P. Pavasant (✉)
Research Unit of Mineralized Tissue, Faculty of Dentistry, Chulalongkorn University,
34 Henri-Dunant road, Pathumwan, Bangkok 10330, Thailand
e-mail: prasit215@gmail.com

© The Author(s) 2017   217
K. Sasaki et al. (eds.), *Interface Oral Health Science 2016*,
DOI 10.1007/978-981-10-1560-1_18

18.1 Introduction

Periodontitis or periodontal disease is one of the most prevalent chronic inflammatory diseases involved in periodontium destruction. The plaque microorganism is the cause of the periodontal disease formation; however, the presence of pathogen alone is not sufficient to trigger the onset of periodontitis. Indeed, the recognition of this pathogen by host immune cells is an essential process in the pathogenesis of periodontal disease [1].

One of the critical recognition components of the host immune response to the pathogenic microorganisms and their products is a family of toll-like receptors (TLRs). The most common TLRs implicated as potent receptors for pathogen-associated periodontitis are TLR-2 and TLR-4 [2, 3]. TLR activation stimulates an intracellular signaling cascade that leads to the activation of transcription factors, such as nuclear factor-kB (NF-kB) and activator protein-1 (AP-1) [4]. Subsequently, the production of pro-inflammatory cytokines that play roles in the induction of osteoclastogenesis leading to bone resorption occurs. Attempts to reduce inflammation could decrease the progression and severity of bone resorption, indicating the importance of inflammatory process in pathogenesis of periodontal disease [5].

The activation of host immune response required antigen-presenting cells (APCs) such as dendritic cells (DCs) to recognize the pathogens and stimulate the adaptive immune response [6–8]. During periodontal inflammation, the increased number of DCs was found related to the severity of inflammation [8]. It has been shown that, at the initiation stage of periodontal inflammation, most of DCs resided in gingival epithelium. However, at the later stage, they moved into the periodontal tissue and activated the differentiation of adaptive immune cells [8, 9].

Interleukin 12 (IL-12) is a potent cytokine that plays an important role in immune response by driving the differentiation of naïve T cells to T helper 1 (Th1) cells and mediating long-term protection against the pathogen [10, 11]. IL-12 is mainly produced by DCs as a result of the induction by antigenic signals, particularly lipopolysaccharide (LPS), and activated T cell signals [12]. Moreover, IL-12 production is further regulated by other cytokines such as interleukin 1beta (IL-1β), a potent osteolytic cytokine [13, 14]. These findings are concordant with the reports from many studies that the level of IL-12 was upregulated and associated with severity of inflammatory diseases [15–17].

IL-12 has been implicated as a potent destructive stimulator in the pathogenesis of several inflammatory diseases that lead to bone destruction [1, 18, 19]. The accumulation of inflammatory cytokines influences bone remodeling by alternating osteoblast and osteoclast activities. In an in vivo study, the injection of exogenous IL-12 resulted in the aggressive joint inflammation [20]. These evidences implicated the destructive role of IL-12 as a trigger of the inflammatory-induced bone destruction.

Although inflammation usually promotes tissue destruction, some inflammatory cytokines can activate immunomodulatory properties of MSCs, a negative mechanism that modulate host immune response. Immunomodulation is an important property of MSCs to promote cell survival under inflammatory environment, since MSCs are susceptible to be destroyed by activated T cells and NK cells [21]. The immunomodulatory property of MSCs occurred via the upregulation of immuno-suppressive molecules such as human leukocyte antigen (HLA) molecules [22, 23] and indoleamine-pyrrole 2,3-dioxygenase (IDO) enzyme [23–25]. HLA molecules attenuate immune cells' activity by signaling via their specific inhibitory receptors [26], while IDO enzyme inhibits immune response by degrading tryptophan, an essential amino acid required for T cell growth, resulted in a decrease of T cell pro-liferation [24]. Therefore, the function of IL-12 may involve in the moderation of host immune response and control survival and function of periodontal ligament cells.

In this study, the presences of IL-12 and IL-12R were detected in all dental tis-sues. The responses of hPDL cells to IL-12, regarding the expression of immuno-modulatory molecules, were investigated. The function of IL-12 in hPDL cells will provide more information toward the understanding of pathogenesis of periodontal disease.

18.2 Materials and Methods

18.2.1 Cell Culture

All experimental protocols were approved by the Ethics Committee of the Faculty of Dentistry, Chulalongkorn University. Non-carious third molars extracted for orthodontic reason were collected for periodontal ligament (PDL) cell isolation as previously described [27]. Cells were cultured in standard medium (Dulbecco's modified Eagle's medium (DMEM) containing 10% fetal bovine serum with L-glutamine (2 mM), penicillin (100 U/ml), streptomycin (100 mg/ml), and ampho-tericin B (5 mg/ml)) and incubated at 37 °C in a humidified atmosphere of 5% CO_2 in air. After the cells reached confluency, they were subcultured at a 1:3 ratio. Cells from the third to fifth passages were used in the experiments. Cell established from at least three different donors were used in each study.

18.2.2 Application of IL-12, IFNγ, IL-1β, and TNFα

Human PDL cells were seeded in 12-well plates (2×10^5 cells/well) for 24 h and then treated with recombinant human IL-12 (p70) (0–10 ng/ml) (Peprotech, Rocky Hill, NJ), recombinant human interferon gamma (IFNγ) (1 ng/ml) (ImmunoTools, Friesoythe, Germany), recombinant human IL-1β (1 ng/ml) (R&D System, Minneapolis, USA), or human recombinant TNFα (1 ng/ml) (Millipore, Darmstadt, Germany) for 1 day. In some experiments, 50 μM of lisofylline (STAT4 inhibitor)

(Cayman chemical, Ann Arbor, MI, USA) or 10 μM of NF-kB inhibitor (Millipore) was added into the culture medium 30 min prior to the IL-12 treatment.

18.2.3 Reverse Transcription-Polymerase Chain Reaction (RT-PCR)

Trizol reagent (Molecular Research Center, Cincinnati, OH) was used for total cellular RNA extraction. Total amount of RNA was quantified using a NanoDrop 2000 Spectrophotometer (Thermo Scientific, Wilmington, DE). One microgram of total RNA per sample was converted to complementary DNA by reverse transcriptase (Promega, Madison, WI). Subsequently, polymerase chain reactions (PCR) were performed with specific primers for IL-12Rβ_2, IL-1β, TNFα, and GAPDH. The sequences of used primers were GAPDH (NM002046.3), forward 5′ (TGAAGGTCGGAGTCAACGGAT-3′), reverse 5′ (TCACACCCATGACGAACATGG-3′); IL-12Rβ_2 (NC018912.2), forward 5′ (CAGCACATCTCCCTTTCTGTTTTC-3′), reverse 5′ (ACTTTAAGGCTTGAAGCCTCACC-3′); IL-1β (NM000576.2), forward 5′ (GGAGCAACAAGTGGTGTTCT-3′), reverse 5′ (AAAGTCCAGGCTATAGCCGT-3′); and TNFα (NM000594.2), forward 5′ (AAGCCTGTAGCCCATGTTGT-3′), reverse 5′ (CAGATAGATGGGCTCATACC-3′). Conventional PCR was performed using Taq polymerase (Taq DNA Polymerase, Invitrogen, Brazil) in a DNA thermal cycler (BiometraGmH, Göttingen, Germany). The products were electrophoresed on a 2 % agarose gel and visualized using ethidium bromide (EtBr; Bio-Rad, Hercules, CA) fluorostaining.

18.2.4 Real-Time Polymerase Chain Reaction (Real-Time PCR)

Real-time PCR was performed in a MJ Mini™ Thermal Cycler (Bio-Rad) using the LightCycler 480 SYBR Green I Master kit (Roche diagnostic) according to the manufacturer's specifications. Gene expression levels were normalized to the GAPDH expression. Then relative gene expression was calculated by CFX Manager™ software (Bio-Rad). The results were shown as fold-change values relative to the control group. The sequence of primers used in the experiment was shown in Table 18.1.

Table 18.1 Primer sequence for real-time PCR

Primer	Forward/reverse	Sequence (5′–3′)	Sequence ID
qGAPDH	F	CACTGCCAACGTGTCAGTGGTG	NM 002046.4
	R	GTAGCCCAGGATGCCCTTGAG	
IDO	F	CAT CTG CAA ATC GTG ACT AAG	NG_028155.1
	R	GTT GGG TTA CAT TAA CCT TCC TT	
HLA-G	F	GCT GTG ATC ACT GGA GCT GT	NM_002127.5
	R	ACT CTT GCC TCT CAG TCC CA	
HLA-A	F	AAG AGG AGA CAC GGA ACA CC	NM_001242758.1
	R	TCG CAG CCA ATC ATC CAC TG	

18.2.5   Statistical Analyses

Data were reported as mean ± standard deviation (SD). Statistical significance was assessed by the Kruskal-Wallis test. The differences at $p < 0.05$ were considered as a statistical significant difference.

18.3   Results and Discussions

18.3.1   Expressions of IL-12 and IL-12 Receptor in Periodontal Tissue Increased During Periodontal Inflammation

The results in Fig. 18.1a showed the expression of IL-12 and IL-12 receptor (IL-12R) in dental tissues. Total RNA was extracted from healthy gingiva (Gin), periodontal ligament (PDL), and dental pulp (Pulp) and subjected to RT-PCR analysis. The expression of IL-12 and its receptor were found in all three dental tissues. In PDL cells, the expression of IL-12R, but not IL-12, could be detected (data not shown, manuscript in preparation).

The expression of IL-12 in dental tissue might be associated with DCs, the major cell types that are responsible for IL-12 production [28]. It has been shown that, during gingivitis, DCs were resided in gingival tissues. These cells penetrated into the periodontal tissue when the disease progresses into periodontitis [6–9]. This data suggested the distribution of DCs within all of the dental soft tissues. In addition, the expression of IL-12Rβ_2 detected in all three dental tissues suggested the presence of IL-12 responsive cells in the healthy dental tissues.

The expression of both IL-12 and IL-12Rβ_2 was significantly increased in periodontal tissue isolated from inflammatory lesion (Fig. 18.1b). The upregulation of IL-12 suggested the increased number of IL-12 producing cells, possibly DCs, during periodontal inflammation. These data were in agreement with the report showing the increased number of DCs in the tissue from healthy to gingivitis and to

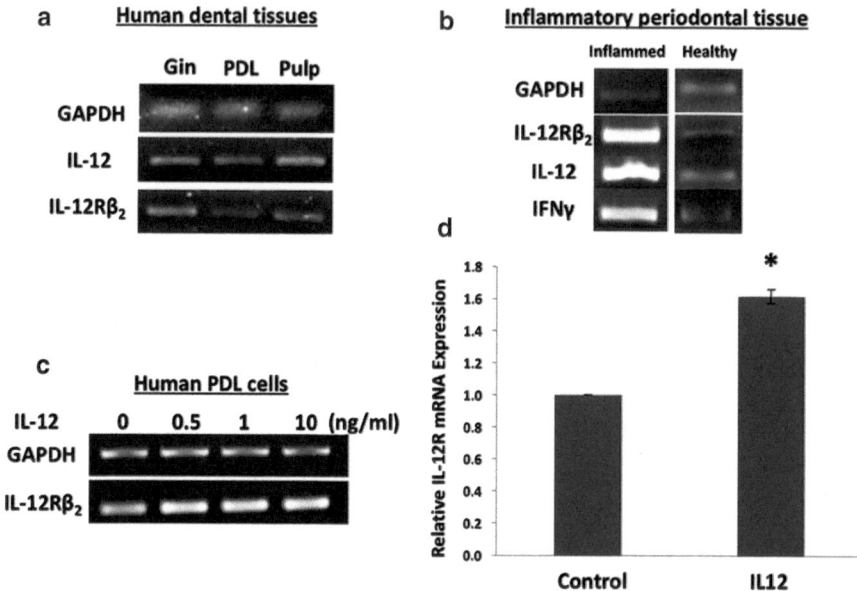

Fig. 18.1 The expression of IL-12 and IL-12R in dental tissues. Dental tissues, including gingiva (Gin), periodontal ligament (PDL), and dental pulp (Pulp), were obtained from patient with informed consent. Total RNA was extracted and subjected to RT-PCR analysis. The results showed all three types of tissue expressed in both IL-12 and IL-12R expression (**a**). The expression level of both IL-12 and IL-12R was higher in the inflammatory PDL tissue than the healthy tissue (**b**). IL-12 could also induce IL-12R expression in hPDL cells (**c**). The relative expression of IL-12R after induced with 1 ng/ml of IL-12 was shown as a graph in (**d**). The pictures were the representative of triplicated experiments and * indicated the statistical significance ($p < 0.05$)

periodontitis [8]. In case of IL-12R, it is possible that the upregulation of IL-12Rβ_2 represents the autocrine regulation of IL-12 for promoting the cellular response during periodontal inflammation.

Figure 18.1c, d showed a significant increase of IL-12Rβ_2 expression after IL-12 treatment in hPDL cells, supporting the concept of autocrine feedback loop. This loop has been demonstrated previously in the in vivo study reported by Thibodeaux et al. [29], showing the increased expression of both IL-12Rβ_1 and β_2 following the systemic administration of recombinant IL-12.

The production of IFNγ following IL-12 activation has been demonstrated to be a potent cytokine that regulated the expression of IL-12R [30]. Thus, the presence of IFNγ found in inflamed periodontal tissue (Fig. 18.1b) might enhance the expression of IL-12R during periodontal inflammation.

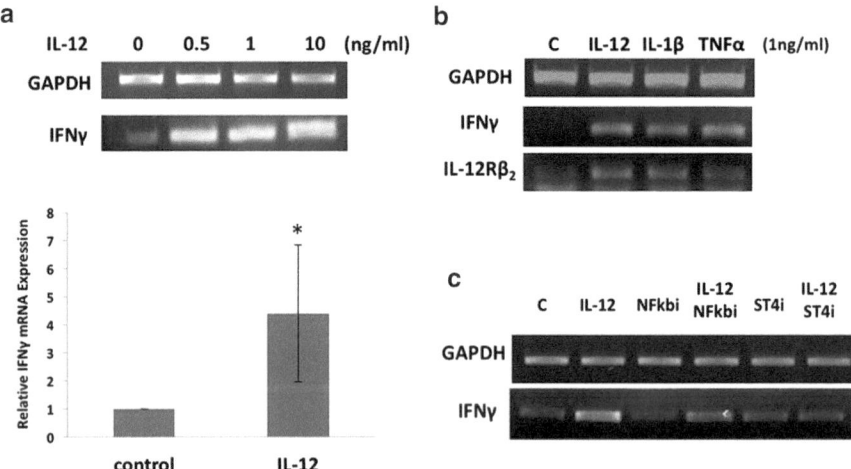

Fig. 18.2 IL-12-induced IFNγ expression in HPDL cells. Human PDL cells were treated with 0.5–10 ng/ml of IL-12 for 24 h. RT-PCR analysis showed the upregulation of IFNγ in a dose-dependent manner (**a**). The graph below showed the relative expression of IFNγ when cells were treated with 1 ng/ml of IL-12. * indicated the significant difference when compared to the control ($p < 0.05$). **b** showed the inductive effect of 1 ng/ml of IL-12, IL-1β, and TNFα on the expressions of IFNγ and IL-12R. The inductive effect of IL-12 on IFNγ was inhibited by NF-kB and STAT4 inhibitors as determined by RT-PCR (**c**)

18.3.2   IL-12-Mediated IFNγ Expression in hPDL Cells via STAT4 and NF-kB Signaling Pathways

Next, we monitored the expression levels of IFNγ, a classical target of IL-12 signaling [28, 31, 32] after IL-12 treatment in hPDL cells. Figure 18.2a showed that IL-12 significantly increased IFNγ expression within 24 h. However, other destructive cytokines, including IL-1β and TNFα, could also induce the expression of IFNγ and IL-12R (Fig. 18.2b). It is possible that the inductive effect of IL-12 might occur via other IL-12-induced cytokines. Nevertheless, both IL-1β and TNFα could promote the response of PDL cells to IL-12 via the increased expression of IL-12Rβ2.

Normally, IL-12 signaling is generated via STAT4 [33]. In addition, STAT4-deficient mice had an identical phenotype to IL-12-deficient mice [34] supporting the role of STAT4 in IL-12 signaling pathway. However, in DCs, NF-kB has been proposed as a major signaling molecule mediated in IL-12 activation [35]. Thus, the role of STAT4 and NF-kB in IL-12-induced IFNγ in hPDL cells by means of inhibitors was examined. The results in Fig. 18.2c showed that both STAT4 and NF-kB inhibitors significantly reduced the IL-12-induced IFNγ expression, suggesting the involvement of both STAT4 and NF-kB in IL-12-induced IFNγ in hPDL cells.

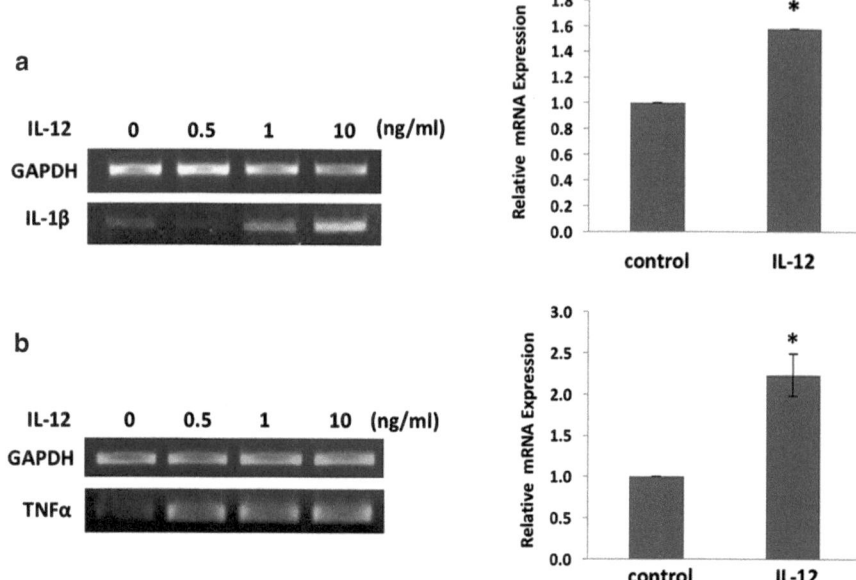

Fig. 18.3 IL-12 induced the expression of IL-1β and TNFα in HPDL cells. Human PDL cells were treated with 0.5–10 ng/ml of IL-12. The upregulation of IL-1β (**a**) and TNFα (**b**) was detected dose dependently. Graphs on the right showed the relative expression of IL-1β and TNFα after being treated with 1 ng/ml of IL-12 as compared to the control. * indicated the statistic difference ($p < 0.05$)

18.3.3 IL-12-Modulated Periodontal Inflammatory Response and Immunomodulatory Property of hPDL Cells

To investigate the modulatory effect of IL-12 on periodontal inflammatory response, the expression levels of IL-1β and TNFα in hPDL cells following IL-12 treatment were monitored. After 24 h incubation, a significant increase in IL-1β and TNFα mRNA expressions was detected, as shown in Fig. 18.3a, b. IL-12-induced TNFα expression was also reported in mouse BV-2 microglial cells [36]. Moreover, the study by Kim et al. [37] demonstrated the upregulation of IL-1β following IL-12 administration in joint tissue. These evidences support the possible role of IL-12 on the enhancement of inflammatory response during periodontal disease.

Indeed, inflammatory cytokines also function as a key inducer to activate the immunomodulatory properties of mesenchymal stem cells (MSCs) [22]. In physiological condition, immunomodulatory property of MSCs is kept inactive and will be activated by inflammatory environment [22]. It is possible that during inflammation, the presence of pro-inflammatory cytokines could also serve as a potent activator to induce the immunomodulatory property of MSCs [38, 39].

The immunomodulatory properties of MSCs are the ability to express immunosuppressive molecules, such as human leukocyte antigen (HLA) [22, 23] and

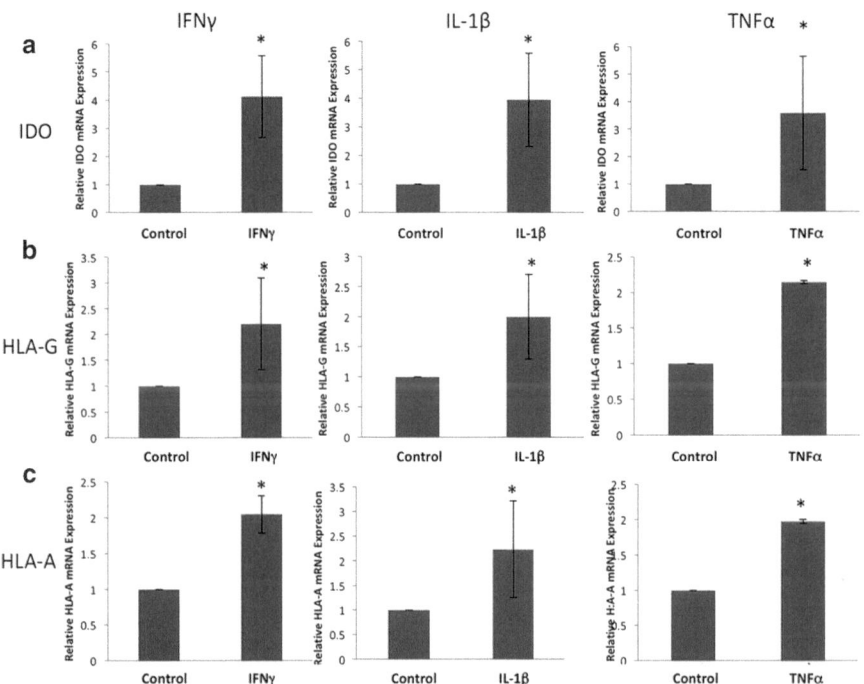

Fig. 18.4 Upregulation of IDO, HLA-G, and HLA-A by IFNγ, IL-1β, and TNFα. Human PDL cells were treated with 1 ng/ml of IFNγ, IL-1β, and TNFα for 24 h. Real-time PCR analysis indicated that all three molecules could significantly induce the expression of IDO, HLA-G, and HLA-A. * indicated the statistical difference ($p < 0.05$)

indoleamine-pyrrole-2,3dioxygenase (IDO) enzyme [23–25]. HLA genes are the gene encoding the major histocompatibility complex (MHC) in human. HLA functions by binding to their inhibitory receptors expressed on various types of immune cells and subsequently attenuating the immunological activation and inhibits the inflammatory response [40, 41]. IDO is a catalytic enzyme that catalyzes and degrades tryptophan, a crucial amino acid for T cell growth, leading to the inhibition of T cell proliferation [24, 42]. HPDL cells have been shown to be able to secrete IDO enzyme upon activation [43, 44]. Therefore, the upregulation of HLA-G, HLA-A, and IDO in this study may participate in the survival of PDLSCs as well as the reduction of immune responses.

The results in Fig. 18.4a–c indicated that 1 ng/ml of exogenous IFNγ, IL-1β, and TNFα could induce IDO, HLA-G, and HLA-A as judged by real-time RT-PCR. Taken together with the results that IL-12 could induce IFNγ, IL-1β, and TNFα, these data strongly suggested the important role of IL-12 in modulating host immune response during periodontal inflammation.

18.4   Conclusion

Our data indicated the ability of hPDL cells to respond to IL-12. The IL-12-induced IL-12R expression might be an autocrine mechanism of IL-12 to promote its cellular response in hPDL cells. IL-12 also stimulated the expression of IL-1β and TNFα, indicated the pro-inflammatory activity of IL-12 in periodontal inflammation. Moreover, the increase in inflammatory cytokine production induced by IL-12 was demonstrated as the factors to activate the immunomodulatory property of hPDL cells, a mechanism to protect the cells from inflammatory environment. Thus, the increased level of IL-12 in periodontal inflammation may play roles in regulating periodontal tissue destruction and survival of PDL stem cells.

Acknowledgments This work was supported by Research Chair Grant 2012, the National Science and Technology Development Agency (NSTDA), Thailand. BI was supported by H.M. King Bhumipol Adulyadej's 72nd Birthday Anniversary Scholarship and Royal Golden Jubilee Scholarship from the Thailand Research Fund.

References

1. Graves DT, Oates T, Garlet GP. Review of osteoimmunology and the host response in endodontic and periodontal lesions. J Oral Microbiol. 2011;3:5304. doi:10.3402/jom.v3i0.5304.
2. Gelani V, Fernandes AP, Gasparoto TH, Garlet TP, Cestari TM, Lima HR, et al. The role of toll-like receptor 2 in the recognition of Aggregatibacter actinomycetemcomitans. J Periodontol. 2009;80(12):2010–9. doi:10.1902/jop.2009.090198.
3. Lima HR, Gelani V, Fernandes AP, Gasparoto TH, Torres SA, Santos CF, et al. The essential role of toll like receptor-4 in the control of Aggregatibacter actinomycetemcomitans infection in mice. J Clin Periodontol. 2010;37(3):248–54. doi:10.1111/j.1600-051X.2009.01531.x.
4. Mahanonda R, Pichyangkul S. Toll-like receptors and their role in periodontal health and disease. Periodontology 2000. 2007;43:41–55. doi:10.1111/j.1600-0757.2006.00179.x.
5. Williams RC, Jeffcoat MK, Kaplan ML, Goldhaber P, Johnson HG, Wechter WJ. Flurbiprofen: a potent inhibitor of alveolar bone resorption in beagles. Science. 1985;227(4687):640–2.
6. Cirrincione C, Pimpinelli N, Orlando L, Romagnoli P. Lamina propria dendritic cells express activation markers and contact lymphocytes in chronic periodontitis. J Periodontol. 2002;73(1):45–52. doi:10.1902/jop.2002.73.1.45.
7. Cutler CW, Jotwani R, Palucka KA, Davoust J, Bell D, Banchereau J. Evidence and a novel hypothesis for the role of dendritic cells and Porphyromonas gingivalis in adult periodontitis. J Periodontal Res. 1999;34(7):406–12.
8. Jotwani R, Palucka AK, Al-Quotub M, Nouri-Shirazi M, Kim J, Bell D, et al. Mature dendritic cells infiltrate the T cell-rich region of oral mucosa in chronic periodontitis: in situ, in vivo, and in vitro studies. J Immunol. 2001;167(8):4693–700.
9. Cutler CW, Jotwani R. Antigen-presentation and the role of dendritic cells in periodontitis. Periodontol 2000. 2004;35:135–57. doi:10.1111/j.0906-6713.2004.003560.x.
10. Hsieh CS, Macatonia SE, Tripp CS, Wolf SF, O'Garra A, Murphy KM. Development of TH1 CD4+ T cells through IL-12 produced by Listeria-induced macrophages. Science. 1993;260(5107):547–9.
11. Murphy EE, Terres G, Macatonia SE, Hsieh CS, Mattson J, Lanier L, et al. B7 and interleukin 12 cooperate for proliferation and interferon gamma production by mouse T helper clones that are unresponsive to B7 costimulation. J Exp Med. 1994;180(1):223–31.

12. Schulz O, Edwards AD, Schito M, Aliberti J, Manickasingham S, Sher A, et al. CD40 triggering of heterodimeric IL-12 p70 production by dendritic cells in vivo requires a microbial priming signal. Immunity. 2000;13(4):453–62.
13. Kawashima N, Stashenko P. Expression of bone-resorptive and regulatory cytokines in murine periapical inflammation. Arch Oral Biol. 1999;44(1):55–66.
14. Wesa AK, Galy A. IL-1 beta induces dendritic cells to produce IL-12. Int Immunol. 2001;13(8):1053–61.
15. Ebrahimi AA, Noshad H, Sadreddini S, Hejazi MS, Mohammadzadeh Sadigh Y, Eshraghi Y, et al. Serum levels of TNF-alpha, TNF-alphaRI, TNF-alphaRII and IL-12 in treated rheumatoid arthritis patients. Iran J Immunol: IJI. 2009;6(3):147–53. doi:IJIv6i3A5.
16. Sanchez-Hernandez PE, Zamora-Perez AL, Fuentes-Lerma M, Robles-Gomez C, Mariaud-Schmidt RP, Guerrero-Velazquez C. IL-12 and IL-18 levels in serum and gingival tissue in aggressive and chronic periodontitis. Oral Dis. 2011;17(5):522–9. doi:10.1111/j.1601-0825.2011.01798.x.
17. Tsai IS, Tsai CC, Ho YP, Ho KY, Wu YM, Hung CC. Interleukin-12 and interleukin-16 in periodontal disease. Cytokine. 2005;31(1):34–40. doi:10.1016/j.cyto.2005.02.007.
18. Queiroz-Junior CM, Silva MJ, Correa JD, Madeira MF, Garlet TP, Garlet GP, et al. A controversial role for IL-12 in immune response and bone resorption at apical periodontal sites. Clin Dev Immunol. 2010;2010:327417. doi:10.1155/2010/327417.
19. Takayanagi H. Osteoimmunology and the effects of the immune system on bone. Nat Rev Rheumatol. 2009;5(12):667–76. doi:10.1038/nrrheum.2009.217.
20. Leung BP, McInnes IB, Esfandiari E, Wei XQ, Liew FY. Combined effects of IL-12 and IL-18 on the induction of collagen-induced arthritis. J Immunol. 2000;164(12):6495–502.
21. Hoogduijn MJ, Popp F, Verbeek R, Masoodi M, Nicolaou A, Baan C, et al. The immunomodulatory properties of mesenchymal stem cells and their use for immunotherapy. Int Immunopharmacol. 2010;10(12):1496–500. doi:10.1016/j.intimp.2010.06.019.
22. Krampera M, Cosmi L, Angeli R, Pasini A, Liotta F, Andreini A, et al. Role for interferon-gamma in the immunomodulatory activity of human bone marrow mesenchymal stem cells. Stem Cells. 2006;24(2):386–98. doi:10.1634/stemcells.2005-0008.
23. Noone C, Kihm A, English K, O'Dea S, Mahon BP. IFN-gamma stimulated human umbilical-tissue-derived cells potently suppress NK activation and resist NK-mediated cytotoxicity in vitro. Stem Cells Dev. 2013;22(22):3003–14. doi:10.1089/scd.2013.0028.
24. Meisel R, Zibert A, Laryea M, Gobel U, Daubener W, Dilloo D. Human bone marrow stromal cells inhibit allogeneic T-cell responses by indoleamine 2,3-dioxygenase-mediated tryptophan degradation. Blood. 2004;103(12):4619–21. doi:10.1182/blood-2003-11-3909.
25. Tipnis S, Viswanathan C, Majumdar AS. Immunosuppressive properties of human umbilical cord-derived mesenchymal stem cells: role of B7-H1 and IDO. Immunol Cell Biol. 2010;88(8):795–806. doi:10.1038/icb.2010.47.
26. Rajagopalan S, Long EO. A human histocompatibility leukocyte antigen (HLA)-G-specific receptor expressed on all natural killer cells. J Exp Med. 1999;189(7):1093–100.
27. Osathanon T, Ritprajak P, Nowwarote N, Manokawinchoke J, Giachelli C, Pavasant P. Surface-bound orientated Jagged-1 enhances osteogenic differentiation of human periodontal ligament-derived mesenchymal stem cells. J Biomed Mater Res A. 2013;101(2):358–67. doi:10.1002/jbm.a.34332.
28. Ma X, Trinchieri G. Regulation of interleukin-12 production in antigen-presenting cells. Adv Immunol. 2001;79:55–92.
29. Thibodeaux DK, Hunter SE, Waldburger KE, Bliss JL, Trepicchio WL, Sypek JP, et al. Autocrine regulation of IL-12 receptor expression is independent of secondary IFN-gamma secretion and not restricted to T and NK cells. J Immunol. 1999;163(10):5257–64.
30. Wu CY, Gadina M, Wang K, O'Shea J, Seder RA. Cytokine regulation of IL-12 receptor beta2 expression: differential effects on human T and NK cells. Eur J Immunol. 2000;30(5):1364–74.
31. Schroder K, Hertzog PJ, Ravasi T, Hume DA. Interferon-gamma: an overview of signals, mechanisms and functions. J Leukoc Biol. 2004;75(2):163–89. doi:10.1189/jlb.0603252.

32. Kubin M, Kamoun M, Trinchieri G. Interleukin 12 synergizes with B7/CD28 interaction in inducing efficient proliferation and cytokine production of human T cells. J Exp Med. 1994;180(1):211–22.

33. Morinobu A, Gadina M, Strober W, Visconti R, Fornace A, Montagna C, et al. STAT4 serine phosphorylation is critical for IL-12-induced IFN-gamma production but not for cell proliferation. Proc Natl Acad Sci U S A. 2002;99(19):12281–6. doi:10.1073/pnas.182618999.

34. Trinchieri G. Interleukin-12 and the regulation of innate resistance and adaptive immunity. Nat Rev Immunol. 2003;3(2):133–46. doi:10.1038/nri1001.

35. Grohmann U, Belladonna ML, Bianchi R, Orabona C, Ayroldi E, Fioretti MC, et al. IL-12 acts directly on DC to promote nuclear localization of NF-kappaB and primes DC for IL-12 production. Immunity. 1998;9(3):315–23.

36. Jana M, Dasgupta S, Saha RN, Liu X, Pahan K. Induction of tumor necrosis factor-alpha (TNF-alpha) by interleukin-12 p40 monomer and homodimer in microglia and macrophages. J Neurochem. 2003;86(2):519–28.

37. Kim HS, Chung DH. TLR4-mediated IL-12 production enhances IFN-gamma and IL-1beta production, which inhibits TGF-beta production and promotes antibody-induced joint inflammation. Arthritis Res Ther. 2012;14(5):R210. doi:10.1186/ar4048.

38. English K, Barry FP, Field-Corbett CP, Mahon BP. IFN-gamma and TNF-alpha differentially regulate immunomodulation by murine mesenchymal stem cells. Immunol Lett. 2007;110(2):91–100. doi:10.1016/j.imlet.2007.04.001.

39. Li W, Ren G, Huang Y, Su J, Han Y, Li J, et al. Mesenchymal stem cells: a double-edged sword in regulating immune responses. Cell Death Differ. 2012;19(9):1505–13. doi:10.1038/cdd.2012.26.

40. Amiot L, Vu N, Samson M. Biology of the immunomodulatory molecule HLA-G in human liver diseases. J Hepatol. 2015;62(6):1430–7. doi:10.1016/j.jhep.2015.03.007.

41. Murphy B, Krensky AM. HLA-derived peptides as novel immunomodulatory therapeutics. J Am Soc Nephrol. 1999;10(6):1346–55.

42. Haddad R, Saldanha-Araujo F. Mechanisms of T-cell immunosuppression by mesenchymal stromal cells: what do we know so far? Biomed Res Int. 2014;2014:216806. doi:10.1155/2014/216806.

43. Moon JS, Cheong NR, Yang SY, Kim IS, Chung HJ, Jeong YW, et al. Lipopolysaccharide-induced indoleamine 2,3-dioxygenase expression in the periodontal ligament. J Periodontal Res. 2013;48(6):733–9. doi:10.1111/jre.12063.

44. Seo BM, Miura M, Gronthos S, Bartold PM, Batouli S, Brahim J, et al. Investigation of multipotent postnatal stem cells from human periodontal ligament. Lancet. 2004;364(9429):149–55. doi:10.1016/S0140-6736(04)16627-0.

Open Access This chapter is distributed under the terms of the Creative Commons Attribution 4.0 International License (http://creativecommons.org/licenses/by/4.0/), which permits use, duplication, adaptation, distribution and reproduction in any medium or format, as long as you give appropriate credit to the original author(s) and the source, provide a link to the Creative Commons license and indicate if changes were made.

The images or other third party material in this chapter are included in the work's Creative Commons license, unless indicated otherwise in the credit line; if such material is not included in the work's Creative Commons license and the respective action is not permitted by statutory regulation, users will need to obtain permission from the license holder to duplicate, adapt or reproduce the material.

Chapter 19
Development and Performance of Low-Cost Beta-Type Ti-Based Alloys for Biomedical Applications Using Mn Additions

Pedro F. Santos, Mitsuo Niinomi, Huihong Liu, Masaaki Nakai, Ken Cho, Takayuki Narushima, Kyosuke Ueda, Naofumi Ohtsu, Mitsuhiro Hirano, and Yoshinori Itoh

Abstract The microstructures, mechanical properties, and biocompatibility of various low-cost β-type Ti-Mn alloys fabricated by both cold crucible levitation melting (CCLM) and metal injection molding (MIM) were investigated after solution treatment. Mn was chosen as a potential low-cost alloying element for Ti. Among the alloys fabricated by both methods, Ti-9Mn shows the best combination of tensile strength and elongation, and their performances are mostly comparable to or superior to those of Ti-6Al-4V (Ti-64) ELI. However, alloys fabricated by MIM show a higher O and C content, along with precipitated Ti carbides and pores, which all cause the ductility of the alloys fabricated by MIM to be lower than that of the alloys fabricated by CCLM. Furthermore, the cell viability and metallic ion release ratios of the alloys fabricated by CCLM are comparable to those of commercially pure Ti, making this alloy promising for biomedical applications. The Young's modulus of the alloys is also lower than that of Ti-64 ELI (which is of approximately 110 GPa), which can possibly reduce the stress shielding effect in implanted patients.

P.F. Santos (✉) • T. Narushima • K. Ueda
Department of Materials Processing, Tohoku University, 6-6-02 Aza Aoba, Aramaki, Aoba-ku, Sendai, Miyagi 980-8579, Japan
e-mail: pedro@imr.tohoku.ac.jp

M. Niinomi
Institute for Materials Research, Tohoku University,
2-1-1 Katahira, Aoba-ku, Sendai, Miyagi 980-8577, Japan

Graduate School of Science and Technology, Meijyo University,
1-501, Shiogamaguchi, Tempaku-ku, Nagoya, Aichi 468-8502, Japan

Graduate School of Engineering, Osaka University,
2-1 Yamadagaoka, Suita, Osaka 565-0871, Japan

Institute of Materials and Systems for Sustainability, Nagoya University,
Furo-cho, Chikusa-ku, Nagoya, Aichi 464-8603, Japan

H. Liu
Joining and Welding Research Institute, Osaka University,
11-1 Mihogaoka, Suita, Osaka 567-0047, Japan

© The Author(s) 2017
K. Sasaki et al. (eds.), *Interface Oral Health Science 2016*,
DOI 10.1007/978-981-10-1560-1_19

229

Keywords Ti-Mn alloys • β phase • Mechanical properties • Metallic biomaterials • Low-cost Ti alloy

19.1   Introduction

Among the most widely used metallic biomaterials, titanium (Ti) and its alloys are the most suitable because of their high specific strength, good corrosion resistance, and biocompatibility [1–3]. However, the most widely used Ti-based alloys in biomedical applications were not designed for this kind of application and can show issues. Commercially pure Ti (CP-Ti) has insufficient mechanical properties for some biomedical applications, whereas Ti-6Al-4V (Ti-64) ELI has the necessary mechanical properties, but contains Al and V, which can be released as ions from the alloy and are associated to health issues in the human body [2, 4–8]. Therefore, recent efforts have been made in recent years to produce new and more biocompatible β-type Ti alloys, designed specifically for biomedical applications [3]. β-type Ti alloys can achieve the highest specific strength when compared with α- or ($\alpha+\beta$)-type Ti alloys [9]. Furthermore, β stabilizing alloying elements can contribute to reduce the Young's modulus of the alloys [10, 11]. The lower Young's modulus is useful because metallic biomaterials usually have a Young's modulus much higher than that of the human bone, which can end up causing the stress shielding effect [12]. However, many of the newly developed alloys contain scarce and high-cost elements such as Nb and Ta, which may be difficult to obtain in the future because of the limited amounts of their natural deposits [13, 14].

Considering the abovementioned factors (i.e., biocompatibility, balance of mechanical properties, availability, and cost), Mn was selected as the primary alloying element for Ti-Mn system alloys due to its β stabilizing effect, lower cytotoxicity, higher availability, and lower cost when compared to other popular alloying elements [15, 16]. Some of the binary Ti-Mn alloys fabricated using a cold crucible

M. Nakai
Department of Mechanical Engineering, Faculty of Science and Engineering,
Kindai University, 3-4-1 Kowakae, Higashiosaka, Osaka 577-8502, Japan

K. Cho
Graduate School of Engineering, Osaka University,
2-1 Yamadaoka, Suita, Osaka 565-0871, Japan

N. Ohtsu • M. Hirano
Instrument Analysis Center, Kitami Institute of Technology,
165 Koen-Cho, Kitami, Hokkaido 090-8507, Japan

Y. Itoh
Hamamatsu Technical Support Center, Industrial Research Institute of Shizuoka Prefecture,
1-3-3 Shinmiyakoda, Kita-ku, Hamamatsu, Shizuoka 431-2103, Japan

levitation melting (CCLM) showed a performance comparable or superior to that of the Ti-64 ELI for biomedical applications [15]. Furthermore, Ti-Mn alloys containing up to 13 mass% Mn showed cytotoxicity levels comparable to that of CP-Ti, but the alloy containing 18 mass% Mn showed cytotoxicity [15]. Thereafter, further cost reduction was considered by fabricating Ti-Mn alloys using a metal injection molding (MIM), a powder metallurgy near-net shape fabrication method [16]. The alloys fabricated by MIM showed some mechanical properties comparable to those of the alloys fabricated by CCLM, but because of both high porosity and high amount of interstitial impurities – inherent to the fabrication method – there was a drastic decrease in ductility compared to the alloys fabricated using CCLM [16]. Therefore, it is necessary to improve the ductility of the alloys.

19.2   Experimental Procedures

19.2.1   Ingots Preparation

For the first part of this study, ingots of Ti-(6, 9, 13, and 18 mass%)Mn were fabricated by CCLM. The ingots were initially subjected to a homogenization treatment for 21.6 ks at 1,273 K followed by ice water quenching and then to hot forging and hot rolling, both at 1,173 K. Finally, in order to retain the β phase in the alloys, the plates were subjected to solution treatment for 3.6 ks at 1,173 K in vacuum, followed by ice water quenching. Hereafter, the alloys fabricated by CCLM will be identified by the subscript "LM," as in Ti-Mn$_{LM}$.

For the second part of this study, ingots of Ti-(8, 9, 12, 13, 15, and 17 mass%)Mn were fabricated by MIM. Sintering was conducted in vacuum at 1,373 K for 28.8 ks. Finally, in order to retain the β phase in the alloys, the plates were subjected to solution treatment for 3.6 ks at 1,173 K in vacuum, followed by ice water quenching. Further details on the specimen preparation methods are described elsewhere [16]. Hereafter, the alloys fabricated by MIM will be identified by the subscript "MIM," as in Ti-Mn$_{MIM}$.

19.2.2   Microstructure Characterization

For all parts of this study, the material characterization was performed by chemical composition analysis, optical microscopy (OM), X-ray diffractometry (XRD), and transmission electron microscopy (TEM). Furthermore, electron probe microanalysis (EPMA) was carried out on the alloys fabricated by MIM. The experimental conditions and other details are described elsewhere [15, 16].

19.2.3   Mechanical Properties Evaluation

For all parts of this study, the mechanical property investigation was conducted by means of Vickers hardness tests, Young's modulus measurements, and tensile tests. Furthermore, compressive tests were carried out on the alloys fabricated by MIM. The experimental conditions and other details are described elsewhere [15, 16].

19.2.4   Biocompatibility Evaluation

The biocompatibility of the alloys fabricated by CCLM was evaluated by means of immersion tests in simulated body fluids (SBF), followed by X-ray photoelectron spectroscopy (XPS) of the surface of SBF-immersed samples, and also cytotoxicity tests using MC3T3-E1 cells. The experimental conditions and other details are described elsewhere [15].

19.3   Results and Discussion

19.3.1   Ti-Mn Binary Alloys Fabricated by CCLM

Table 19.1 shows the chemical compositions of Ti-(6-18)Mn$_{LM}$. The lower amount of Mn content in comparison to the nominal value is caused by the evaporation of Mn during melting processes [17]. Mn loss of up to 25 % was expected [17]. However, the observed difference (2–10 %) is considerably smaller in the Ti-Mn$_{LM}$, thanks to the shorter melting time required by this method.

Figures 19.1 and 19.2 show the optical micrographs and XRD profiles of the Ti-(6-18)Mn$_{LM}$, respectively. As shown in Fig. 19.2, only diffraction peaks attributed to β planes are detected in the XRD profiles of Ti-(9-18)Mn$_{LM}$. However, diffraction peaks that can be attributed to ω and α or α' phases are also detected in the XRD profile of Ti-6Mn$_{LM}$. A concentration of approximately 6.3 mass% Mn is required to fully retain the β phase upon quenching [4]. However, the Mn content of Ti-6Mn$_{LM}$ used for this study is 5.60 mass%. Furthermore, the presence of the athermal ω phase has been confirmed by the TEM observations. Figure 19.3a–d shows

Table 19.1 Chemical compositions of Ti-Mn alloys fabricated by CCLM (mass%)

Alloy	Element					
	Ti	Mn	O	C	N	Nominal Mn
Ti-6Mn$_{LM}$	bal.	5.60	0.0736	0.0036	0.0045	6.0
Ti-9Mn$_{LM}$	bal.	9.47	0.0727	0.0031	0.0036	10.0
Ti-13Mn$_{LM}$	bal.	12.65	0.0747	0.0024	0.0043	14.0
Ti-18Mn$_{LM}$	bal.	17.65	0.0591	0.0073	0.0021	18.0

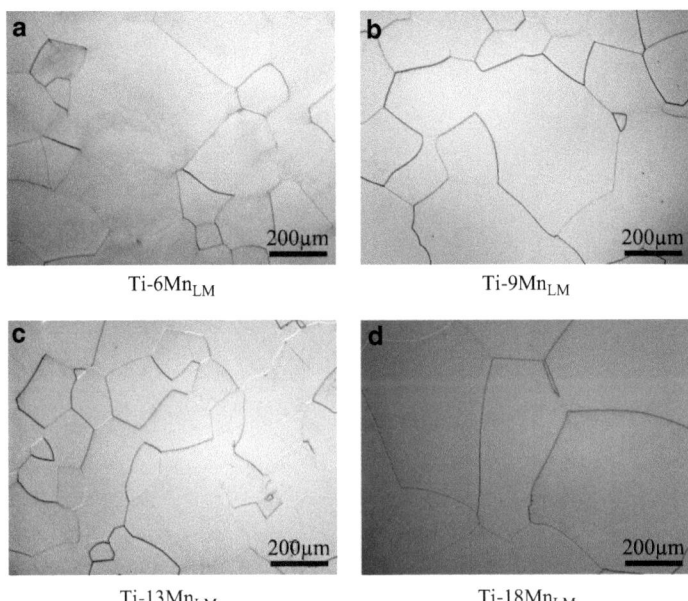

Fig. 19.1 Typical optical micrographs of (**a**) Ti-6Mn$_{LM}$, (**b**) Ti-9Mn$_{LM}$, (**c**) Ti-13Mn$_{LM}$, and (**d**) Ti-18Mn$_{LM}$

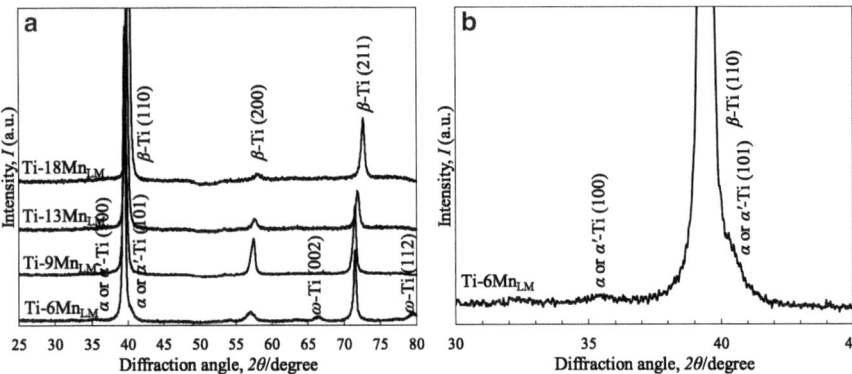

Fig. 19.2 (**a**) XRD profiles of Ti-Mn$_{LM}$ alloys and (**b**) magnified section of XRD profile of Ti-6Mn$_{LM}$

the selected area electron diffraction (SAED) patterns, where the spots or streaks attributed to the athermal ω phase decrease in intensity with increasing Mn content. The bright particles in the corresponding dark field (DF) images (Fig. 19.3e–h) represent the athermal ω phase particles. Ti-6Mn$_{LM}$ contains the largest amount of particles, as it decreases with increasing Mn content. The decrease in the volume fraction of the athermal ω phase, observed through the DF images of the Ti-(6–18)

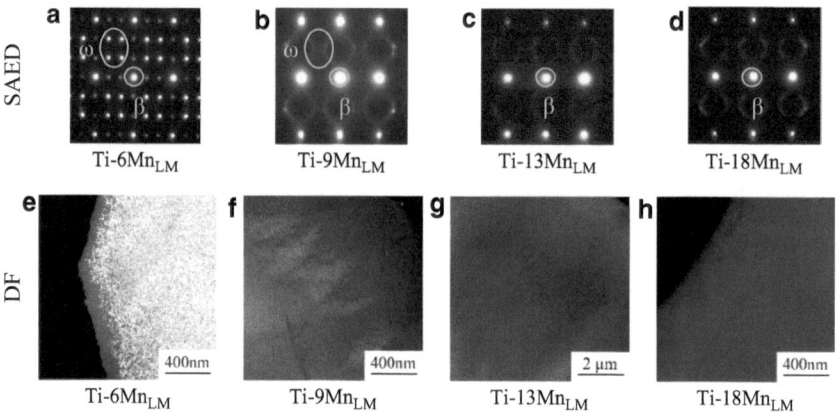

Fig. 19.3 Typical SAED patterns viewed from $[110]_\beta$, and corresponding DF images of diffraction spots or streaks of ω phase of (**a**) and (**e**) Ti-6Mn$_{LM}$, (**b**) and (**f**) Ti-9Mn$_{LM}$, (**c**) and (**g**) Ti-13Mn$_{LM}$, and (**d**) and (**h**) Ti-18Mn$_{LM}$

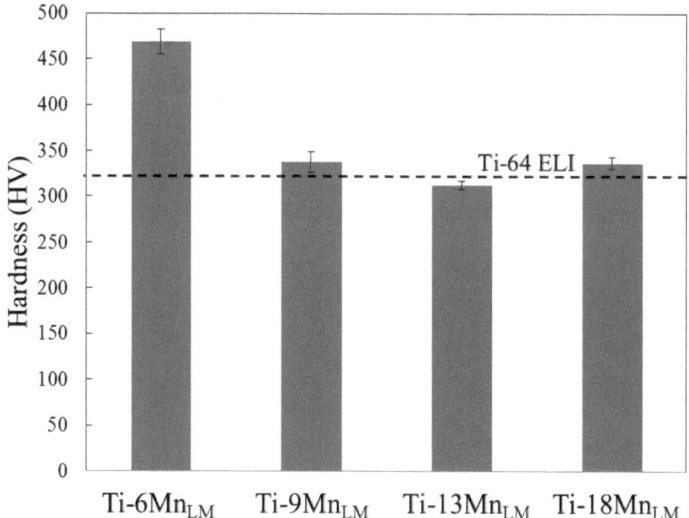

Fig. 19.4 Vickers hardness of Ti-Mn$_{LM}$ alloys compared to that of Ti-64 ELI alloy (*dashed line*)

Mn$_{LM}$, is attributed to increasing β phase stability obtained by increasing the Mn content. Because of the higher β phase stability of the Ti-13Mn$_{LM}$ and Ti-18Mn$_{LM}$, the ω phase cannot be observed in the DF images of these alloys (Fig. 19.3g, h).

Figures 19.4 and 19.5 show the Vickers hardness and Young's modulus of the Ti-(6–18)Mn$_{LM}$ alloys, respectively. The hardness of Ti-9Mn$_{LM}$ and Ti-13Mn$_{LM}$ is comparable to that of Ti-64 ELI reported in the literature (approximately 325 HV) [5], indicated by the dashed line in Fig. 19.4. Ti-(9–18)Mn$_{LM}$ shows a lower value

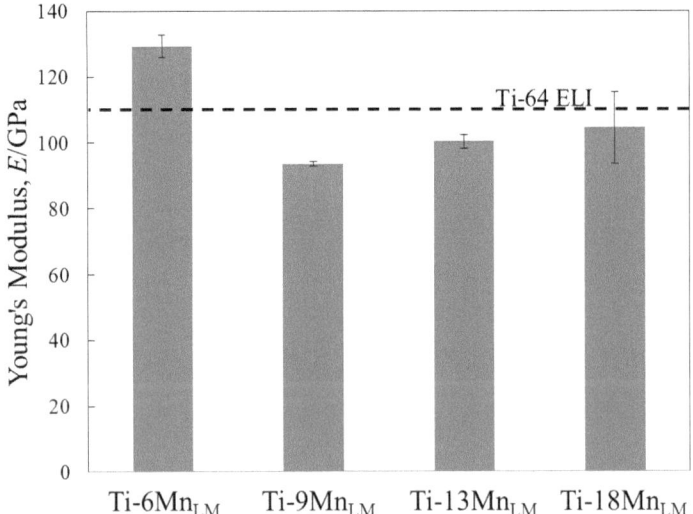

Fig. 19.5 Young's moduli of Ti-Mn$_{LM}$ alloys compared to that of Ti-64 ELI (*dashed line*)

of Young's modulus in comparison to that reported for Ti-64 ELI (110 GPa), indicated by the dashed line in Fig. 19.5 [5]. The main factors that affect both the Vickers hardness and Young's modulus measurements of Ti-(6–18)Mn$_{LM}$ are the solid solution hardening effects of Mn and the presence and amount of the athermal ω phase in the alloys. The increase in Mn leads to an increase in both the Vickers hardness and Young's modulus of the alloys by solid solution hardening and its effect on the lattice, respectively [15, 16, 18]. On the other hand, the decrease in the volume fraction of the athermal ω phase with increasing Mn leads to a decrease in both Vickers hardness and Young's modulus [4, 19].

Figure 19.6 shows the tensile properties of Ti-(9–18)Mn$_{LM}$ along with that of Ti-64 ELI obtained from literature [5]. The tensile properties of Ti-6Mn were not obtained because the specimens suffered premature fracture due to the high amount of athermal ω phase. Ti-9Mn$_{LM}$ and Ti-13Mn$_{LM}$ show ultimate tensile strength (UTS, σ_B), 0.2 % proof stress ($\sigma_{0.2}$), and elongation comparable to those reported for Ti-64 ELI (σ_B: 965 MPa, $\sigma_{0.2}$: 869 MPa, elongation: 15 %) [5]. However, the results shown by Ti-9Mn$_{LM}$ (σ_B: 1,048 MPa, $\sigma_{0.2}$: 1,023 MPa, and elongation: 19 %) make the use of this alloy in biomedical applications more promising than either Ti-13Mn$_{LM}$ or Ti-64 ELI.

Figure 19.7 shows the ratio of the amount of released Mn ions to the sum of released Ti and Mn ions to a 1 % lactic acid solution, as a function of Mn content in the Ti-Mn alloys released from the alloys. The ratio of released Mn ions increases in an almost linear relationship with the Mn content observed in the alloys. Concerning the toxic effect of Mn ions released in the human body, adult patients that displayed symptoms related to Mn intoxication were parenterally administered

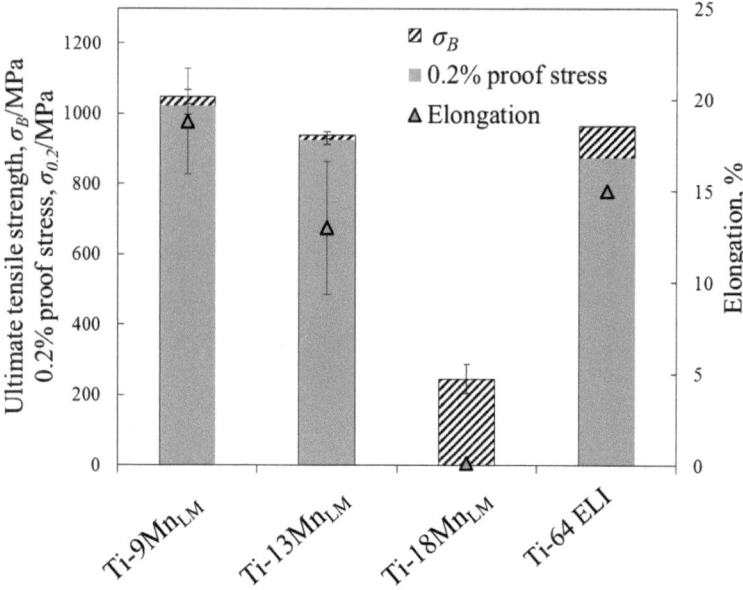

Fig. 19.6 Tensile properties of Ti-Mn$_{LM}$ alloys along with that of Ti-64 ELI (*rightmost column*)

Fig. 19.7 Ratio of amount of released Mn ions to the sum of released Ti and Mn ions as a function of Mn content in the Ti-Mn$_{LM}$ alloys

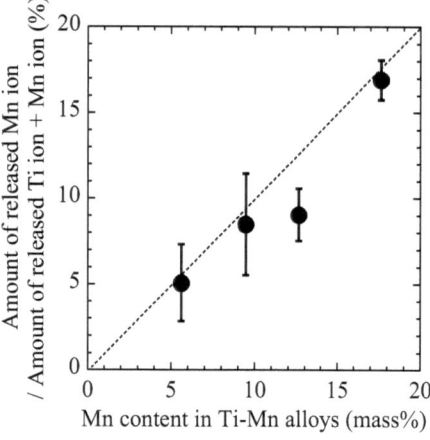

a daily dosage of at least 0.1 mg of Mn and usually more [20, 21]. The Ti-Mn$_{LM}$ alloys immersed in a 1 % lactic acid solution, which is a solution recommended for accelerated immersion tests [22], showed a maximum amount of released Mn ions of approximately 0.9 μg/cm^2 (Ti-18Mn) over a period of 7 days. Because the amount released is orders of magnitude lower than the amount administered during parenteral nutrition, it is believed that the amount of released Mn ions will not reach the levels known to cause symptoms related to Mn intoxication by itself, even in an

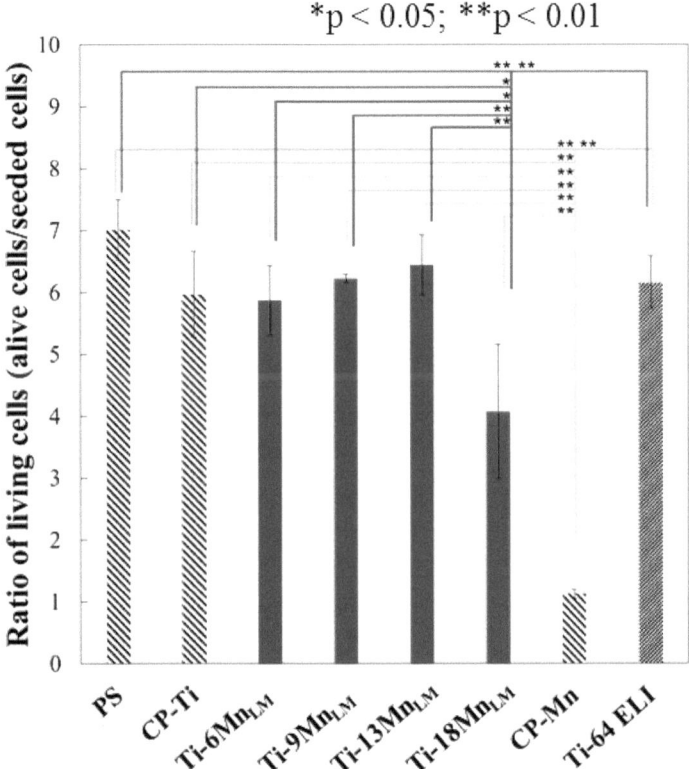

Fig. 19.8 MC3T3-E1 cell viability ratios for the CP-Ti, Ti-Mn$_{LM}$ alloys, and CP-Mn after 72 h of incubation

implant with considerable surface area. Furthermore, XPS analysis [15] revealed that the immersion tests using 1 % lactic acid caused the passive oxide layer of the alloys to become thinner, but the oxide layer regenerated.

Figure 19.8 shows the results obtained for the cytotoxicity test. After counting the living cells following a 72 h incubation period, significant differences are observed between Ti-18Mn$_{LM}$, CP-Mn, and the other materials. The cytotoxicity of Ti-(6–13)Mn$_{LM}$ is comparable to that of CP-Ti. The cell proliferations of Ti-18Mn$_{LM}$ and CP-Mn are lower than those of the other alloys. The incubation time of the 72 h test leads to a higher amount of released Mn ions from the alloy. It is supposed that the amount of released Mn ions from Ti-18Mn$_{LM}$ can inhibit cell proliferation [15].

19.3.2   Ti-Mn Binary Alloys Fabricated by MIM

Table 19.2 shows the chemical compositions of Ti-(8–17)Mn$_{MIM}$ fabricated by MIM. The O content of Ti-(8–17)Mn$_{MIM}$ increases with increasing Mn content, from approximately 0.23–0.32 mass%. Most of the O present in the alloys originates

Table 19.2 Chemical compositions of Ti-Mn alloys$_{MIM}$ fabricated by MIM (mass%)

Alloy	Element					
	Ti	Mn	O	C	N	Nominal Mn
Ti-8Mn$_{MIM}$	bal.	7.56	0.233	0.0606	0.0064	8.0
Ti-9Mn$_{MIM}$	bal.	9.29	0.244	0.0654	0.0081	10.0
Ti-12Mn$_{MIM}$	bal.	11.5	0.252	0.0604	0.0096	12.0
Ti-13Mn$_{MIM}$	bal.	13.3	0.270	0.0512	0.0081	14.0
Ti-15Mn$_{MIM}$	bal.	15.4	0.275	0.0655	0.0123	16.0
Ti-17Mn$_{MIM}$	bal.	17.1	0.316	0.0512	0.0075	18.0

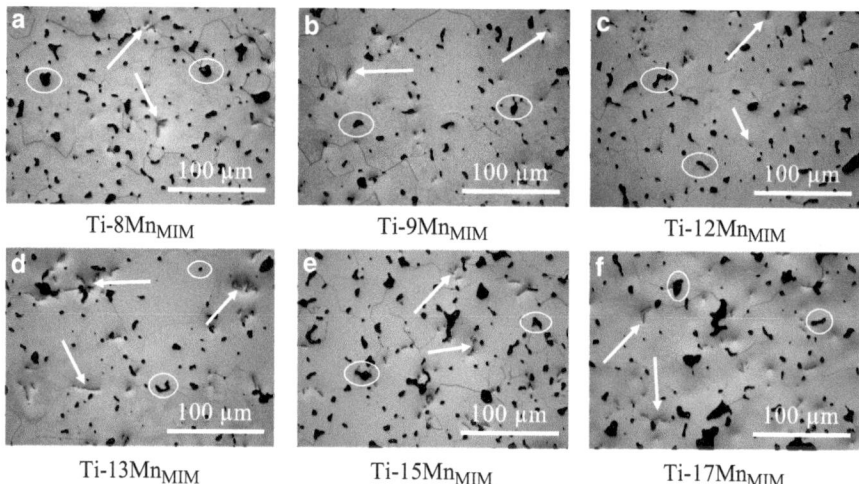

Fig. 19.9 Typical optical micrographs of (**a**) Ti-8Mn$_{MIM}$, (**b**) Ti-9Mn$_{MIM}$, (**c**) Ti-12Mn$_{MIM}$, (**d**) Ti-13Mn$_{MIM}$, (**e**) Ti-15Mn$_{MIM}$, and (**f**) Ti-17Mn$_{MIM}$. *Ellipses* indicate some of the pores and *arrows* indicate some of the precipitates

from the Ti and Mn powders, which contain 0.16 and 0.77 mass% O, respectively. The C content of the alloys is of approximately 0.06 mass%. This is likely due to C pickup during the debinding process [23]. Furthermore, Mn evaporation is also observed to occur during fabrication by MIM, as the Mn content of the alloys is lower than the nominal content [16].

Figures 19.9 and 19.10 show the optical micrographs and XRD profiles of the Ti-(8–17)Mn$_{MIM}$, respectively. Small closed pores, large interconnected pores, and elongated precipitates can be observed in each alloy. Both the pores (ellipses) and precipitates (arrows) are mostly located at the grain boundaries. Only the diffraction peaks of the attributed to β planes are observable in the XRD profiles of Ti-(8–17) Mn$_{MIM}$. Again, the presence of the athermal ω phase has been confirmed by the TEM observations. Figure 19.11 shows the SAED patterns and DF images of the diffraction spots or streaks of the ω phase of Ti-(8–17)Mn$_{MIM}$. Clear diffraction spots produced by the athermal ω phase can be observed in the SAED patterns of

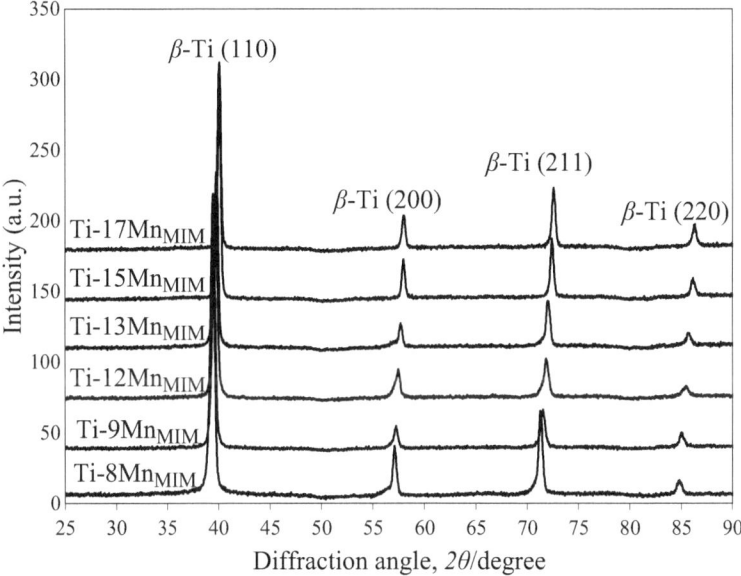

Fig. 19.10 XRD profiles of Ti-Mn$_{MIM}$ alloys

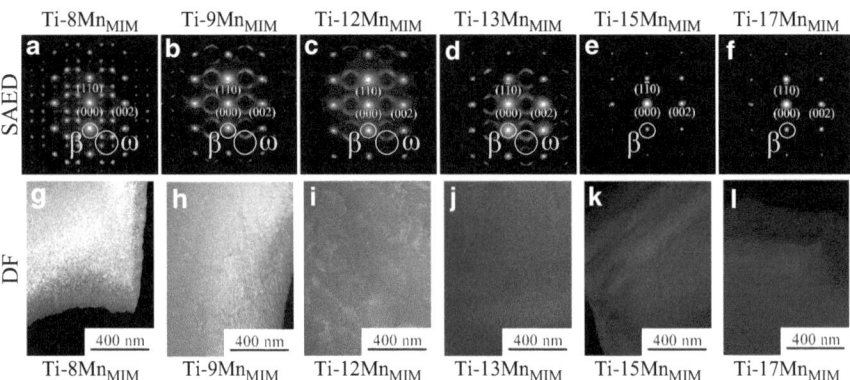

Fig. 19.11 Typical SAED patterns viewed from [110]$_\beta$, and corresponding DF images of the diffraction spots or streaks of ω phase of (**a**) and (**g**) Ti-8Mn$_{MIM}$, (**b**) and (**h**) Ti-9Mn$_{MIM}$, (**c**) and (**i**) Ti-12Mn$_{MIM}$, (**d**) and (**j**) Ti-13Mn$_{MIM}$, (**e**) and (**k**) Ti-15Mn$_{MIM}$, and (**f**) and (**l**) Ti-17Mn$_{MIM}$

Ti-8Mn$_{MIM}$ (Fig. 19.11a). The DF image of the ω spots (Fig. 19.11g) evidences a high-volume fraction of the athermal ω phase of Ti-8Mn$_{MIM}$. The intensity of the spots attributed to the athermal ω phase and the apparent volume fraction of the athermal ω particles gradually decrease with increasing Mn content. It is observed that Ti-18Mn$_{LM}$ still shows very diffuse streaks associated to the athermal ω phase, while Ti-(15–17)Mn$_{MIM}$ don't show spots or streaks. This is likely caused by the

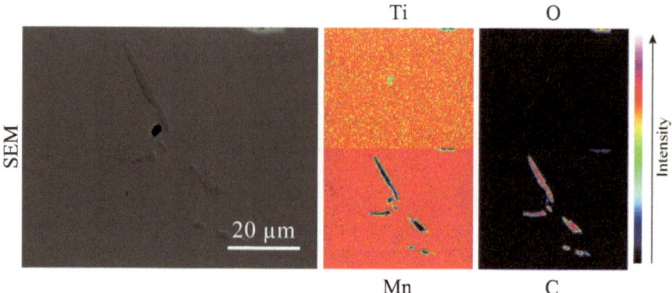

Fig. 19.12 SEM image and EPMA elemental mapping images of Ti, Mn, O, and C in a precipitate in a Ti-17Mn$_{MIM}$ alloy specimen

higher O content of the MIM alloys. O has been reported to inhibit the formation of ω phase in Ti alloys [24]. Figure 19.12 shows the results of the EPMA analysis of the precipitates. The precipitates are low in Mn and high in C. The precipitates are thus identified as Ti carbides.

The volume fraction and average diameter of both pores and carbides in the alloys have been estimated from the optical micrographs using an image analysis software [16]. There is no significant variation of either the volume fraction and average diameter of both pores and precipitates among Ti-(8–17)Mn$_{MIM}$. However, the shapes of the pores in Ti-(8–17)Mn$_{MIM}$ are irregular, although there are no significant differences among the pore morphologies of the alloys. The irregular pore shape is most likely due to the fact that the Mn powder has a more irregular shape compared to the Ti powder [16]. The high C levels facilitate the formation of carbides. The morphology of the carbides does not particularly vary among Ti-(8–17)Mn$_{MIM}$.

Figures 19.13 and 19.14 show the Vickers hardness and Young's modulus of the Ti-(8–17)Mn$_{MIM}$ alloys, respectively. The hardness does not significantly vary among the alloys, and they are comparable to that of Ti-64 ELI, indicated by the dashed line in Fig. 19.13 [5]. The Young's moduli of the alloys are all lower than that for annealed Ti-64 ELI, indicated by the dashed line in Fig. 19.14 [5]. Besides the parameters already discussed for the alloys fabricated by CCLM, such as the amount of athermal ω phase and the effects of increasing Mn, which can affect both hardness and Young's modulus measurements, other parameters must be considered for the alloys fabricated by MIM [15, 16]. The presence of the pores, which cause localized stress concentration, tends to decrease the hardness of the alloys [23, 25]. Conversely, the presence of carbides, which can cause precipitation strengthening, tends to increase the hardness [23, 25]. However, because there is little variation of the volume fraction and average diameter of both pores and carbides, these parameters do not cause significant variation of the measured hardness and Young's modulus among the alloys. Furthermore, higher O contents increase the solid solution hardening effect, which in turn increases the hardness [25–27]. These opposing effects balance each other so that there is little hardness variation among the alloys.

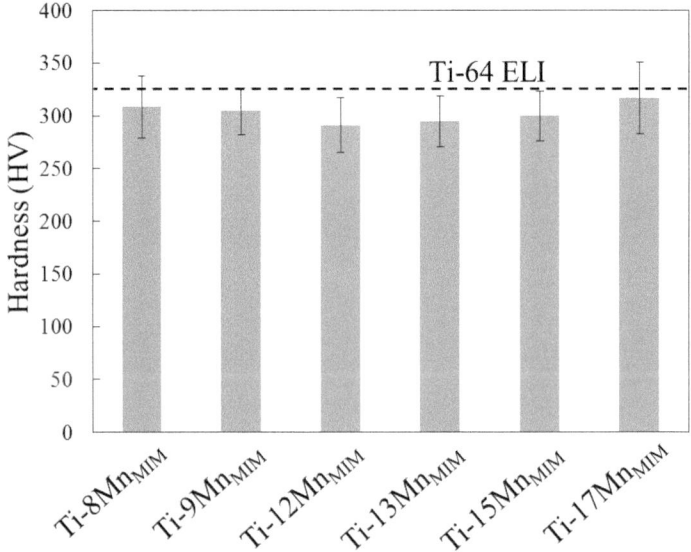

Fig. 19.13 Comparison of Vickers hardness of Ti-Mn$_{MIM}$ alloys with that of Ti-64 ELI (*dashed line*)

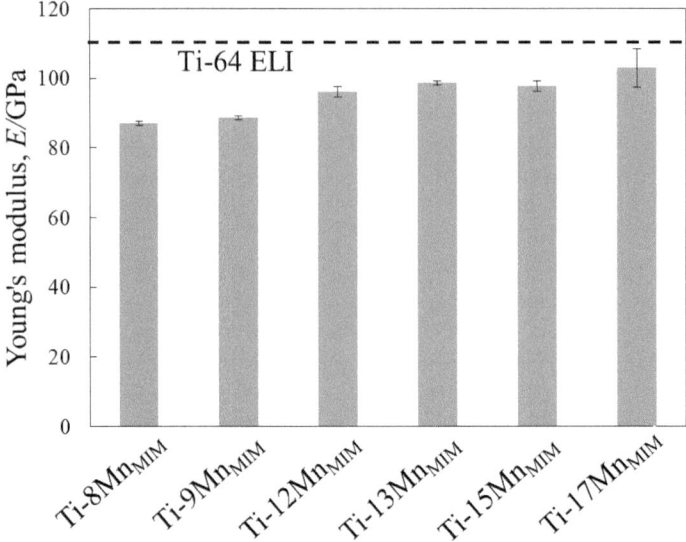

Fig. 19.14 Comparison of Young's moduli of Ti-Mn$_{MIM}$ alloys with that of Ti-64 ELI (*dashed line*)

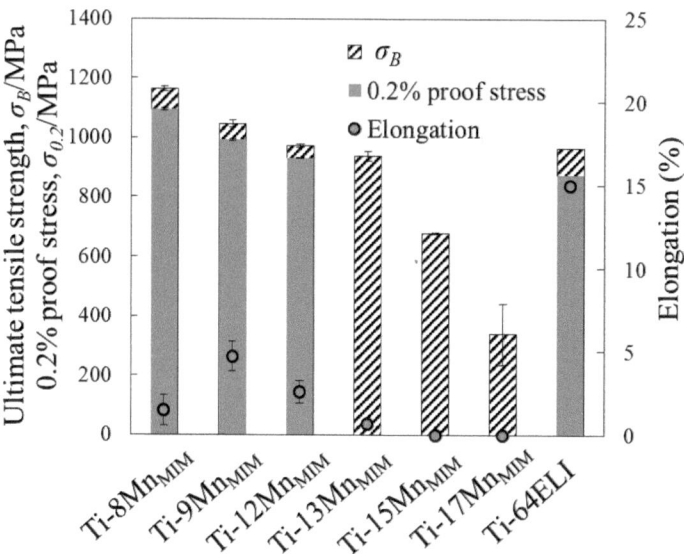

Fig. 19.15 Comparison of tensile properties of Ti-Mn$_{MIM}$ alloys. The properties of Ti-64 ELI are also shown

Figure 19.15 shows the tensile properties of Ti-(8–17)Mn$_{MIM}$ along with those of Ti-64 ELI [5]. Both σ_B and $\sigma_{0.2}$ values for Ti-(8–12)Mn$_{MIM}$ are higher than the corresponding values for Ti-64 ELI [5]. However, the elongation of the Ti-Mn$_{MIM}$ alloys is lower than that of annealed Ti-64 ELI. It is noted that the elongation of Ti-Mn$_{MIM}$, which contains higher O content, pores, and carbides, is lower than that of Ti-Mn$_{LM}$, which do not contain pores and carbides and have lower O content. Thus, the lower elongation of Ti-Mn$_{MIM}$ can be attributed to the combined effects of a higher O content primarily, along with the presence of pores and carbides, which are inherent to the MIM process [23, 28, 29].

Figure 19.16 shows the compressive properties of Ti-(8–17)Mn$_{MIM}$. The compressive 0.2% proof stress ($\sigma_{c0.2}$) and compressive strain (ε_c) values for Ti-(8–17) Mn$_{MIM}$ are higher than those for Ti-64 [4], also shown in Fig. 19.16. The two main parameters that affect the compressive properties of an alloy are the Mn content (which causes solid solution strengthening) and the amount of the ω phase (which causes precipitation strengthening). The higher volume fraction of athermal ω phase in Ti-8Mn$_{MIM}$ and Ti-9Mn$_{MIM}$ causes their $\sigma_{c0.2}$ values to be higher than those of Ti-(12–17)Mn$_{MIM}$. As the volume fraction of the athermal ω phase decreases in Ti-(12–15)Mn$_{MIM}$, the solid solution strengthening effect of the Mn balances the effect of the decreasing ω phase. The increase in compressive strain with increasing Mn content is also due to the decrease in the volume fraction of the athermal ω phase with increasing Mn content.

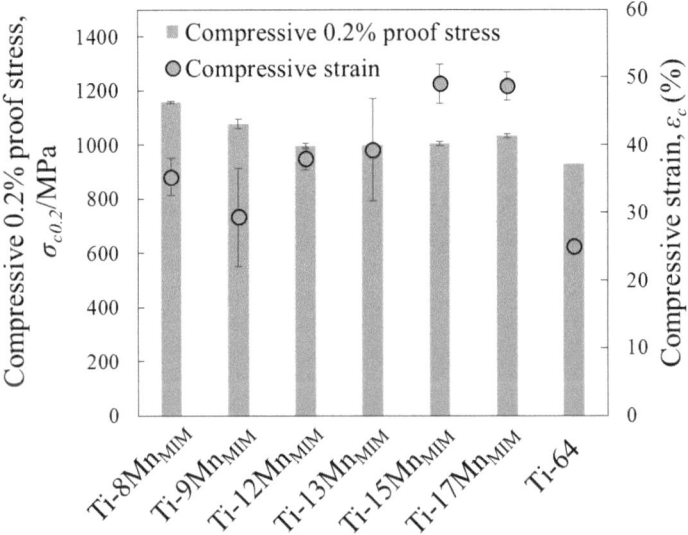

Fig. 19.16 Compressive properties of Ti-Mn$_{MIM}$ alloys and annealed Ti-64

19.4 Conclusions

Mn was selected as a low-cost β-stabilizer alloying element to fabricate Ti alloys of various compositions. The alloys were fabricated by both CCLM and MIM. Microstructural observations, mechanical performance tests, and biocompatibility tests were used to evaluate the mechanical properties and biocompatibility of the fabricated Ti-Mn alloys. Regarding the use of Mn as an alloying element with Ti and the competitive applicability of Ti-Mn alloys, it was possible to conclude that:

1. The alloys are primarily composed of equiaxed β grains. Some of the alloys with lower Mn content also contain the athermal ω phase. The β phase stability increases, thus decreasing the volume fraction of the athermal ω phase, with increasing Mn content. The alloys fabricated by MIM contain high amounts of O and C, which are due to the powders and organic binder used, respectively. Because of the high C content, there is the precipitation of Ti carbide in the alloys fabricated by MIM. Furthermore, the alloys fabricated by MIM also contain pores.

2. The hardness, Young's modulus, and tensile strength of Ti-(9–13)Mn$_{LM}$ are comparable or superior to those of Ti-64 ELI. The hardness, Young's modulus, tensile strength, and compressive properties of Ti-(8–13)Mn$_{MIM}$ are comparable or superior to those of Ti-64 ELI. However, the ductility of the alloys fabricated by MIM is adversely affected by the high oxygen content and the presence of carbides and pores.

3. Among both alloys fabricated by CCLM and those fabricated by MIM, Ti-9Mn shows the best balance between tensile strength and ductility. In particular, the Ti-9Mn$_{LM}$ shows the largest elongation among all the Ti-Mn alloys. Conversely, Ti-9Mn$_{MIM}$ shows the lowest compressive properties among these alloys.
4. Every Ti-Mn$_{LM}$ alloy shows ion release rates consistent with their chemical composition when immersed in SBF. Ti-(6–13)Mn$_{LM}$ shows good cell viability ratios, which are comparable to those of CP-Ti. However, higher Mn concentrations should be avoided because of risks of Mn intoxication.

Acknowledgments This study was supported in part by a Grant-in-Aid for Scientific Research (A) No. 24246111; a Grant-in-Aid for Young Scientists (B) No. 25820367 from the Japan Society for the Promotion of Science (JSPS); the Inter-University Cooperative Research Program "Innovation Research for Biosis-Abiosis Intelligent Interface" from the Ministry of Education, Culture, Sports, Science and Technology (MEXT), Japan; and the Innovative Structural Materials Association (ISMA), Japan.

References

1. Geetha M, Singh AK, Asokamani R, Gogia AK. Ti based biomaterials, the ultimate choice for orthopaedic implants – a review. Prog Mater Sci. 2009;54:397–425. doi:10.1016/j.pmatsci.2008.06.004.
2. Niinomi M. Recent metallic materials for biomedical applications. Metall Mater Trans A. 2002;33:477–86. doi:10.1007/s11661-002-0109-2.
3. Niinomi M. Mechanical biocompatibilities of titanium alloys for biomedical applications. J Mech Behav Biomed Mater. 2008;1:30–42. doi:10.1016/j.jmbbm.2007.07.001.
4. Boyer R, Welsch G, Collings EW. Materials properties handbook: titanium alloys. New York: ASM International; 1993.
5. Niinomi M. Mechanical properties of biomedical titanium alloys. Mater Sci Eng A. 1998;243:231–6. doi:10.1016/S0921-5093(97)00806-X.
6. Perl DP. Relationship of aluminum to Alzheimer's disease. Environ Health Perspect. 1985;63:149–53. doi:10.1289/ehp.8563149.
7. Domingo JL. Vanadium and tungsten derivatives as antidiabetic agents: a review of their toxic effects. Biol Trace Elem Res. 2002;88:97–112. doi:10.1385/BTER:88:2:097.
8. Service PH. Public health statement for aluminum. 2008.
9. Bania PJ. Beta titanium alloys and their role in the titanium industry. JOM. 1994;46:16–9. doi:10.1007/BF03220742.
10. Niinomi M, Hattori T, Morikawa K, Kasuga T, Suzuki A, Fukui H, et al. Development of low rigidity β-type titanium alloy for biomedical applications. Mater Trans. 2002;43:2970–7.
11. Froes FH, Bomberger HB. The beta titanium alloys. JOM. 1985;37:28–37. doi:10.1007/BF03259693.
12. Pilliar RM, Cameron HU, Binnington AG, Szivek J, Macnab I. Bone ingrowth and stress shielding with a porous surface coated fracture fixation plate. J Biomed Mater Res. 1979;13:799–810. doi:10.1002/jbm.820130510.
13. Erdmann L, Graedel TE. Criticality of non-fuel minerals: a review of major approaches and analyses. Environ Sci Technol. 2011;45:7620–30. doi:10.1021/es200563g.
14. U.S. Department of Energy. Critical materials strategy. 2011.
15. Santos PF, Niinomi M, Cho K, Nakai M, Liu H, Ohtsu N, et al. Microstructures, mechanical properties and cytotoxicity of low cost beta Ti–Mn alloys for biomedical applications. Acta Biomater. 2015. doi:10.1016/j.actbio.2015.08.015.

16. Santos PF, Niinomi M, Liu H, Cho K, Nakai M, Itoh Y, et al. Fabrication of low-cost beta-type Ti–Mn alloys for biomedical applications by metal injection molding process and their mechanical properties. J Mech Behav Biomed Mater. 2016;59:497–507. doi:10.1016/j.jmbbm.2016.02.035.

17. You B-D, Lee B-W, Pak J-J. Manganese loss during the oxygen refining of high-carbon ferromanganese melts. Met Mater. 1999;5:497–502. doi:10.1007/BF03026165.

18. Ho WF. Effect of omega phase on mechanical properties of Ti-Mo alloys for biomedical applications. J Med Biol Eng. 2008;28:47–51.

19. Williams JC, Hickman BS, Marcus HL. The effect of omega phase on the mechanical properties of titanium alloys. Metall Trans. 1971;2:1913–19. doi:10.1007/BF02913423.

20. Reynolds N, Blumsohn A, Baxter JP, Houston G, Pennington CR. Manganese requirement and toxicity in patients on home parenteral nutrition. Clin Nutr. 1998;17:227–30.

21. Dickerson RN. Manganese intoxication and parenteral nutrition. Nutrition. 2001;17:689–93. doi:10.1016/S0899-9007(01)00546-9.

22. Okazaki Y, Gotoh E, Manabe T, Kobayashi K. Comparison of metal concentrations in rat tibia tissues with various metallic implants. Biomaterials. 2004;25:5913–20. doi:10.1016/j.biomaterials.2004.01.064.

23. German R. Progress in titanium metal powder injection molding. Materials (Basel). 2013;6:3641–62. doi:10.3390/ma6083641.

24. Paton NE, Williams JC. The influence of oxygen content on the athermal beta-omega transformation. Scr Metall. 1973;7:647–9.

25. Esteban PG, Bolzoni L, Ruiz-Navas EM, Gordo E. PM processing and characterisation of Ti-7Fe low cost titanium alloys. Powder Metall. 2011;54:242–52. doi:10.1179/174329009X457063.

26. Hammond C, Nutting J. The physical metallurgy of superalloys and titanium alloys. Prog Met Phys. 1977;7:65–163. doi:10.1016/0502-8205(58)90004-2.

27. Ogden HR, Jaffee RI. The effects of carbon, oxygen, and nitrogen on the mechanical properties of titanium and titanium alloys. 1955.

28. Liu H, Niinomi M, Nakai M, Cho K, Narita K, Şen M, et al. Mechanical properties and cytocompatibility of oxygen-modified β-type Ti–Cr alloys for spinal fixation devices. Acta Biomater. 2015;12:352–61. doi:10.1016/j.actbio.2014.10.014.

29. Zhao D, Chang K, Ebel T, Qian M, Willumeit R, Yan M, et al. Microstructure and mechanical behavior of metal injection molded Ti-Nb binary alloys as biomedical material. J Mech Behav Biomed Mater. 2013;28:171–82. doi:10.1016/j.jmbbm.2013.08.013.

Open Access This chapter is distributed under the terms of the Creative Commons Attribution 4.0 International License (http://creativecommons.org/licenses/by/4.0/), which permits use, duplication, adaptation, distribution and reproduction in any medium or format, as long as you give appropriate credit to the original author(s) and the source, provide a link to the Creative Commons license and indicate if changes were made.

The images or other third party material in this chapter are included in the work's Creative Commons license, unless indicated otherwise in the credit line; if such material is not included in the work's Creative Commons license and the respective action is not permitted by statutory regulation, users will need to obtain permission from the license holder to duplicate, adapt or reproduce the material.

Chapter 20
Effect of Titanium Surface Modifications of Dental Implants on Rapid Osseointegration

Ting Ma, Xiyuan Ge, Yu Zhang, and Ye Lin

Abstract The initial cellular response to the dental implant is essential for the subsequent tissue regeneration around the foreign implant surface. There are many cells and proteins involved in the integration process which leads to the final osseointegration between implants and peri-implant bone tissue. With regard to materials used in dental implants, titanium is a prevalent biomaterial applied in orthopedic or dental implants due to its premium mechanical and biological properties and osteoconductivity. The roughness and chemical composition of the titanium surface affect the process and rate of the osseointegration of dental implants. Different studies on the effect of roughness and wettability of titanium surface on the process of early events in the osseointegration are reviewed in this article. In addition, in order to accelerate this wound-healing process, varied surface topography and chemical composition have been produced depending on different types of surface modifications. The desirable dental implant surface design caters for the development of implantology for immediate loading and the improvement of long-term stability. An appropriate understanding of the interaction between cells and implant surfaces is essential for the future design of new surface which could enhance the speed and stability of osseointegration of dental implants.

Keywords Osseointegration • Surface modification • Bone-implant interface

T. Ma • Y. Zhang (✉) • Y. Lin
Department of Oral Implantology, Peking University, School
and Hospital of Stomatology, 22th South Avenue, Zhongguancun,
Haidian District, Beijing 100081, People's Republic of China
e-mail: zhang76yu@163.com

X. Ge
Central laboratory, Peking University, School and Hospital of Stomatology,
22th South Avenue, Zhongguancun, Haidian District, Beijing, People's Republic of China

© The Author(s) 2017   247
K. Sasaki et al. (eds.), *Interface Oral Health Science 2016*,
DOI 10.1007/978-981-10-1560-1_20

20.1 Introduction

In the 1960s, Per Ingvar Brånemark and his colleagues accidentally discovered that titanium implants and peri-implant bone tissues could be firmly bonded and there was no fibrous tissue between the interface according to intravital microscopy observation of bone defects around the microcirculation in the experiment, and they named this phenomenon "osseointegration." After continuous revision, in 1985, professor Brånemark redefined "osseointegration" as a direct interface between living bones and the implant surface, and this functional connection can be loaded [1, 2]. Due to its premium mechanical properties, chemical stability and biocompatibility, titanium and its alloys have been widely used in dental and orthopedic implants [3]. When titanium and its alloys are exposed in the air within a very short time, a dense oxide layer will be formed on its surface. This oxide layer does not only boost good biocompatibility but also makes this metallic material biologically inert, and the implant surface treatment is usually referred to the modification on this oxide layer [4]. In recent years, with technological development of implantology, techniques that shortened treatment time and lower failure risk are urgently demanded. In this regard, the osseointegration is affected by the surface characteristics of the titanium implant, such as roughness, wettability, the chemical composition, and so on [5, 6]. Therefore, the osseointegration biological process, the interaction between cells and implant surface, and some commonly used dental implant surface modification methods are reviewed in this article.

20.2 Cellular Events at the Bone-Implant Interface

During the biological process of osseointegration, a great number of different types of cells involved in this process exert an important role. Bone-implant integration is similar to the wound-healing process, which can be divided into four distinct phases: hemostasis phase, inflammatory phase, proliferative phase, and remodeling phase. Once the implant was placed into the prepared site, hemostasis began. Subsequently, growth factors and matrix proteins were activated and released from the injured tissue during the drilling process. Within a few seconds to hours, there was a water layer on the implant surface and extracellular matrix proteins were adsorbed onto the surface. Among these proteins, some of them such as vitronectin and fibronectin have interactions with the following host inflammatory response. Moreover, aggregated fibrinogen matrix facilitates more platelet adhesion and aggregation, and then the release of vasoactive substances and chemokines from the platelets symbolizes the beginning of inflammatory response [7–10] (Fig. 20.1).

The initial stage of the inflammatory phase is activated by the innate immune system. Once adhering to the surface of the implant, neutrophils secrete pro-inflammatory cytokines, collagenase, and other enzymes to remove the foreign bacteria. If the bacteria have not been cleared or a large number of bacteria still exist,

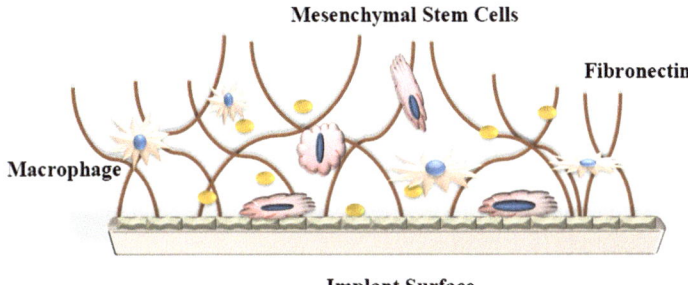

Fig. 20.1 Schematic showing fibrin clot formation on the implant surface and macrophages and MSCs play important roles in this process

the duration of cellular innate immune response will be extended, which would damage the surrounding normal tissue. Therefore, prolonged cellular innate immune response would also impede the wound healing. To sum up, immune and inflammatory responses play an important role during the wound-healing process of foreign implants [9]. In addition, another cell that is adsorbed onto the implant surface is the macrophage. Macrophages play an important role in phagocytosis of necrotic tissue, secretion of fibroblast growth factor, and angiogenic growth factors [11]. As osteoblast precursor cells adhered firmly, with the expression of osteocalcin and alkaline phosphatase, the cells gradually differentiate into mature osteoblasts, followed by continuously woven bone formation. In the process of bone remodeling, the gradual replacement of woven bone with lamellar bone depends on the interactions between osteoblasts and osteoclasts [12]. However, the differences between the natural wound-healing process and dental implant osseointegration lie in the topography, roughness, surface energy, chemical properties, and other elements of the implant surface that play an important role during this process [13, 14].

20.3   Topography and Roughness of Implant Surface

A great number of previous studies suggest that compared to the smooth surface, proper micron roughness serves to enhance osteoblast differentiation and increase bone-implant contact rate in vivo, even in the absence of osteogenic growth factors [15–18]. However, with the increase of surface roughness, the implant surface was also enlarged, which might aggravate bacterial colonization [19]. As a result, the balance between the required biological reaction and elimination of plaque accumulation around the implant needs to be considered [20]. In addition, roughness of micron surface topography to a certain extent could activate specific cell membrane receptors such as integrins, which serves as a communicator in the interactions between the extracellular matrix and the cytoskeleton [21, 22]. Previous studies indicate that surface roughness of natural bones is about 32 nm, and pore size of epithelial basement membrane is about 70–100 nm. Nanoscale roughened material might affect a variety of cells, such as epithelial cells, osteoblasts, fibroblasts, and so on [23]. Some studies show that the nanoscale surface materials are potent in altering the adhesion between the surface and the protein or cells. Moreover, Webster et al. found that nanosurface materials effectively improved the adhesion of vitronectin compared with the microscale surfaces. In addition, the surface of nano-materials promotes osteoblast adhesion and proliferation [24, 25]. Also, Ellingsen et al. found that implants with nanosurface accelerate the bone formation around the implants [26].

20.4 The Surface Wettability

Wettability is another important feature of the material surface. This basic physical parameter can be quantified by the value of the surface contact angle (CA), which is first proposed by Thomas Young in 1805 [27]. Generally, a surface with a contact angle of less than 90° is considered as a hydrophilic one and if the contact angle is equal to 0°, it is considered as a super-hydrophilic surface. On the contrary, the hydrophobic surface is defined as the one with the value of CA more than 90°.

The roughness would also influence the wettability of material surfaces. Previous studies have shown that the surface contact angle of the pure titanium was about 70–90° regardless of surface roughness. However, after a serial surface roughening process including acid etching and sandblasting, the CA of the surface could reach up to 150°. Moreover, it is reported that the CA of the surface rarely exceeds 120° if the material is only treated by chemical methods rather than roughening. Therefore, the hydrophobic surfaces with the CA ranging from 125 to 180° often call for a combination of different treatments [28–30]. Surfaces with different wettability produce diverse biological effects of osseointegration, and the impact can be divided into the following four aspects: protein and biological macromolecular adhesion to the material surface, biological behavior of different cells on the surface, the formation of the bacteria biofilm, and in vivo study of osseointegration.

There are different views on the adhesion of protein on the surfaces with different wettability. Previous studies have shown that the fibronectin adhesion is much more facilitated on a hydrophilic surface than on the hydrophobic surface, and the fibronectin adhering to the hydrophilic surface can maintain a better biological activity which could promote osteoblast adhesion and differentiation [7, 31]. However, Tugulu et al. [32] found that the super-hydrophilic surface treated by diluted alkaline solution would reduce the adhesion of the fibrinogen and thereby reduce the inflammation around the implant, which enhances the potential of the promotion of implant osseointegration in vivo [33]. Besides, the adhesion of other proteins which play important roles in osteogenic differentiation such as vitronectin and type I collagen is also affected by surfaces with diverse characteristics [34].

More recently, some results demonstrate that the expression of genes related to the osteogenic differentiation of mesenchymal stem cells on the super-hydrophilic surface is higher than the level of the expression of the cells on the hydrophobic surface, indicating that the super-hydrophilic surface is more conducive to the osteogenic differentiation of mesenchymal stem cells than the hydrophobic surface [17, 35]. It was also implied that super-hydrophilic surface can serve better to promote osteoblast maturation and mineralization [15, 36]. In 2005, Buser et al. [37] found that the super-hydrophilic surface modified SLA (modSLA) promoted the bone-implant contact rate in the early period (2 and 4 weeks) in vivo, while there was no difference between the experimental group and the control group 8 weeks later after the implantation. In 2007, Schwarz et al. [38] indicated that the formation of collagen fibers was observed on the super-hydrophilic surface modSLA of the implant in the first 4 days after the implantation. Two weeks later, bones formed around the super-hydrophilic surfaces became much denser than those formed on the control surfaces. Another experiment conducted on healthy adult volunteers showed that the implants with super-hydrophilic surface can promote bone integration after 2–4 weeks [39]. In addition, the adhesion and proliferation of the soft tissues around the implant such as epithelial cells and fibroblasts were simultaneously affected by the wettability of the surface. Similar results showed that super-hydrophilic surface was helpful to form the rapid and compact soft tissue seal around the implant [31, 40]. In summary, a large number of studies deliver a conclusion that the hydrophilic surface, especially the super-hydrophilic surface, exhibits preferable advantages in terms of promoting osseointegration in the early stage.

20.5   The Common Methods of Surface Modifications

Since the surface morphology and characteristics of the implant play important roles in the osseointegration process, researches on the surface of the implant are particularly indispensable. The surfaces with relatively similar features can be obtained even through several completely different processes, whereas the same process can fabricate two different types of surfaces only by changing their parameters [41]. Based on the topology and the average roughness value (Sa) of the material surface, Albrektsson and Wennerberg et al. [42] classified implant surfaces into four categories: smooth surface (Sa: 0–0.04 μm), slightly rough surface (Sa: 0.5–1 μm), medium rough surface (Sa: 1–2 μm), and rough surface (Sa: >2 μm). Among them, the medium rough surface exhibited satisfactory clinical results in comparison with smooth surface and rough surface. In another way by some other researchers, the implant surfaces can be classified into three categories based on the surface roughness: large roughness surface (Sa greater than 10 μm), micro-sized roughness surface (Sa: 1–10 μm), and nano-sized roughness surface [16]. Among them, the rough surface demonstrated satisfactory performance in the protein adhesion, the formation of extracellular matrix, the promotion of osteogenic differentiation and osseointegration in vivo, etc. Therefore, the roughening treatment is still prevalent in modifying the implant surface.

The following is the common practical roughening methods, such as plasma spraying, sand blasting, chemical etching, anodic oxidation, electrochemical micro-arc oxidation treatment method, and so on [41]. By virtue of plasma spraying technique, the osteoinductive hydroxyapatite (HA) particles with good biocompatibility and mechanical strength could adhere to the surface and roughen the surface at the same time. Moreover, it has been confirmed that osseointegration can be enhanced by this HA-coated surfaces [43].

Sandblasted and acid-etched (SLA) is a common roughening method employed in titanium dental implant surfaces, which forms the complex surface with considerate roughness and micro-roughness. The method mentioned above has been commercialized. Moreover, this approach has a "gold standard" in the industry: firstly, the surface of the implant is sandblasted by using aluminum oxide particles (250–500 μm), and then the specimen is etched by HCl/H_2SO_4 solution of specific concentration. The average surface roughness (Ra) of the treated material is 1.5 μm [44]. In 1990, Wilke et al. [45] first reported the in vivo application of SLA surface implants and found that the treatment could effectively improve the mechanical stability of the implant. Buser et al. [46] observed that the SLA surface treatment effectively increased the bone contact rate of the implant from small pigs in vivo compared to the titanium plasma spray method (titanium plasma spray, TPS). A body of evidence presented in published data indicates that pits with the diameter of 1–2 μm can be observed being distributed regularly on the SLA surface and that the processing method may raise the short-term bone contact rate of the implant after the implantation. In addition, a large number of studies have shown that the SLA surfaces of in vitro experiments, animal experiments, and clinical practices have achieved good results [44, 47]. Besides, nanoscaled treatments on the implant surface not only affect the topography of the implant surface but also change the chemical properties of the material surface. There are several common nanotechnologies for surface processing, including physical compression method, self-assembled monolayers, chemical treatment method (acid, alkali, and hydrogen peroxide), nanoparticle deposition method, the anodic oxidation method, etc. [23].

There are several surface treatment methods which could enhance the wettability of the surface of the material. For example, Baier et al. [48] found that the use of radio-frequency glow discharge (RFGD) technology could effectively clean and disinfect the surface of inorganic materials, thus increasing the surface energy of the materials and strengthening the adhesion of the cells. It was reported that the superhydrophilic surface could be formed after an atmospheric plasma treatment on titanium foil. In this way, the contact angle could approach 0°, which greatly facilitates osteoblasts to spread on the treated surface [49]. But at the same time, some researchers believe that the chemical or plasma treatment yields satisfactory hydrophilicity that could not last for long in air [50].

However, by roughening and chemical treatment of the implant surface, the osteoconductivity could thus be enhanced. With the development of biochemical techniques, some specific short peptide fragments, proteins or growth factors, and other biologically active substance could be fixed onto the surface of the material so as to promote surface osteoinductivity. Currently used biologically active molecules currently fall into the following categories: cell adhesion molecules, such as those containing arginine-glycine-aspartic acid (RGD) peptide which could promote cell adhesion and extracellular matrix attachment; bone morphogenetic proteins (BMP) such as the application of BMP-2 and BMP-7, which have strong ability to promote bone formation; and growth factors, such as insulin-like growth factors (IGFs), transforming growth factor-β (TGF-β), vascular endothelial growth factor (VEGF), fibroblast growth factor (FGF), and so on [51, 52].

20.6 Polydopamine Surface Modification

Marine mussels adhered to the surface of the metal, rock, and so many different materials even under humid conditions that rely on the main component of mussel adhesive protein (MAPs). The protein contains lysine and L-3,4-dihydroxyphenylalanine (L-3,4-dihydroxyphenylalanine, L-DOPA). Inspired by the adhesion of mussels, a versatile poly dopamine (PDA) has been developed which could adhere to a wide range of organic and inorganic materials, such as metals, glasses, ceramics, and other synthetic polymers. In 2007, Lee et al. [53] found that by soaking the material in an alkaline (pH = 8.5) dopamine solution, a layer of PDA membrane could be formed on the material surface, which also provided a platform for secondary material modification. The two-step method of covalent modification of the fixed type I collagen on the titanium surface by means of PDA coating promotes MC3T3-E1 cell adhesion and early osteogenic differentiation [54]. Lee et al. [55] used 15 short peptides derived from BMP-7 amino acid which is called bone formation peptide1 (BFP1) to modify the polylactic acid-glycolic acid (PLGA) copolymer surface through PDA coating. The modified materials were implanted to reconstruct the skull defects in mice. Eight weeks later, the modified materials exhibited enhanced new bone formation around the bone defects. In another animal experiment, human adipose-derived stem cells (hADSCs) cultured on the PDA and BMP-2-modified PLGA scaffold were implanted in mice skull defects, and the modified materials were also found to effectively enhance new bone formation [56]. It is reported that PDA coating is a biocompatible coating which can be reduced by PDA-modified PLLA of inflammation and immune responses [57].

References

1. Branemark R, Branemark P, Rydevik B, et al. Osseointegration in skeletal reconstruction and rehabilitation: a review. J Rehabil Res Dev. 2001;38:175.
2. Bránemark P, Zarb G, Albrektesson T. Tissue-integrated prostheses. In: Quintessence Publishing Co., Inc., Chicago; 1985.
3. Brunette DM. Principles of cell behavior on titanium surfaces and their application to implanted devices[A]. In: Titanium in medicine. Berlin: Springer; 2001. p. 485–512.

4. Sul Y-T. The significance of the surface properties of oxidized titanium to the bone response: special emphasis on potential biochemical bonding of oxidized titanium implant. Biomaterials. 2003;24:3893–907.
5. Elias CN, Oshida Y, Lima JH, et al. Relationship between surface properties (roughness, wettability and morphology) of titanium and dental implant removal torque. J Mech Behav Biomed Mater. 2008;1:234–42.
6. Ku SH, Park CB. Human endothelial cell growth on mussel-inspired nanofiber scaffold for vascular tissue engineering. Biomaterials. 2010;31:9431–7.
7. Wilson CJ, Clegg RE, Leavesley DI, et al. Mediation of biomaterial-cell interactions by adsorbed proteins: a review. Tissue Eng. 2005;11:1–18.
8. Marx RE. Platelet-rich plasma: evidence to support its use. J Oral Maxillofac Surg. 2004;62:489–96.
9. Terheyden H, Lang NP, Bierbaum S, et al. Osseointegration – communication of cells. Clin Oral Implants Res. 2012;23:1127–35.
10. Jansson E, Tengvall P. In vitro preparation and ellipsometric characterization of thin blood plasma clot films on silicon. Biomaterials. 2001;22:1803–8.
11. Witte MB, Barbul A. Role of nitric oxide in wound repair. Am J Surg. 2002;183:406–12.
12. Zelzer E, McLean W, Ng YS, et al. Skeletal defects in VEGF(120/120) mice reveal multiple roles for VEGF in skeletogenesis. Development. 2002;129:1893–904.
13. Colnot C, Romero DM, Huang S, et al. Molecular analysis of healing at a bone-implant interface. J Dent Res. 2007;86:862–7.
14. Ponche A, Bigerelle M, Anselme K. Relative influence of surface topography and surface chemistry on cell response to bone implant materials. Part 1: physico-chemical effects. Proc Inst Mech Eng H. 2010;224:1471–86.
15. Vlacic-Zischke J, Hamlet SM, Friis T, et al. The influence of surface microroughness and hydrophilicity of titanium on the up-regulation of TGFbeta/BMP signalling in osteoblasts. Biomaterials. 2011;32:665–71.
16. Le Guehennec L, Soueidan A, Layrolle P, et al. Surface treatments of titanium dental implants for rapid osseointegration. Dent Mater. 2007;23:844–54.
17. Olivares-Navarrete R, Hyzy SL, Hutton DL, et al. Direct and indirect effects of microstructured titanium substrates on the induction of mesenchymal stem cell differentiation towards the osteoblast lineage. Biomaterials. 2010;31:2728–35.
18. Cochran DL, Schenk RK, Lussi A, et al. Bone response to unloaded and loaded titanium implants with a sandblasted and acid-etched surface: a histometric study in the canine mandible. J Biomed Mater Res. 1998;40:1–11.
19. An YH, Friedman RJ. Concise review of mechanisms of bacterial adhesion to biomaterial surfaces. J Biomed Mater Res. 1998;43:338–48.
20. Albrektsson T, Wennerberg A. The impact of oral implants – past and future, 1966–2042. J Can Dent Assoc. 2005;71:327.
21. Zhao G, Raines AL, Wieland M, et al. Requirement for both micron- and submicron scale structure for synergistic responses of osteoblasts to substrate surface energy and topography. Biomaterials. 2007;28:2821–9.
22. Keselowsky BG, Wang L, Schwartz Z, et al. Integrin alpha(5) controls osteoblastic proliferation and differentiation responses to titanium substrates presenting different roughness characteristics in a roughness independent manner. J Biomed Mater Res A. 2007;80:700–10.
23. Mendonca G, Mendonca DB, Aragao FJ, et al. Advancing dental implant surface technology – from micron- to nanotopography. Biomaterials. 2008;29:3822–35.
24. Webster TJ, Schadler LS, Siegel RW, et al. Mechanisms of enhanced osteoblast adhesion on nanophase alumina involve vitronectin. Tissue Eng. 2001;7:291–301.
25. Webster TJ, Ergun C, Doremus RH, et al. Specific proteins mediate enhanced osteoblast adhesion on nanophase ceramics. J Biomed Mater Res. 2000;51:475–83.
26. Ellingsen JE, Johansson CB, Wennerberg A, et al. Improved retention and bone-implant contact with fluoride-modified titanium implants. Int J Oral Maxillofac Implants. 2004;19:659–66.

27. Young T. An essay on the cohesion of fluids. Phil Trans R Soc (London). 1805;95:65–87.
28. Rupp F, Scheideler L, Eichler M, et al. Wetting behavior of dental implants. Int J Oral Maxillofac Implants. 2011;26:1256–66.
29. Gittens RA, Olivares-Navarrete R, Cheng A, et al. The roles of titanium surface micro/nanotopography and wettability on the differential response of human osteoblast lineage cells. Acta Biomater. 2013;9:6268–77.
30. Marmur A. Hydro-hygro-oleo-omni-phobic? Terminology of wettability classification. Soft Matter. 2012;8:6867–70.
31. Gittens RA, Scheideler L, Rupp F, et al. A review on the wettability of dental implant surfaces II: biological and clinical aspects. Acta Biomater. 2014;10:2907–18.
32. Tugulu S, Lowe K, Scharnweber D, et al. Preparation of superhydrophilic microrough titanium implant surfaces by alkali treatment. J Mater Sci Mater Med. 2010;21:2751–63.
33. Tang L, Wu Y, Timmons RB. Fibrinogen adsorption and host tissue responses to plasma functionalized surfaces. J Biomed Mater Res. 1998;42:156–63.
34. Salasznyk RM, Williams WA, Boskey A, et al. Adhesion to vitronectin and collagen I promotes osteogenic differentiation of human mesenchymal stem cells. BioMed Res Int. 2004;2004:24–34.
35. Wall I, Donos N, Carlqvist K, et al. Modified titanium surfaces promote accelerated osteogenic differentiation of mesenchymal stromal cells in vitro. Bone. 2009;45:17–26.
36. Zhao G, Schwartz Z, Wieland M, et al. High surface energy enhances cell response to titanium substrate microstructure. J Biomed Mater Res A. 2005;74:49–58.
37. Buser D, Broggini N, Wieland M, et al. Enhanced bone apposition to a chemically modified SLA titanium surface. J Dent Res. 2004;83:529–33.
38. Schwarz F, Herten M, Sager M, et al. Histological and immunohistochemical analysis of initial and early osseous integration at chemically modified and conventional SLA titanium implants: preliminary results of a pilot study in dogs. Clin Oral Implants Res. 2007;18:481–8.
39. Lang NP, Salvi GE, Huynh-Ba G, et al. Early osseointegration to hydrophilic and hydrophobic implant surfaces in humans. Clin Oral Implants Res. 2011;22:349–56.
40. Scheideler L, Rupp F, Wendel HP, et al. Photocoupling of fibronectin to titanium surfaces influences keratinocyte adhesion, pellicle formation and thrombogenicity. Dent Mater. 2007;23:469–78.
41. Annunziata M, Guida L. The effect of titanium surface modifications on dental implant osseointegration. Front Oral Biol. 2015;17:62–77.
42. Albrektsson T, Wennerberg A. Oral implant surfaces: part 1 – review focusing on topographic and chemical properties of different surfaces and in vivo responses to them. Int J Prosthodont. 2004;17:536–43.
43. Darimont G, Cloots R, Heinen E, et al. In vivo behaviour of hydroxyapatite coatings on titanium implants: a quantitative study in the rabbit. Biomaterials. 2002;23:2569–75.
44. Szmukler-Moncler S, Perrin D, Ahossi V, et al. Biological properties of acid etched titanium implants: effect of sandblasting on bone anchorage. J Biomed Mater Res B Appl Biomater. 2004;68:149–59.
45. Wilke H, Claes L, Steinemann S. The influence of various titanium surfaces on the interface shear strength between implants and bone. Adv Biomater. 1990;9:309–14.
46. Buser D, Schenk R, Steinemann S, et al. Influence of surface characteristics on bone integration of titanium implants. A histomorphometric study in miniature pigs. J Biomed Mater Res. 1991;25:889–902.
47. Cochran DL, Buser D, Ten Bruggenkate CM, et al. The use of reduced healing times on ITI® implants with a sandblasted and acid-etched (SLA) surface. Clin Oral Implants Res. 2002;13:144–53.
48. Baier R, Meyer A, Natiella J, et al. Surface properties determine bioadhesive outcomes: methods and results. J Biomed Mater Res. 1984;18:337–55.
49. Duske K, Koban I, Kindel E, et al. Atmospheric plasma enhances wettability and cell spreading on dental implant metals. J Clin Periodontol. 2012;39:400–7.

50. Rupp F, Axmann D, Ziegler C, et al. Adsorption/desorption phenomena on pure and Teflon® AF-coated titania surfaces studied by dynamic contact angle analysis. J Biomed Mater Res. 2002;62:567–78.
51. Kim T-I, Jang J-H, Kim H-W, et al. Biomimetic approach to dental implants. Curr Pharm Des. 2008;14:2201–11.
52. Avila G, Misch K, Galindo-Moreno P, et al. Implant surface treatment using biomimetic agents. Implant Dent. 2009;18:17–26.
53. Lee H, Dellatore SM, Miller WM, et al. Mussel-inspired surface chemistry for multifunctional coatings. Science. 2007;318:426–30.
54. Yu X, Walsh J, Wei M. Covalent immobilization of collagen on titanium through polydopamine coating to improve cellular performances of MC3T3-E1 cells. RSC Adv. 2013;4:7185–92.
55. Lee YJ, Lee JH, Cho HJ, et al. Electrospun fibers immobilized with bone forming peptide-1 derived from BMP7 for guided bone regeneration. Biomaterials. 2013;34:5059–69.
56. Ko E, Yang K, Shin J, et al. Polydopamine-assisted osteoinductive peptide immobilization of polymer scaffolds for enhanced bone regeneration by human adipose-derived stem cells. Biomacromolecules. 2013;14:3202–13.
57. Hong S, Kim KY, Wook HJ, et al. Attenuation of the in vivo toxicity of biomaterials by polydopamine surface modification. Nanomedicine. 2011;6:793–801.

Open Access This chapter is distributed under the terms of the Creative Commons Attribution 4.0 International License (http://creativecommons.org/licenses/by/4.0/), which permits use, duplication, adaptation, distribution and reproduction in any medium or format, as long as you give appropriate credit to the original author(s) and the source, provide a link to the Creative Commons license and indicate if changes were made.

The images or other third party material in this chapter are included in the work's Creative Commons license, unless indicated otherwise in the credit line; if such material is not included in the work's Creative Commons license and the respective action is not permitted by statutory regulation, users will need to obtain permission from the license holder to duplicate, adapt or reproduce the material.

Chapter 21
Development of Powder Jet Deposition Technique and New Treatment for Discolored Teeth

Kuniyuki Izumita, Ryo Akatsuka, Akihiko Tomie, Chieko Kuji,
Tsunemoto Kuriyagawa, and Keiichi Sasaki

Abstract The powder jet deposition (PJD) process is for creation of a hydroxyapatite (HA) layer on human teeth. To develop the PJD device, the layer-forming properties have improved. Created HA layers with a new handpiece-type PJD device demonstrate excellent material properties in vitro. Titanium dioxide (TiO_2) is known to cause whitening because of the selective reflection of the light. Therefore, we assessed the possibility of using the creation of TiO_2-HA layers with the new PJD device as a new treatment for discolored teeth. In this study, the microstructural and mechanical properties of TiO_2-HA layers were evaluated under the same conditions of the previous study. These properties were evaluated before and after 500 cycles of thermal cycling (5–55 °C). Furthermore, the CIE $L^*a^*b^*$ color system was used for color measurement and $\varDelta E^*$ values for color differences were calculated. The maximum thickness of the TiO_2-HA layers was about 60 μm. There were no significant differences in thickness, hardness, or bonding strength before and after thermal cycling. The layers showed an increased L^* parameter and a decreased b^* parameter, and the color difference $\varDelta E^*$ was approximately 6.7 units. Creation of TiO_2-HA layers by PJD might be a valuable new treatment for discolored teeth.

Keywords Powder jet deposition • Hydroxyapatite • Titanium dioxide • Discolored teeth • Thermal cycle

K. Izumita (✉) • R. Akatsuka
Division of Advanced Prosthetic Dentistry, Graduate School of Dentistry, Tohoku University,
4-1 Seiryo-machi, Aoba-ku, Sendai, Miyagi 980-8575, Japan
e-mail: k-izumita@dent.tohoku.ac.jp

A. Tomie • C. Kuji • T. Kuriyagawa
Department of Nanomechanics, Graduate School of Engineering, Tohoku University,
Sendai, Miyagi, Japan

K. Sasaki
Division of Advanced Prosthetic Dentistry, Graduate School of Dentistry, Tohoku University,
4-1 Seiryo-machi, Aoba-ku, Sendai, Japan

© The Author(s) 2017
K. Sasaki et al. (eds.), *Interface Oral Health Science 2016*,
DOI 10.1007/978-981-10-1560-1_21

21.1   Introduction

The biomaterials used for dental restorative treatments have quite different chemical compositions and mechanical properties from tooth substances which are enamel and dentin. Even if the adhesion technique is properly applied, these differences in properties between restorative materials and tooth substance can give rise to different mechanical, thermal, or chemical stresses in the oral cavity and can lead to clinical problems such as drop-off of the restorative material from the tooth or secondary caries [1]. In our previous studies, we presented the possibility of applying a layer of HA as a new restorative material with chemical, compositional, and mechanical properties corresponding to the tooth substance [2, 3]. A newly developed advanced technology called powder jet deposition (PJD) was used to create the HA layers. The PJD technique was originated in abrasive jet machining (AJM), which is one of the precise processing methods for hard-brittle materials through an average particle size of approximately 10–20 μm, and it is also a well-established or developed noncontact mechanical removal process. When smaller particles of ceramic or metal, approximately 1–3 μm in diameter, were blasted onto ceramic substrate by the AJM process, the particles were deposited on the substrate. This is called the PJD technique. The PJD technique is performed at room temperature and at atmospheric pressure, which enables it to be used with HA particles for the creation of HA layers on human teeth [1].

White and well-aligned teeth are the most important aspect of a smile; discolored teeth can influence self-esteem and professional relationships [4, 5]. Tooth discoloration is classified as an intrinsic or extrinsic depending on its cause [4–7]. Discoloration caused by extrinsic factors, including chromogens derived from coffee, cigarettes, or dental plaque, is generally removed by professional tooth polishing [4–8]. Treatment of teeth with intrinsic discoloration, which may be drug related (tetracycline) or caused by pulp tissue remnants after endodontic therapy [9], is done by bleaching with radical agents or tooth restoration with dental materials such as composite resin. Several studies [10, 11] have reported side effects of tooth bleaching agents, such as tooth hypersensitivity, irritation of dental pulp, and decreased hardness of enamel. These observations suggest the need for a new treatment for discolored teeth including the creation of an interface layer with material similar to the tooth and no adverse tooth substance and pulpal response.

In daily life, TiO_2 is a common additive in many foods, pigments, and other consumer products used. TiO_2-based ingredients are mostly used as a coloring agent [12]. According to a study carried out to determine the influence of TiO_2 particles in a concentration of 0.10–0.25 %, the particles induced a whiter tooth color and simulated the opalescence of human enamel [13]. We therefore proposed the possibility of applying creation of TiO_2-HA layers on the human tooth surface by noble PJD process as a new treatment for discolored teeth.

21.2 The Initial Development of PJD Process and Evaluation of HA Layers

The first previous study aimed to create HA layers by PJD with manipulating the blasting nozzle above human enamel and to evaluate the microstructural properties of HA layers in vitro. The caries-free human molars extracted for orthodontic treatment with the informed consent of the patients were used. For each specimen, the crown of the molar was severed using a diamond blade cutter and fixed onto an aluminum scanning electron microscope (SEM) stage using an epoxy resin-based adhesive agent. A flat enamel surface perpendicular to the enamel rods was prepared by polishing each specimen with a silicon carbide grinding wheel, using diamond pastes, to obtain an identical surface. The specimens were treated by PJD, using HA particles synthesized by Sangi Co., Ltd. (Tokyo, Japan). A dental PJD device developed by Kuriyagawa and Sendai Nikon Co., Ltd. (Miyagi, Japan) was used to create the HA layers on the enamel substrate of specimens. The PJD device with a blasting nozzle was about 3 cm long, 4 cm wide, and 4 cm tall. In this system, when the high-speed solenoid value is opened, the difference in cross-sectional area across the particle feed tube generates negative pressure at the particle charging port located in the middle of particle feed tube. The negative pressure causes particles to be taken up into the tube. The high-speed solenoid value is controlled by a personal computer and can be opened and closed at a frequency of approximately 100 Hz. The particles are carried to the mixing chamber, where they are mixed with a continuous flow of accelerating gas and then blasted from the nozzle. The details of experimental conditions are shown in Table 21.1. The specimens were fixed on a stage, with their enamel surface perpendicular to the nozzle. The layers were polished using diamond polishing paste (Dia Polisher Paste; GC, Tokyo, Japan) and a felt wheel for 30 s with intermittent pressure of 5,000 g [2]. To evaluate the microstructure, the cross section of the HA layers was observed by SEM (JSM-6500F, JEOL, Tokyo, Japan). Three-dimensional profiles were obtained and surface thickness evaluated using a three-dimensional noncontact measurement system (NH-3;

Table 21.1 The experimental conditions

The first, second, and present study shows captions "29.2," "29.3," and "29.4," respectively			
Parameters	The first study	The second study	The present study
Substrate	Enamel	Enamel	Enamel
Particles	HA	HA	TiO_2-HA
Size of particles [μm]	4.7	3.0 ± 1.0	3.0 ± 1.0
Accelerating gas pressure [MPa]	0.5	0.5	0.5
Feed gas pressure [MPa]	–	0.5	0.5
Blasting angle [°]	90	90	90
Blasting time[s]	–	30	30
Gap between nozzle and substrate [mm]	1	1	1

Mitaka Kohki, Tokyo, Japan). The HA particles were densely packed, and there were no obvious gaps between the HA layers and the enamel substrate. The maximum and average thickness was approximately 40 μm and 30 μm, respectively.

21.3 New Handpiece-Type PJD Device and Evaluation of Thermal Stress in HA Layers

The aims of the second previous study were to use the PJD device, newly developed for use with dental handpieces, to create the HA layer, and to evaluate the microstructural and mechanical properties of the HA layer, in particular the effects of thermal stress. To apply the HA layers, a newly developed PJD device, of similar size to normal dental handpieces, was used (Fig. 21.1). The HA particles were mixed in the camber of the PJD device with a continuous flow of accelerating gas (air), and then blasted form the nozzle onto the enamel substrate at room temperature (25 °C) and atmospheric pressure (1 atm). Detailed experimental conditions of this study are shown in Table 21.1. For thermal-cycling procedure, a computer-controlled two-temperature thermal-cycler (Thermal-cycling K-179; Tokyo-Giken, Tokyo, Japan) was used. Two water baths were maintained at 5 °C and 55 °C. Each cycle consisted of 20 s immersion in each water bath and a travel time of 10 s. The water baths were constantly stirred with stirrers, and the variation in the temperature of each water bath was within 1 °C of the set temperature [2, 14]. The specimens were first immersed in distilled water at 37 °C for 1 day. After the immersion, the specimens underwent 500 thermal cycles [2]. To evaluate their microstructure, the cross section of the HA layers was observed by SEM. The three-dimensional profiles, including surface thickness, of the layers were evaluated using a three-dimensional noncontact measurement system. To evaluate the mechanical properties of the layers, micro-Vickers hardness was measured using a dynamic microhardness tester (FM-ARS 9000; Future-Tech, Kawasaki, Japan). A load of 100 gf was applied for 5 s using a pyramid-shaped die; the depth of the impression was used to

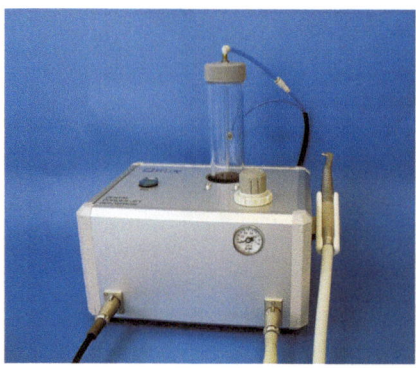 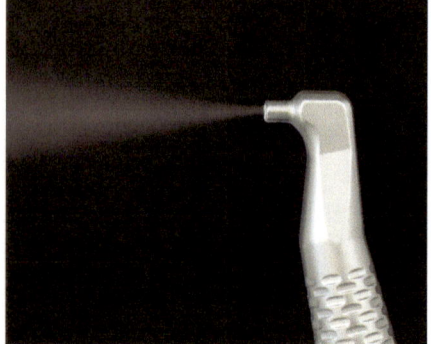

Fig. 21.1 New handpiece-type PJD device

calculate the hardness of the specimens [2, 15]. Furthermore, the bonding strength of the HA layers to the enamel substrates was evaluated using a micro-tensile test (Romulus; Quad Group, Spokane, WA, USA) [2, 16]. An epoxy pre-coated alumina stud was placed perpendicularly onto the surface of the specimen. The area of the surface coated with the epoxy glue was approximately 2.7 mm in diameter. After curing at 150 °C for 1 h, the specimen was set into the testing machine and gripped. The study was pulled until destruction of the specimen, and the bonding strength was determined as the maximum load recorded [2]. The microstructural properties (SEM images, three-dimensional profiles, and surface thickness) and the mechanical properties (micro-Vickers hardness and bonding strength) were evaluated before and after the abovementioned thermal-cycling process. The HA particles in the deposited layer were densely packed, and the surface of the HA layer was unchanged after thermal cycling. The maximum and average thickness were approximately 60 μm and 50 μm, respectively. There were also no significant differences in the hardness and the bonding strength of HA layer before and after thermal cycling.

21.4   New Handpiece-Type PJD Device and Evaluation of Thermal Stress in TiO_2-HA Layers

These previous studies have led to an understanding of the excellent properties of HA layers created by PJD process. However, the microstructural and mechanical properties of TiO_2-HA layers have not yet been clarified. The objectives of the present study are to evaluate the microstructural and mechanical properties of TiO_2-HA layers before and after thermal-cycling tests under the same conditions of the second precious study. To apply the TiO_2-HA layers, a newly developed PJD device was used. The specimens were treated by with PJD, using HA-TiO_2 particles synthesized by Sangi Co., Ltd. (Tokyo, Japan), in which TiO_2 was added as a color regulator [17]. The microstructural properties of TiO_2-HA layers were evaluated from SEM and the three-dimensional profiles: surface thickness. The mechanical properties were evaluated from micro-Vickers hardness and bonding strength. These properties were evaluated before and after the abovementioned 500 cycles of thermal cycling. TiO_2-HA particles were also densely packed; the maximum and average thickness were approximately 60 μm and 50 μm, respectively. There were no significant differences in thickness, hardness, or bonding strength before and after thermal cycling.

21.5   Color Differences of TiO_2-HA Layers

In the previous study, the degree to which TiO_2-HA layers influence tooth whitening has not been well documented. The objectives of the present study were also to evaluate the color differences of TiO_2-HA layers. The color was measured using the

CIE $L^*a^*b^*$ color measurement system on the spectrophotometer (CMS-35FS/C; Murakami Color Research Laboratory, Tokyo, Japan) [17]. The CIE $L^*a^*b^*$ system is composed of three coordinates: L^* (lightness, from 0 = black to 100 = white), a^* (from –a = green to +a = red), and b (from –b = blue to + b = yellow). The area of irradiation and color measurement were φ6 mm and φ3 mm, respectively, and the illuminating and viewing configuration used was CIE diffuse/8° geometry [16, 18]. Each specimen was chromatically measured three times under different conditions – (1) before and (2) after creation of the TiO_2-HA layer – and the average values were calculated. The total color differences between the abovementioned three conditions were calculated according to the following equation:

$$\Delta E^* = \sqrt{\left(\Delta L^*\right)^2 + \left(\Delta a^*\right)^2 + \left(\Delta b^*\right)^2}$$

Color differences between conditions (1) and (2) were calculated as ΔE which evaluates the degree of whitening.

The L^*, a^*, and b^* color parameters and color differences of the three measured conditions are listed in Table 21.3. The L^* color parameter significantly increased between conditions (1) and (2). The a^* color parameter did not significantly change between any of the conditions. In contrast, the b^* color parameter significantly decreased between conditions (1) and (2) ($p > 0.05$, Steel-Dwass multiple comparison test). From these data, the calculated total color difference was ΔE (Table 21.4).

21.6 Discussions

In our previous and present study, SEM images clearly showed that the HA and TiO_2-HA particles were densely packed in the layers and there were no visible pores or cracks between layer and enamel substrate. Cross-sectional SEMs reported in a previous study on the microstructure of a plasma-sprayed 50-μm-thick HA coating on titanium alloy revealed the presence of cracks in the titanium alloy and along the HA coating-substrate interface [19]. Such cracks may be directly related to the coating bonding strength [20] and may affect the long-term stability of HA coatings [21, 22]. Although the substrate was different, it could be considered that HA layers with good material properties are more effectively created using the PJD process than the abovementioned plasma spray-coating technique. From the three-dimensional viewing of HA and TiO_2-HA layers created with new handpiece-type PJD device, the maximum and the average layer thickness were approximately 60 μm and 50 μm, respectively. Using of a new developed PJD device enabled us to create HA and TiO_2-HA layers on the enamel substrate that was thicker than the previously reported layers created by the initial development PJD device. The PJD process was improved in two ways. First, smaller particles were used than in the previous studies. Second, experimental conditions, such as the accelerating pressure and the feed pressure, could be adjusted for the new PJD device for using in dental handpieces. As a result, it was considered that the kinetic energy of particles was effectively imparted to give good adhesion properties of the particles. In case of using new PJD device, the thickness of TiO_2-HA layers was almost equal to that of HA layers. Therefore, it was expected that the TiO_2-HA layers would show almost the same good microstructural properties and high stability as those of the previous HA layers.

The hardness of the TiO_2-HA layers was in the range of approximately 350–390 Hv. These results are almost the same as those obtained for the HA layers created on enamel under the same PJD experimental conditions in the second previous study [2]. On the other hand, the bonding strength of the TiO_2-HA layers was approximately 15 MPa. The micro-tensile bonding strength test used in present study has advantages including more economical use of teeth, better control of regional differences, and better stress distribution at the true interface [23]. The bonding strength of the HA layers created in the second previous study [2] was almost the same as that of the TiO_2-HA layers in the present study. Furthermore, several studies [24–26] have reported bonding strengths of composite resin to enamel substrate of 5.9–22.2 MPa; these values are similar to the 15 MPa bonding strength measured for the present TiO_2-HA layers. Therefore, it was confirmed that the present PJD process could create the HA and TiO_2-HA layers with good material properties on enamel substrates. In addition, our results suggested that addition of TiO_2 particles would not influence the creation of layers by the PJD process and the material properties of HA layers, since the TiO_2-HA layers showed almost the same excellent material properties as the HA layers.

Thermal-cycling procedures simulate the frequent changes in intraoral temperature induced by eating, drinking, and breathing [27–29]. Thermal stresses are related to mechanical stresses, as differential thermal changes can induce crack propagation through bonded interfaces, and the changing gap dimensions are associated with gap volume changes which pump pathogenic oral fluids in and out of the gaps [30]. According to a review of literature, the sequence of temperatures 35 °C, 15 °C, 35 °C, and then 45 °C, with a corresponding dwell sequence of 28 s, 2 s, 28 s, and 2 s, is suggested to be sufficiently clinically relevant [30]. However, the present study adopted thermal cycling between 5 and 55 °C for 500 cycles based on the International Organization for Standardization (ISO) specifications to examine the durability of dental materials [16] in order to evaluate under the same conditions as for HA layers in the second previous study [2]. The HA and TiO_2-HA layers maintained their three-dimensional morphology even after thermal cycling, as determined from three-dimensional views and SEM images. Table 21.2 showed that there were no significant differences in micro-Vickers hardness and bonding strength before and after the thermal-cycling procedure.

The color difference results of the TiO_2-HA layers are shown in Tables 21.3 and 21.4. Between conditions (1) and (2), the L^* color parameter significantly increased, the b^* color parameter significantly decreased, and the a^* parameter was not changed. These results indicate that the created TiO_2-HA layers brightened the color of the specimens; their color became slightly bluer. According to the study of evaluation of whitening effects in vitro, whitening occurs mainly by increasing the L^* param-

Table 21.2 Mean values (and standard deviation) of the micro-Vickers hardness and the bonding strength ($n = 10$)

The first, second, and present study shows captions "29.2," "29.3," and "29.4," respectively		The first study	The second study	The present study
Micro-Vickers hardness	Before thermal cycling	–	391.6(9.80)	371.3(20.7)
	After thermal cycling	–	401.3(9.74)	365.6(27.5)
Bonding strength	Before thermal cycling	–	15.6(5.6)	15.7(1.5)
	After thermal cycling	–	14.8(5.9)	15.5(3.8)

Table 21.3 Mean values (and standard deviation) of CIE L*, a*, b* color parameters for the TiO_2-HA layers ($n = 5$)

	L*	a*	b*
(1) Before creation of layers	71.2(1.2)	−2.3(0.3)	−1.4(0.2)
(2) After creation of layers	78.1[a](1.1)	−2.3(0.3)	−3.9[a](0.4)

[a]The mean difference is significant at the 0.05 level

Table 21.4 Mean values (and standard deviation) for color difference ΔE^* of the TiO_2-HA layers ($n = 5$)

ΔE	6.7(1.1)

eter (higher L^*), reducing the yellowness (lower b^*), and, to a lesser extent, reducing the redness (lower a^*) [5, 31]. In addition, subjective responses to whiteness improvement are significantly correlated with changes in the b^* parameter, and the yellow-blue shift is of primary perceptual importance in tooth-whitening procedures [4]. The reason for the observed color change might be related to selective absorption and reflection of incident light by the TiO_2 particles [21]. From the data of L^*, a^*, and b^* color parameters, the calculated total color difference ΔE, which evaluates the degree of whitening, was approximately 6.7 units. Several reports have been published about the threshold levels of total color difference that can be visually perceived [22, 32]. According to a study on intraoral determination of acceptability and perceptibility tolerances for shade mismatch, the color difference in the test denture which 50 % of dentists could perceive was 2.6 units, while the value at which 50 % of dentists would remake due to color mismatch was 5.5 units [32]. This indicates that although the present TiO_2-HA layers significantly whitened the enamel substrates beyond the perceptible threshold level, their color was mismatched with the optical properties of natural teeth.

In conclusion, the new handpiece-type PJD device could create a thick HA and TiO_2-HA layer on human enamel substrate, and these layers created demonstrated and maintained excellent microstructural and mechanical properties comparable to those of HA layers in our first previous study even after thermal cycling, which simulated the oral environment. Furthermore, it was confirmed that TiO_2-HA layers could whiten the tooth surface and showed high color stability. However, we found that the whiteness of the layers mismatched the natural tooth color to too great an extent, suggesting that adjustment of the HA-TiO_2 particles' TiO_2 content may be necessary to match the natural tooth color. Stated further, if the creation of layers can be adjusted to match the color which patients desires, this raises the possibility that creation TiO_2-HA layers by the PJD process may be used clinically as a new treatment for tooth discoloration. Our future research will assess whether layers demonstrating excellent material properties can be created using particles prepared by adjusting the TiO_2 concentration.

21.7 Conclusion

It was concluded that the new handpiece-type PJD process could create thick HA and TiO_2-HA layers on enamel substrates, and these layers demonstrated excellent microstructural and mechanical properties comparable to those of HA layers created by the initial developmental PJD device, even after thermal-cycling procedure. Furthermore, TiO_2-HA layers could whiten the tooth color and showed high color stability. Creation of TiO_2-HA layers by the PJD process might be a valuable new treatment for discolored teeth.

References

1. Akatsuka R, Sasaki K, Zahmaty MS, Noji M, Anada T, Suzuki O, et al. Characteristics of hydroxyapatite layers formed on human enamel with the powder jet deposition technique. J Biomed Mater Res B Appl Biomater. 2011;98(2):210–6.
2. Akatsuka R, Ishihata H, Noji M, Matsumura K, Kuriyagawa T, Sasaki K. Effect of hydroxyapatite layers formed by powder jet deposition on dentin permeability. Eur J Oral Sci. 2012;120:558–62.
3. Akatsuka R, Matsumura K, Noji M, Kuriyagawa T, Sasaki K. Evaluation of thermal stress in hydroxyapatite layers fabricated by powder jet deposition. Eur J Oral Sci. 2013;121:504–7.
4. Joiner A. Tooth colour: a review of the literature. J Dent. 2004;32:3–12.
5. Lima FG, Rotta TA, Penso S, Meireles SS, Demarco FF. In vitro evaluation of the whitening effect of mouth rinses containing hydrogen peroxide. Braz Oral Res. 2012;26(3):269–74.
6. Dahl JE, Pallesen U. Tooth bleaching-a critical review of the biological aspects. Crit Rev Oral Biol Med. 2003;14:292–304.
7. Hattab FN, Qudeimat MA, Al-Rimawi HS. Dental discoloration: an overview. J Esthet Dent. 1999;11:291–310.
8. Watts A, Addy M. Tooth discoloration and staining: a review of the literature. Br Dent J. 2001;190:309–16.
9. Plotino G, Buono L, Grande NM, Pameijer CH, Somma F. Nonvital tooth bleaching: a review of the literature and clinical procedures. J Endod. 2008;34(4):394–407.

10. Hannig C, Zech R, Henze E, Dorr-Tolui R, Attin T. Determination of peroxides in saliva-kinetics of peroxide release into saliva during home-bleaching with Whitestrips and Vivastyle. Arch Oral Biol. 2003;48:559–66.
11. Azer SS, Hague AL, Johnston WM. Effect of bleaching on tooth discoloration from food colourant in vitro. J Dent. 2011;39 Suppl 3:e52–6.
12. Periasamy VS, Athinarayanan J, Al-Hadi AM, Juhaimi FA, Alshatwi AA. Effects of titanium dioxide nanoparticles isolated from confectionery products on the metabolic stress pathway in human lung fibroblast cells. Arch Environ Contam Toxicol. 2015;68(3):521–33.
13. Yu B, Ahn JS, Lim JI, Lee YK. Influence of TiO_2 nanoparticles on the optical properties of resin composites. Dent Mater. 2009;25(9):1142–7.
14. Weir MD, Moreau JL, Levine ED, Strassler HE, Chow LC, Xu HH. Nanocomposite containing CaF(2) nanoparticles: thermal cycling, wear and long-term water-aging. Dent Mater. 2012;28(6):642–52.
15. International organization for standardization. (1994) Guidance on testing of adhesion to tooth structure. ISO TR 11450. Geneva, Switzerland: ISO, 1–14.
16. CIE (Commission International de l'Eclairage). Colorimetry, technical report. CIE Pub. No.15, 2nd ed. Vienna: Bureau Central de la CIE; 1986 (corrected reprint 1996).
17. Akatsuka R, Izumita K, Nishikawa A, Kayaba C, Kadota S, Hoshi T, et al. Exploratory trial to evaluate the hydroxyapatite layer formed by a new dental treatment system. Open J Stomatol. 2015;5:281–6.
18. Park MY, Lee YK, Lim BS. Influence of fluorescent whitening agent on the fluorescent emission of resin composites. Dent Mater. 2007;23(6):731–5.
19. Garcia-Sanz FJ, Mayor MB, Arias JL, Pou J, Leon B, Perez-Amor M. Hydroxyapatite coatings: a comparative study between plasma-spray and pulsed laser deposition techniques. J Mater Sci Mater Med. 1997;8(12):861–5.
20. Quirynen M, Bollen CM. The influence of surface roughness and surface-free energy on supra- and subgingival plaque formation in man. A review of the literature. J Clin Periodontol. 1995;22(1):1–14.
21. Power JM. Restorative dental materials. 12th ed. St. Louis: Mosby; 2006. p. 35–42.
22. Ruyter IE, Nilner K, Moller B. Color stability of dental composite resin materials for crown and bridge veneers. Dent Mater. 1987;3(5):246–51.
23. Van Meerbeek B, Peumans M, Poitevin A, Mine A, Van Ende A, Neves A, et al. Relationship between bond-strength tests and clinical outcomes. Dent Mater. 2010;26(2):e100–21.
24. Eminkahyagil N, Gokalp S, Korkmaz Y, Baseren M, Karabulut E. Sealant and composite bond strength to enamel with antibacterial/self-etching adhesives. Int J Paediatr Dent. 2005;15:274–81.
25. Kameyama A, Kato J, Aizawa K, Suemori T, Nkazawa Y, Ogata T, et al. Tensile bond strength of one-step self-etch adhesives to Er:YAG laser-irradiated and non-irradiated enamel. Dent Mater J. 2008;27:386–91.
26. Turkun M, Kaya AD. Effect of 10 % sodium ascorbate on the shear bond strength of composite resin to bleached bovine enamel. J Oral Rehabil. 2004;31(12):1184–91.
27. Asaka Y, Amano S, Rikuta A, Kurokawa H, Miyazaki M, Platt JA, et al. Influence of thermal cycling on dentin bond strengths of single-step self-etch adhesive systems. Oper Dent. 2007;32(1):73–8.
28. Saboa VP, Silva FC, Nato F, Mazzoni A, Cadenaro M, Mazzotti G, et al. Analysis of differential artificial ageing of the adhesive interface produced by a two-step etch-and-rinse adhesive. Eur J Oral Sci. 2009;117(5):618–24.

29. Krejci I, Planinic M, Stavridakis M, Bouillaguet S. Resin composite shrinkage and marginal adaptation with different pulse-delay light curing protocols. Eur J Oral Sci. 2005;113(6):531–6.
30. Gale MS, Darvell BW. Thermal cycling procedures for laboratory testing of dental restorations. J Dent. 1999;27(2):89–99.
31. International Commission of Illumination. Recommendations on uniform colour spaces, color difference equations and psychometric colour terms. CIE Publication 1978;15 Suppl 2.
32. Douglas RD, Steinhauer TJ, Wee AG. Intraoral determination of the tolerance of dentists for perceptibility and acceptability of shade mismatch. J Prosthet Dent. 2007; 97(4):200–8.

Open Access This chapter is distributed under the terms of the Creative Commons Attribution 4.0 International License (http://creativecommons.org/licenses/by/4.0/), which permits use, duplication, adaptation, distribution and reproduction in any medium or format, as long as you give appropriate credit to the original author(s) and the source, provide a link to the Creative Commons license and indicate if changes were made.

The images or other third party material in this chapter are included in the work's Creative Commons license, unless indicated otherwise in the credit line; if such material is not included in the work's Creative Commons license and the respective action is not permitted by statutory regulation, users will need to obtain permission from the license holder to duplicate, adapt or reproduce the material.

Chapter 22
Osteogenetic Effect of Low-Magnitude High-Frequency Loading and Parathyroid Hormone on Implant Interface in Osteoporosis

Aya Shibamoto, Toru Ogawa, Masayoshi Yokoyama, Joke Duyck, Katleen Vandamme, Ignace Naert, and Keiichi Sasaki

Abstract Osteoporosis could potentially complicate oral implant treatment because of disease-specific characteristics such as the abnormal bone condition, poor healing ability caused by bisphosphonates (BPs), and bisphosphonate-related osteonecrosis of the jaw (BRONJ). These problems must be resolved to ensure that oral implant treatment is successful in osteoporotic patients.

As a novel therapeutic option for increasing the success rate of oral implantation in patients with osteoporosis, we focused on parathyroid hormone (PTH) and low-magnitude high-frequency (LMHF) loading. Compared to BPs, which inhibit osteoclastic bone resorption and suppress bone turnover, PTH stimulates osteoblastic bone formation and promotes bone turnover. Intermittent PTH administration is a new class of anabolic therapy for the treatment of severe osteoporosis. LMHF loading, which elicits a positive effect on skeleton, has been proposed as a nonpharmacological and adjunctive intervention in the treatment of osteoporosis. Previous investigation reported that both intermittent PTH administration and LMHF loading have an independent osteogenic effect on peri-implant bone healing and implant osseointegration. In addition, our recent study reveals their combined therapy acts locally and synergistically on peri-implant bone healing process, strengthening osseointegration.

Therefore, this can be a new therapeutic option for oral implant treatment in osteoporotic patients without any problems.

A. Shibamoto • T. Ogawa (✉) • M. Yokoyama
Division of Advanced Prosthetic Dentistry, Tohoku University Graduate School of Dentistry,
4-1 Seiryo-machi, Aoba-ku, Sendai, Miyagi 980-8575, Japan
e-mail: ogat-thk@umin.ac.jp

J. Duyck • K. Vandamme • I. Naert
Department of Oral Health Sciences, Prosthetic Dentistry, BIOMAT – Biomaterials,
Katholieke Universiteit Leuven, Leuven, Belgium

K. Sasaki
Division of Advanced Prosthetic Dentistry, Tohoku University Graduate School of Dentistry,
4-1 Seiryo-machi, Aoba-ku, Sendai, Japan

© The Author(s) 2017
K. Sasaki et al. (eds.), *Interface Oral Health Science 2016*,
DOI 10.1007/978-981-10-1560-1_22

Keywords Oral implant • High-frequency loading • Parathyroid hormone • Osseointegration

22.1 Background

Oral implants are a well-accepted and predictable treatment option for the rehabilitation of partially and completely edentulous patients. The osseointegrated implants' success depends on the mechanical support of the host bone in achieving primary stability and the biological process of bone adaptation and regeneration in achieving secondary stability [1]. Nevertheless, the extended life expectancy in today's society has expanded the indications for oral implantation in elderly patients with systemic diseases.

Osteoporosis is a metabolic bone disorder characterized by low bone mass and microarchitectural deterioration of the bone, leading to enhanced bone fragility and a consequent increase in fracture risk [2]. Regardless of the disease characteristics, osteoporosis is not considered an absolute contraindication for oral implant treatment [3, 4]. However, some studies have reported implant failure because of a lack of primary stability and difficulty in achieving osseointegration in patients with osteoporosis [5, 6]. In addition, bisphosphonates (BPs), which are antiresorptive agents and are widely used as the first-choice therapy for osteoporosis, could be a risk factor for implant failure. Kasai et al. [7] compared the success rate of oral implants placed in female patients taking oral BPs with a control group not taking BPs. The BP group had an 86 % success rate, while the control group had a 95 % success rate. BPs also are known to induce BP-related osteonecrosis of the jaw (BRONJ), a serious side effect in patients undergoing invasive oral surgery [8]. There are no universally accepted prevention or treatment protocols for BRONJ [9, 10].

Therefore, to treat osteoporotic patients with oral implants successfully, it is necessary to overcome problems associated with the characteristics of osteoporosis, BPs, and BRONJ. As a novel therapeutic option for increasing the success rate of oral implantation in patients with osteoporosis, we focused on teriparatide [hPTH(1-34)] and low-magnitude high-frequency (LMHF) loading. The aim of this review was to evaluate the single and combined effects of LMHF loading and PTH treatment on peri-implant bone healing and implant osseointegration in osteoporosis.

22.2 Teriparatide [hPTH(1-34)]

22.2.1 PTH as a Therapy for Osteoporosis

Teriparatide [hPTH(1-34)] is an analog of human parathyroid hormone (PTH) containing the amino acid sequence 1–34. It is a new class of anabolic agents acting on the skeleton and should be considered as an alternative to existing antiresorptive agents for the treatment of severe osteoporosis and intractable fractures [11]. Indeed, intermittent systemic administration of hPTH(1-34) reduces the risk of fractures

[12, 13] by improving bone microarchitecture and enhancing overall bone mass [14, 15]. Black et al. [16] reported that hPTH(1-34) exceeded BPs in increasing bone mineral density. In contrast with BPs, which inhibit osteoclastic bone resorption and decrease the bone remodeling rate, PTH stimulates osteoblastic bone formation through an increase in the bone remodeling rate [17]. However, the clinical problems of PTH are as follows: PTH is significantly more expensive than antiresorptive agents [11], PTH is administered by subcutaneous injection [11], and the duration of administration is limited to <2 years based on the induction of osteosarcoma in a rat model of carcinogenicity [18, 19].

22.2.2   Effect of Intermittent PTH Administration on Peri-implant Bone

Recent studies have also reported that intermittent hPTH(1-34) administration promotes peri-implant bone healing in animal and clinical models [20–22]. Our present study also confirmed that intermittent hPTH(1-34) administration has a potent osteogenic capability in stimulating implant osseointegration in ovariectomized (OVX) rats as described in Sect. 30.4. Although there are clinical problems associated with PTH, replacement of BPs with PTH is expected to improve bone density and quality in surgical sites, promote peri-implant bone formation, and prevent the development of BRONJ.

22.3   Low-Magnitude High-Frequency (LMHF) Loading

22.3.1   Effect of LMHF Loading on Skeletal Tissue

LMHF loading, which is generally defined as an LM of $<1\,g$ ($1\,g = 9.98$ m/s^2) and HF of 20–90 Hz, elicits a positive effect on the skeleton (Fig. 22.1) [23, 24]. Numerous studies have evidenced that LMHF loading, applied by means of whole-body vibration (WBV), stimulates bone formation and fracture healing [25–28]. WBV loading has already been used clinically as a nonpharmacological intervention in the treatment of osteoporosis [29–34].

22.3.2   Effect of LMHF Loading on Peri-implant Bone

Regarding titanium implant osseointegration, which has similarities with bone fracture healing, the coauthors' previous studies showed that LMHF loading has an osteogenetic effect on peri-implant bone [35–37]. In particular, Ogawa et al. [35, 38] confirmed that the specific parameters of a loading regimen, such as the

Fig. 22.1 Osteogenetic
effect of LMHF loading.
Montages of
photomicrographs of the
proximal sheep femur used
for static
histomorphometric
evaluation after 1 year of
exposure (20 min per day)
to a 0.3 g, 30-Hz
mechanical stimulus. There
was 32 % more trabecular
bone in the proximal femur
of experimental animals
(**a**) compared with
age-matched controls (**b**)
($P < 0.04$) [23]

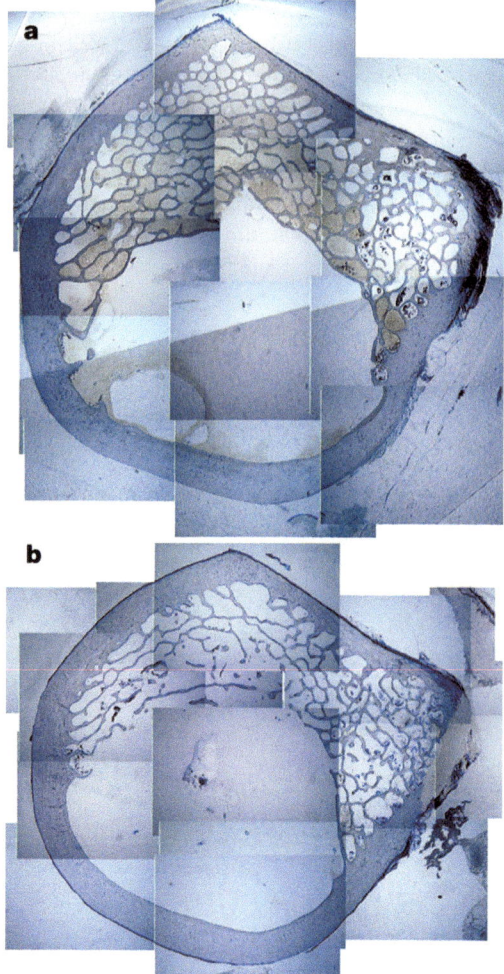

duration, session distribution, frequency, and amplitude of loading, play an important role in the impact of LMHF loading on the bone (Fig. 22.2). Additionally, the
application of LMHF loading reportedly enhances bone-implant osseointegration in
OVX rats [39–41], which was observed in our present study as described in Sect.
30.4.

Fig. 22.2 Effect of LMHF loading on peri-implant bone healing and implant osseointegration. Representative images of the test (loaded) and control (unloaded) group from the 1-week healing period (**a**) and the 4-week healing period (**b**). *Scale bars*: 1 mm. After 4 weeks of healing, the bone neoformation and cortical bone width were much greater in the test group than in the control [35]

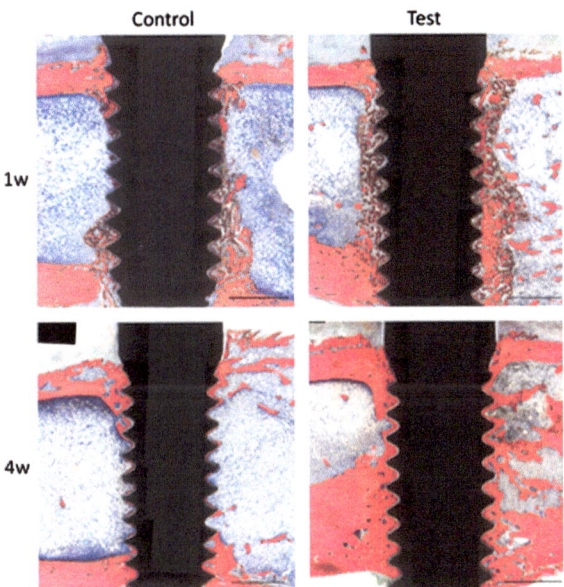

22.4   Effect of LMHF Loading and Intermittent PTH on Peri-implant Bone

Both LMHF loading and intermittent hPTH(1-34) administration have an independent osteogenic effect on peri-implant bone healing and implant osseointegration. However, there are no reports on the impact of their combined therapy on peri-implant bone. It seems likely that combined therapy would act synergistically on the bone healing process and strengthen bone-implant osseointegration. Additionally, the potential synergistic effect may shorten the healing period, thereby relieving the clinical problems associated with PTH.

Our recent study compared the osteogenic impact of LMHF loading and intermittent hPTH(1-34) administration on peri-implant bone healing and implant osseointegration in an osteoporosis model and evaluated their combined effect on these processes. Thirteen-week-old ovariectomized rats ($n = 88$) were divided into three groups: each group of rats received PTH (40 µg/kg, 5 days/week), alendronate (15 µg/kg, 2 days/week), and saline (volume-matched vehicle control), respectively. After 3 weeks, a titanium implant was inserted in both tibiae. Again, each group was subdivided into two groups: with or without LMHF loading via whole-body vibration (WBV, 50 Hz at $0.5\,g$, 15 min/day, 5 days/week). The rats were sacrificed 1 or 4 weeks after implant installation. Peri-implant bone healing and implant osseointegration were assessed using removal torque tests (RT value) and micro-CT analyses (relative gray (RG) value, water = 0 and implant = 100). The data were analyzed by three-factor ANOVA (loading, drug, healing period) followed with a Tukey-HSD test ($\alpha = 0.05$). RT value was significantly influenced by all three factors

($P<0.01$). In particular for PTH-WBV group, these values were highest in all groups after 4 weeks of healing. In the cortical bone, RG value was significantly influenced by the loading ($P<0.01$). In the trabecular bone, on the other hand, RG value was significantly influenced by the drug ($P<0.01$). The RG values of the PTH-treated groups were significantly higher than those of other drug-treated groups ($P<0.01$). The results reveal that LMHF loading and PTH act locally and synergistically on bone healing process, thereby strengthening implant osseointegration. Interestingly, a previous study reported that the combination of ALN and LMHF loading did not lead to a synergistic reaction influencing the bone healing response [41]. Similar to the present study, no obvious positive effect was found in the ALN and ALN+WBV groups. This might be because ALN inhibits osteoclastic bone activity, which is required in the process of bone adaptation and therefore of implant osseointegration. The results also indicate that PTH combined with LMHF loading has a bone-stimulating effect superior to that of ALN and LMHF loading.

22.5   Conclusion

There were four main findings in this review. In osteoporosis model:

- LMHF loading has an osteogenetic effect on the peri-implant bone.
- Intermittent hPTH(1-34) administration has a potent osteogenic capability in stimulating implant osseointegration.
- Two treatment modalities act locally on the bone healing process. The cortical bone was influenced by LMHF loading. The trabecular bone was influenced by PTH.
- The combined application of LMHF loading and PTH synergistically stimulates implant osseointegration. Additionally, PTH combined with LMHF loading has a bone-stimulating effect superior to that of ALN and LMHF loading.

Therefore, this can be a new therapeutic option for oral implant treatment in osteoporotic patients without problems such as failure of osseointegration, delayed healing, or BRONJ.

Conflicts of Interest   The authors report no conflicts of interest.

References

1. Meredith N. Assessment of implant stability as a prognostic determinant. Int J Prosthodont. 1998;11(5):491–501.
2. Kanis JA. Assessment of fracture risk and its application to screening for postmenopausal osteoporosis: synopsis of a WHO report. WHO Study Group. Osteoporos Int. 1994;4(6):368–81.

3. Hwang D, Wang HL. Medical contraindications to implant therapy: part I: absolute contraindications. Implant Dent. 2006;15(4):353–60. doi:10.1097/01.id.0000247855.75691.03.
4. Hwang D, Wang HL. Medical contraindications to implant therapy: part II: relative contraindications. Implant Dent. 2007;16(1):13–23. doi:10.1097/ID.0b013e31803276c8.
5. Alsaadi G, Quirynen M, Komarek A, van Steenberghe D. Impact of local and systemic factors on the incidence of oral implant failures, up to abutment connection. J Clin Periodontol. 2007;34(7):610–7. doi:10.1111/j.1600-051X.2007.01077.x.
6. Ozawa S, Ogawa T, Iida K, Sukotjo C, Hasegawa H, Nishimura RD, et al. Ovariectomy hinders the early stage of bone-implant integration: histomorphometric, biomechanical, and molecular analyses. Bone. 2002;30(1):137–43.
7. Kasai T, Pogrel MA, Hossaini M. The prognosis for dental implants placed in patients taking oral bisphosphonates. J Calif Dent Assoc. 2009;37(1):39–42.
8. Russell RG, Xia Z, Dunford JE, Oppermann U, Kwaasi A, Hulley PA, et al. Bisphosphonates: an update on mechanisms of action and how these relate to clinical efficacy. Ann N Y Acad Sci. 2007;1117:209–57. doi:10.1196/annals.1402.089.
9. Khan AA, Morrison A, Hanley DA, Felsenberg D, McCauley LK, O'Ryan F, et al. Diagnosis and management of osteonecrosis of the jaw: a systematic review and international consensus. J Bone Miner Res. 2015;30(1):3–23. doi:10.1002/jbmr.2405.
10. Yoneda T, Hagino H, Sugimoto T, Ohta H, Takahashi S, Soen S, et al. Bisphosphonate-related osteonecrosis of the jaw: position paper from the Allied Task Force Committee of Japanese Society for Bone and Mineral Research, Japan Osteoporosis Society, Japanese Society of Periodontology, Japanese Society for Oral and Maxillofacial Radiology, and Japanese Society of Oral and Maxillofacial Surgeons. J Bone Miner Metab. 2010;28(4):365–83. doi:10.1007/s00774-010-0162-7.
11. Hodsman AB, Bauer DC, Dempster DW, Dian L, Hanley DA, Harris ST, et al. Parathyroid hormone and teriparatide for the treatment of osteoporosis: a review of the evidence and suggested guidelines for its use. Endocr Rev. 2005;26(5):688–703. doi:10.1210/er.2004-0006.
12. Neer RM, Arnaud CD, Zanchetta JR, Prince R, Gaich GA, Reginster JY, et al. Effect of parathyroid hormone (1-34) on fractures and bone mineral density in postmenopausal women with osteoporosis. N Engl J Med. 2001;344(19):1434–41. doi:10.1056/nejm200105103441904.
13. Aspenberg P, Genant HK, Johansson T, Nino AJ, See K, Krohn K, et al. Teriparatide for acceleration of fracture repair in humans: a prospective, randomized, double-blind study of 102 postmenopausal women with distal radial fractures. J Bone Miner Res. 2010;25(2):404–14. doi:10.1359/jbmr.090731.
14. Hodsman AB, Kisiel M, Adachi JD, Fraher LJ, Watson PH. Histomorphometric evidence for increased bone turnover without change in cortical thickness or porosity after 2 years of cyclical hPTH(1-34) therapy in women with severe osteoporosis. Bone. 2000;27(2):311–8.
15. Jiang Y, Zhao JJ, Mitlak BH, Wang O, Genant HK, Eriksen EF. Recombinant human parathyroid hormone (1-34) [teriparatide] improves both cortical and cancellous bone structure. J Bone Miner Res. 2003;18(11):1932–41. doi:10.1359/jbmr.2003.18.11.1932.
16. Black DM, Greenspan SL, Ensrud KE, Palermo L, McGowan JA, Lang TF, et al. The effects of parathyroid hormone and alendronate alone or in combination in postmenopausal osteoporosis. N Engl J Med. 2003;349(13):1207–15. doi:10.1056/NEJMoa031975.
17. Rosen CJ, Bilezikian JP. Clinical review 123: anabolic therapy for osteoporosis. J Clin Endocrinol Metab. 2001;86(3):957–64. doi:10.1210/jcem.86.3.7366.
18. Tashjian Jr AH, Chabner BA. Commentary on clinical safety of recombinant human parathyroid hormone 1-34 in the treatment of osteoporosis in men and postmenopausal women. J Bone Miner Res. 2002;17(7):1151–61. doi:10.1359/jbmr.2002.17.7.1151.
19. Vahle JL, Sato M, Long GG, Young JK, Francis PC, Engelhardt JA, et al. Skeletal changes in rats given daily subcutaneous injections of recombinant human parathyroid hormone (1-34) for 2 years and relevance to human safety. Toxicol Pathol. 2002;30(3):312–21.
20. Shirota T, Tashiro M, Ohno K, Yamaguchi A. Effect of intermittent parathyroid hormone (1-34) treatment on the bone response after placement of titanium implants into the tibia of ovariectomized rats. J Oral Maxillofac Surg. 2003;61(4):471–80. doi:10.1053/joms.2003.50093.

21. Kuchler U, Luvizuto ER, Tangl S, Watzek G, Gruber R. Short-term teriparatide delivery and osseointegration: a clinical feasibility study. J Dent Res. 2011;90(8):1001–6. doi:10.1177/0022034511407920.

22. Almagro MI, Roman-Blas JA, Bellido M, Castaneda S, Cortez R, Herrero-Beaumont G. PTH [1-34] enhances bone response around titanium implants in a rabbit model of osteoporosis. Clin Oral Implants Res. 2013;24(9):1027–34. doi:10.1111/j.1600-0501.2012.02495.x.

23. Rubin C, Turner AS, Bain S, Mallinckrodt C, McLeod K. Anabolism. Low mechanical signals strengthen long bones. Nature. 2001;412(6847):603–4. doi:10.1038/35088122.

24. Judex S, Gupta S, Rubin C. Regulation of mechanical signals in bone. Orthod Craniofac Res. 2009;12(2):94–104. doi:10.1111/j.1601-6343.2009.01442.x.

25. Omar H, Shen G, Jones AS, Zoellner H, Petocz P, Darendeliler MA. Effect of low magnitude and high frequency mechanical stimuli on defects healing in cranial bones. J Oral Maxillofac Surg. 2008;66(6):1104–11. doi:10.1016/j.joms.2008.01.048.

26. Hwang SJ, Lublinsky S, Seo YK, Kim IS, Judex S. Extremely small-magnitude accelerations enhance bone regeneration: a preliminary study. Clin Orthop Relat Res. 2009;467(4):1083–91. doi:10.1007/s11999-008-0552-5.

27. Sehmisch S, Galal R, Kolios L, Tezval M, Dullin C, Zimmer S, et al. Effects of low-magnitude, high-frequency mechanical stimulation in the rat osteopenia model. Osteoporos Int. 2009;20(12):1999–2008. doi:10.1007/s00198-009-0892-3.

28. Shi HF, Cheung WH, Qin L, Leung AH, Leung KS. Low-magnitude high-frequency vibration treatment augments fracture healing in ovariectomy-induced osteoporotic bone. Bone. 2010;46(5):1299–305. doi:10.1016/j.bone.2009.11.028.

29. Russo CR, Lauretani F, Bandinelli S, Bartali B, Cavazzini C, Guralnik JM, et al. High-frequency vibration training increases muscle power in postmenopausal women. Arch Phys Med Rehabil. 2003;84(12):1854–7.

30. Rubin C, Recker R, Cullen D, Ryaby J, McCabe J, McLeod K. Prevention of postmenopausal bone loss by a low-magnitude, high-frequency mechanical stimuli: a clinical trial assessing compliance, efficacy, and safety. J Bone Miner Res. 2004;19(3):343–51. doi:10.1359/jbmr.0301251.

31. Verschueren SM, Roelants M, Delecluse C, Swinnen S, Vanderschueren D, Boonen S. Effect of 6-month whole body vibration training on hip density, muscle strength, and postural control in postmenopausal women: a randomized controlled pilot study. J Bone Miner Res. 2004;19(3):352–9. doi:10.1359/jbmr.0301245.

32. Iwamoto J, Takeda T, Sato Y, Uzawa M. Effect of whole-body vibration exercise on lumbar bone mineral density, bone turnover, and chronic back pain in post-menopausal osteoporotic women treated with alendronate. Aging Clin Exp Res. 2005;17(2):157–63.

33. Gusi N, Raimundo A, Leal A. Low-frequency vibratory exercise reduces the risk of bone fracture more than walking: a randomized controlled trial. BMC Musculoskelet Disord. 2006;7:92. doi:10.1186/1471-2474-7-92.

34. von Stengel S, Kemmler W, Engelke K, Kalender WA. Effects of whole body vibration on bone mineral density and falls: results of the randomized controlled ELVIS study with postmenopausal women. Osteoporos Int. 2011;22(1):317–25. doi:10.1007/s00198-010-1215-4.

35. Ogawa T, Possemiers T, Zhang X, Naert I, Chaudhari A, Sasaki K, et al. Influence of whole-body vibration time on peri-implant bone healing: a histomorphometrical animal study. J Clin Periodontol. 2011;38(2):180–5. doi:10.1111/j.1600-051X.2010.01637.x.

36. Ogawa T, Zhang X, Naert I, Vermaelen P, Deroose CM, Sasaki K, et al. The effect of whole-body vibration on peri-implant bone healing in rats. Clin Oral Implants Res. 2011;22(3):302–7. doi:10.1111/j.1600-0501.2010.02020.x.

37. Zhang X, Torcasio A, Vandamme K, Ogawa T, van Lenthe GH, Naert I, et al. Enhancement of implant osseointegration by high-frequency low-magnitude loading. PLoS One. 2012;7(7), e40488. doi:10.1371/journal.pone.0040488.

38. Ogawa T, Vandamme K, Zhang X, Naert I, Possemiers T, Chaudhari A, et al. Stimulation of titanium implant osseointegration through high-frequency vibration loading is enhanced when applied at high acceleration. Calcif Tissue Int. 2014;95(5):467–75. doi:10.1007/s00223-014-9896-x.

39. Akca K, Sarac E, Baysal U, Fanuscu M, Chang TL, Cehreli M. Micro-morphologic changes around biophysically-stimulated titanium implants in ovariectomized rats. Head Face Med. 2007;3:28. doi:10.1186/1746-160X-3-28.
40. Chen B, Li Y, Xie D, Yang X. Low-magnitude high-frequency loading via whole body vibration enhances bone-implant osseointegration in ovariectomized rats. J Orthop Res. 2012;30(5):733–9. doi:10.1002/jor.22004.
41. Chatterjee M, Hatori K, Duyck J, Sasaki K, Naert I, Vandamme K. High-frequency loading positively impacts titanium implant osseointegration in impaired bone. Osteoporos Int. 2015;26(1):281–90. doi:10.1007/s00198-014-2824-0.

Open Access   This chapter is distributed under the terms of the Creative Commons Attribution 4.0 International License (http://creativecommons.org/licenses/by/4.0/), which permits use, duplication, adaptation, distribution and reproduction in any medium or format, as long as you give appropriate credit to the original author(s) and the source, provide a link to the Creative Commons license and indicate if changes were made.

The images or other third party material in this chapter are included in the work's Creative Commons license, unless indicated otherwise in the credit line; if such material is not included in the work's Creative Commons license and the respective action is not permitted by statutory regulation, users will need to obtain permission from the license holder to duplicate, adapt or reproduce the material.